Product Recovery in Bioprocess Technology

BOOKS IN THE BIOTOL SERIES

The Molecular Fabric of Cells
Infrastructure and Activities of Cells

Techniques used in Bioproduct Analysis
Analysis of Amino Acids, Proteins and Nucleic Acids
Analysis of Carbohydrates and Lipids

Principles of Cell Energetics
Energy Sources for Cells
Biosynthesis and the Integration of Cell Metabolism

Genome Management in Prokaryotes
Genome Management in Eukaryotes

Crop Physiology
Crop Productivity

Functional Physiology
Cellular Interactions and Immunobiology
Defence Mechanisms

Bioprocess Technology: Modelling and Transport Phenomena
Operational Modes of Bioreactors

In vitro Cultivation of Micro-organisms
In vitro Cultivation of Plant Cells
In vitro Cultivation of Animal Cells

Bioreactor Design and Product Yield
Product Recovery in Bioprocess Technology

Techniques for Engineering Genes
Strategies for Engineering Organisms

Principles of Enzymology for Technological Applications
Technological Applications of Biocatalysts
Technological Applications of Immunochemicals

Biotechnological Innovations in Health Care

Biotechnological Innovations in Crop Improvement
Biotechnological Innovations in Animal Productivity

Biotechnological Innovations in Energy and Environmental Management

Biotechnological Innovations in Chemical Synthesis

Biotechnological Innovations in Food Processing

Biotechnology Source Book: Safety, Good Practice and Regulatory Affairs

BIOTECHNOLOGY BY OPEN LEARNING

Product Recovery in Bioprocess Technology

PUBLISHED ON BEHALF OF:

Open universiteit and **Thames Polytechnic**

Valkenburgerweg 167
6401 DL Heerlen
Nederland

Avery Hill Road
Eltham, London SE9 2HB
United Kingdom

Butterworth-Heinemann Ltd
Linacre House, Jordan Hill, Oxford OX2 8DP

 PART OF REED INTERNATIONAL BOOKS

OXFORD LONDON BOSTON
MUNICH NEW DELHI SINGAPORE SYDNEY
TOKYO TORONTO WELLINGTON

First published 1992

British Library Cataloguing in Publication Data
A catalogue record for this book is
available from the British Library

Library of Congress Cataloguing in Publication Data
A catalogue record for this book is
available from the Library of Congress

ISBN 0 7506 1510 9

Composition by Thames Polytechnic
Printed and bound in Great Britain by
Thomson Litho Ltd, East Kilbride, Scotland

The Biotol Project

The BIOTOL team

OPEN UNIVERSITEIT, THE NETHERLANDS
Dr M. C. E. van Dam-Mieras
Professor W. H. de Jeu
Professor J. de Vries

THAMES POLYTECHNIC, UK
Professor B. R. Currell
Dr J. W. James
Dr C. K. Leach
Mr R. A. Patmore

This series of books has been developed through a collaboration between the Open universiteit of the Netherlands and Thames Polytechnic to provide a whole library of advanced level flexible learning materials including books, computer and video programmes. The series will be of particular value to those working in the chemical, pharmaceutical, health care, food and drinks, agriculture, and environmental, manufacturing and service industries. These industries will be increasingly faced with training problems as the use of biologically based techniques replaces or enhances chemical ones or indeed allows the development of products previously impossible.

The BIOTOL books may be studied privately, but specifically they provide a cost-effective major resource for in-house company training and are the basis for a wider range of courses (open, distance or traditional) from universities which, with practical and tutorial support, lead to recognised qualifications. There is a developing network of institutions throughout Europe to offer tutorial and practical support and courses based on BIOTOL both for those newly entering the field of biotechnology and for graduates looking for more advanced training. BIOTOL is for any one wishing to know about and use the principles and techniques of modern biotechnology whether they are technicians needing further education, new graduates wishing to extend their knowledge, mature staff faced with changing work or a new career, managers unfamiliar with the new technology or those returning to work after a career break.

Our learning texts, written in an informal and friendly style, embody the best characteristics of both open and distance learning to provide a flexible resource for individuals, training organisations, polytechnics and universities, and professional bodies. The content of each book has been carefully worked out between teachers and industry to lead students through a programme of work so that they may achieve clearly stated learning objectives. There are activities and exercises throughout the books, and self assessment questions that allow students to check their own progress and receive any necessary remedial help.

The books, within the series, are modular allowing students to select their own entry point depending on their knowledge and previous experience. These texts therefore remove the necessity for students to attend institution based lectures at specific times and places, bringing a new freedom to study their chosen subject at the time they need and a pace and place to suit them. This same freedom is highly beneficial to industry since staff can receive training without spending significant periods away from the workplace attending lectures and courses, and without altering work patterns.

Contributors

AUTHORS

Ir J. Krijgsman, Principal Scientist and Research Project Manager, Gist-brocades, Delft and Delft University of Technology, Delft, The Netherlands

EDITOR

Dr R. O. Jenkins, Leicester Polytechnic, Leicester, UK

SENIOR BIOPROCESS TECHNOLOGY ADVISOR

Professor Ir K. Ch. A. M. Luyben, Delft University of Technology, Delft, The Netherlands

SCIENTIFIC AND COURSE ADVISORS

Dr M. C. E. van Dam-Mieras, Open universiteit, Heerlen, The Netherlands

Dr C. K. Leach, Leicester Polytechnic, Leicester, UK

ACKNOWLEDGEMENTS

Grateful thanks are extended, not only to the author, editor and course advisors, but to all those who have contributed to the development and production of this book. They include Mrs A. Allwright, Miss K. Brown, Mrs S. Connor, Mrs A. van Galen, Mr L. van Geest, Mr M. de Kok, Miss J. Skelton and Professor R. Spier. Special thanks for the support of Gist-brocades, Delft.

The development of the BIOTOL project is funded by **COMETT, The European Community Action Programme for Education and Training for Technology**. Additional support is provided by the Open universiteit of The Netherlands and by Thames Polytechnic.

Project Manager: Dr J. W. James

Contents

How to use an open learning text

An open learning text presents to you a very carefully thought out programme of study to achieve stated learning objectives, just as a lecturer does. Rather than just listening to a lecture once, and trying to make notes at the same time, you can with a BIOTOL text study it at your own pace, go back over bits you are unsure about and study wherever you choose. Of great importance are the self assessment questions (SAQs) which challenge your understanding and progress and the responses which provide some help if you have had difficulty. These SAQs are carefully thought out to check that you are indeed achieving the set objectives and therefore are a very important part of your study. Every so often in the text you will find the symbol Π, our open door to learning, which indicates an activity for you to do. You will probably find that this participation is a great help to learning so it is important not to skip it.

Whilst you can, as a open learner, study where and when you want, do try to find a place where you can work without disturbance. Most students aim to study a certain number of hours each day or each weekend. If you decide to study for several hours at once, take short breaks of five to ten minutes regularly as it helps to maintain a higher level of overall concentration.

Before you begin a detailed reading of the text, familiarise yourself with the general layout of the material. Have a look at the contents of the various chapters and flip through the pages to get a general impression of the way the subject is dealt with. Forget the old taboo of not writing in books. There is room for your comments, notes and answers; use it and make the book your own personal study record for future revision and reference.

At intervals you will find a summary and list of objectives. The summary will emphasise the important points covered by the material that you have read and the objectives will give you a check list of the things you should then be able to achieve. There are notes in the left hand margin, to help orientate you and emphasise new and important messages.

BIOTOL will be used by universities, polytechnics and colleges as well as industrial training organisations and professional bodies. The texts will form a basis for flexible courses of all types leading to certificates, diplomas and degrees often through credit accumulation and transfer arrangements. In future there will be additional resources available including videos and computer based training programmes.

Preface

The remarkable advances in biotechnology in the past few decades can be attributed to a combination of increased knowledge of how biological systems function and on developments in the process technology associated with production and harvesting of the products of biotransformations. Central to the successful application of the opportunities arising from the new knowledge of biological phenomena are the issues of bioproduct recovery. This text is about the recovery of bioproducts.

The recovery of the desired products from bioreactor outflow is greatly influenced by the nature of the products themselves, the size of the market, the need to achieve market and legally required specification standards and on the market value of the product. The achievement of preparing a product with desired characteristics in a commercially viable manner is, therefore, not a simple matter. It not only requires knowledge of the technical and scientific principles involved in product recovery, but that selection of appropriate strategies be made to achieve commercial and specification targets. This texts aims to provide the knowledge needed to understand, select and develop strategies for the recovery of products from bioreactors and to develop awareness of the issues relating to the final formulation of products. This experience will enable readers to contribute to the commercial realisation of bioproduct manufacture.

We have been fortunate with our author:editor team in bringing together both real life industrial experience of the key processes and experience of sound teaching and learning principles. This combination has prepared a text which not only reflects the forefront of commercial downstream processing, but which also provides easy to read, sound instruction in the strategies and techniques employed in this vitally important aspect of bioprocess technology. To achieve this, in-text activities have been incorporated into the text to facilitate learning and readers are provided with opportunities to check their own learning through the inclusion of self assessment questions.

The text begins by putting downstream processing into a commercial context. It then goes on to examine the starting point of downstream processing, the fermentation broth. Consideration is then given of the processes available for the release of intracellular components and the primary processes available for separating solids from solid-liquid suspensions. Subsequent chapters deal with concentrating and purification of products. Product formulation and market research are essential for the successful commercialisation of biotechnology. These topics are dealt with in Chapter 8. The final chapter deals with the more recently introduced separation processes and discusses the integration of downstream and upstream processes. The text is supported by appendices which provide a summary of the symbols used within the text and useful physical constants. Suggestions for further reading are also included.

Scientific and Course Advisors: Dr M. C. E. van Dam-Mieras
Dr C. K. Leach

Downstream processing in biotechnology

Downstream processing in biotechnology

1.1 Introduction

Central in biotechnology are living cells making desired (and often sophisticated) products. These are formed in a complicated and often quite dilute mixture in a bioreactor. 'Downstream' from this reactor we then have to concentrate and purify the desired product. Downstream processing is the principle subject of this book.

economic product recovery

Downstream processing is not the most glamorous part of biotechnology. A glimpse in any introductory book on the subject will show you that. Chances are that it will not take up more than a few percent of the text. This can be understood - at least partly. Nowadays we know infinitely more about microbiology, genetics and biochemistry than we did a few decennia ago. This has opened our eyes to the possibilities for making products and applications which were formerly undreamt of. The development of these new applications always starts in the laboratory using small scale analytical and preparative techniques. However, in the further development of biotechnological products processing usually becomes very important. As the scale increases, economic ways of conducting fermentation and product recovery become essential. Production scales are larger and changes in the process are more difficult to implement. In this stage downstream processing may require half or more of the cost price. It is therefore worthwhile to try to conceive a proper downstream scheme earlier in the process development. This is difficult: it requires that not only process engineers, but also laboratory development personnel have some idea of which steps are economical and easily scaled up. On the other hand it must be said that small improvements on a production scale can greatly improve the economics of a process.

1.2 Different sectors in biotechnology

In biotechnology, the concept of recovery and product purification is different for the different market sectors. To those manufacturers trying to develop a new drug (a therapeutic enzyme, for example), recovery and purification are just laboratory techniques, only to be used on a preparatory scale. In the production of bulk antibiotics and enzymes, where a high stage yield and a low cost price are very important, a typical chemical engineering approach is more common.

market volume and selling price

In biotechnology we may distinguish a number of market sectors based on market volume and selling price.

In Figure 1.1 the selling price is plotted as function of the concentration in the starting material (blood plasma, fermentation broth, etc).

The products in the upper left corner in particular look very attractive. Selling prices up to 10^9 US $ kg^{-1} strongly appeal to one's imagination. The total market volume however is rather low. The world demand for factor VIII, for example, is about 0.1-1 kg a year. The world market for penicillin is 1.5 10^7 kg a year with a total turnover of 4.10^8 US $ a year, equalling the total turnover of factor VIII.

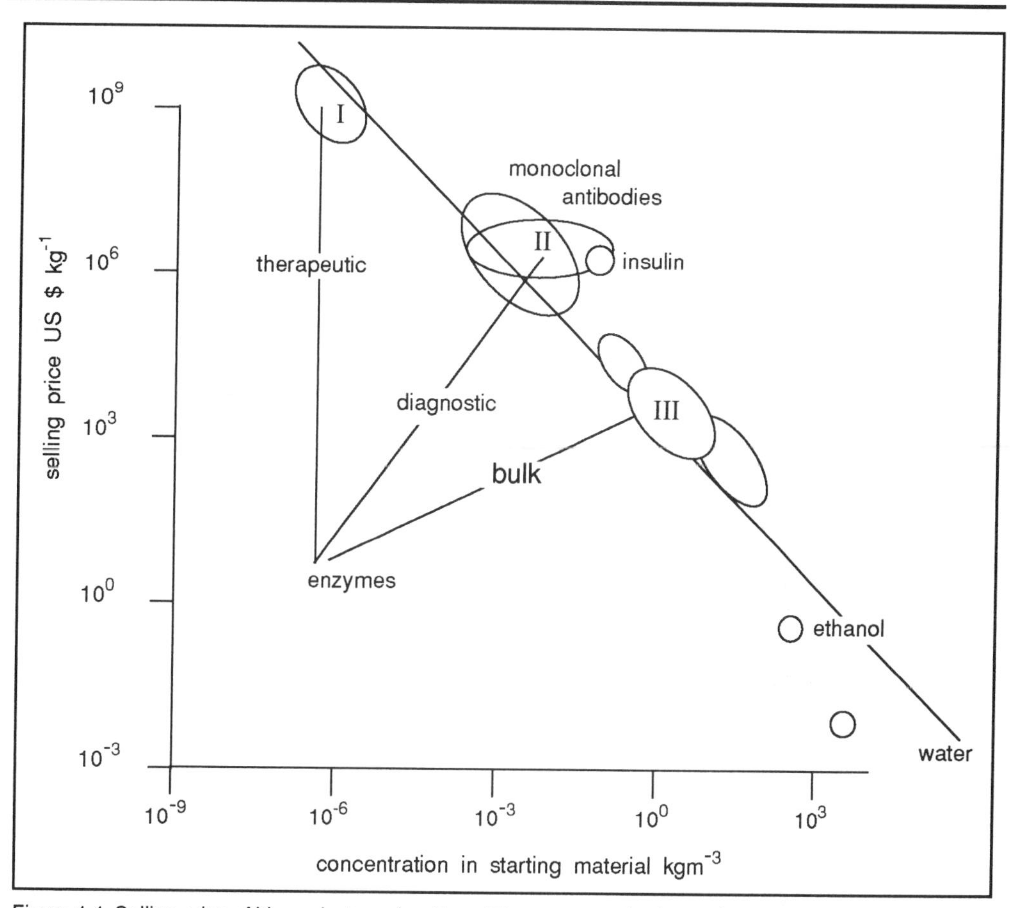

Figure 1.1 Selling price of bioproducts as function of the concentration in starting material.

added value influences net profit

It would seem that the product of selling price and market volume for the different products is roughly constant. However, the added value and, consequently, the net profit may differ greatly.

three market sectors

Three different market sectors can be distinguished:

- Sector I can be characterised as the therapeutic protein sector, with products such as factor VIII and urokinase.
- Sector II is the area of research into diagnostic enzymes and monoclonal antibodies. Other products in this sector are: insulin, luciferase and glycerophosphate dehydrogenase.
- Sector III is characterised by industrial bio-bulk products ranging from antibiotics, proteases, amylases and organic acids to ethanol.

Table 1.1 gives some typical characteristics of bioprocesses in the different sectors.

As can be seen from the table, downstream processing may differ quite substantially from one sector to the other.

Characteristics	Sector I	Sector II	Sector III
Volumes	0.1-10^2 kg y^{-1}	10^3-10^5 kg y^{-1}	10^6 - 10^9 kg y^{-1}
Organism	recombinant-DNA	partly recombinant-DNA	natural producers
Product purity	very high	high/very high	relatively low
Recovery yield	subordinate importance	of minor importance	highly important
Cost price	fraction	20-50% determined by raw materials	50-90% determined by raw materials
Technology	affinity chromatography, preparative electrophoresis	adsorption chromatography, membranes	filtration / extraction / adsorption / precipitation / evaporation / membranes

Table 1.1 Characteristics of bioprocesses in market sectors.

Many of the products in sectors I and II are manufactured under Good Manufacturing Practice (GMP). The process is carefully designed and validated according to the requirements of the national authorities or the Food and Drug Administration (FDA). But in these sectors many processes are being run under suboptimal conditions because early product registration and the wish to be the first on the market are more important than optimal processing.

Many of the processes in sector III are not restricted by these stringent rules and in this field the typical chemical engineering approach works very well.

SAQ 1.1

	Enzyme A	Enzyme B
Selling price (US$ kg^{-1})	5×10^4	100
Total market volume (kg y^{-1})	5×10^2	2×10^6
Added value (US$ kg^{-1})	12×10^3	10

1) Determine the total turnover for each enzyme.

2) Determine the annual net profit for each enzyme.

3) To which market sector do each of these enzymes belong?

4) Which of these enzymes is likely to have the highest concentration in the starting material (eg fermentation broth)? Give a reason for your choice.

1.3 Characterisation of biomolecules

Because biomolecules differ greatly in nature, different separation principles are required for their recovery and purification.

biomolecules as products

Their relative molecular masses vary from approximately 60 to over 2,000,000. Generally, biomolecules are rather unstable and their stability depends on many different factors such as:

- pH;
- temperature;
- ionic strength;
- type of solvent used;
- presence of surfactant;
- metal ions, etc.

In addition, many biomolecules are sensitive to shear and are hydrophobic. Finally, they are sometimes present in very low concentrations, as can be seen in Figure 1.1. These characteristics of biomolecules strongly influence the characteristics of bioprocesses used for their production.

1.4 Characterisation of bioprocesses

One of the most striking characteristics of bioprocesses is the difference in production scale. As can be seen from Table 1.1, the difference is a factor 10^{10}.

Table 1.2 shows some typical characteristics of bioprocesses.

- almost exclusively batch
- small scale relative to chemical industry
- multifunctional equipment
- very flexible and easy to extend
- equipment sterilisable
- suited for containment production
- validated equipment in case of pharmaceutical production

Table 1.2 Characteristics of bioprocesses.

recovery processes should be flexible and easy to extend

Recovery processes in particular should be easy to extend. When starting a new product, generally the expression levels are very low. During production, the fermentation process will be optimised, and medium and strain selection will continue. The result is a strong increase in the expression level. A classical example is penicillin.

In forty years the concentration has increased from 0.4 to approximately 80 mol m^{-3}, an increase of a factor 200! The process should be sufficiently flexible to handle strong fluctuations in fermentation behaviour. These fluctuations are the result of changes in raw materials or strains used during the fermentation. Also, the behaviour patterns of micro-organisms may be fairly unpredictable.

1.5 Recovery in modern versus classical biotechnology

low number of unit operations and high purity

Modern biotechnology is characterised by the way in which organisms are conditioned to make their products. In modern biotechnology, products are made by organisms that do not produce these by nature, but have been manipulated by recombinant DNA (rDNA) techniques in order to achieve this.

The recovery is restricted to a limited number of process steps of high resolution. The degree of purification is high to very high.

Typical characteristics of modern biotechnology processes are given in Table 1.3.

- production on a small scale, 0.1-10m^3
- extension of the variety of organisms used for production (plant and mammalian cells)
- use of non classical fermenters such as:
  - airlift;
  - membrane reactor;
  - immobilised cell reactor.
- sometimes continuous or equipped with *in situ* recovery system

Table 1.3 Characteristics of contemporary bioprocesses.

The separation of biomolecules is largely focused on biospecific interactions and properties such as:

- molecular weight (mass);
- charge distribution;
- hydrophobicity;
- immunogenic structure;
- structure.

In classical biotechnology, however, the product is made by the natural producers (molds for penicillin, bacteria for organic acids and proteases, etc).

high number of unit operations

In many cases a large sequence of classical unit operations of low resolution is necessary to achieve the required purity.

Most of the unit operations are relatively well described mathematically and scale up parameters are relatively easy to determine in pilot plant experiments.

The yield is an important factor, as is minimising the costs of raw materials and labour.

Figure 1.2 shows a typical flow sheet for the production of a food grade enzyme. Most striking is the large number of unit operations required to obtain a high purity and a proper formulation (liquid, powder, tablet, granule, etc.).

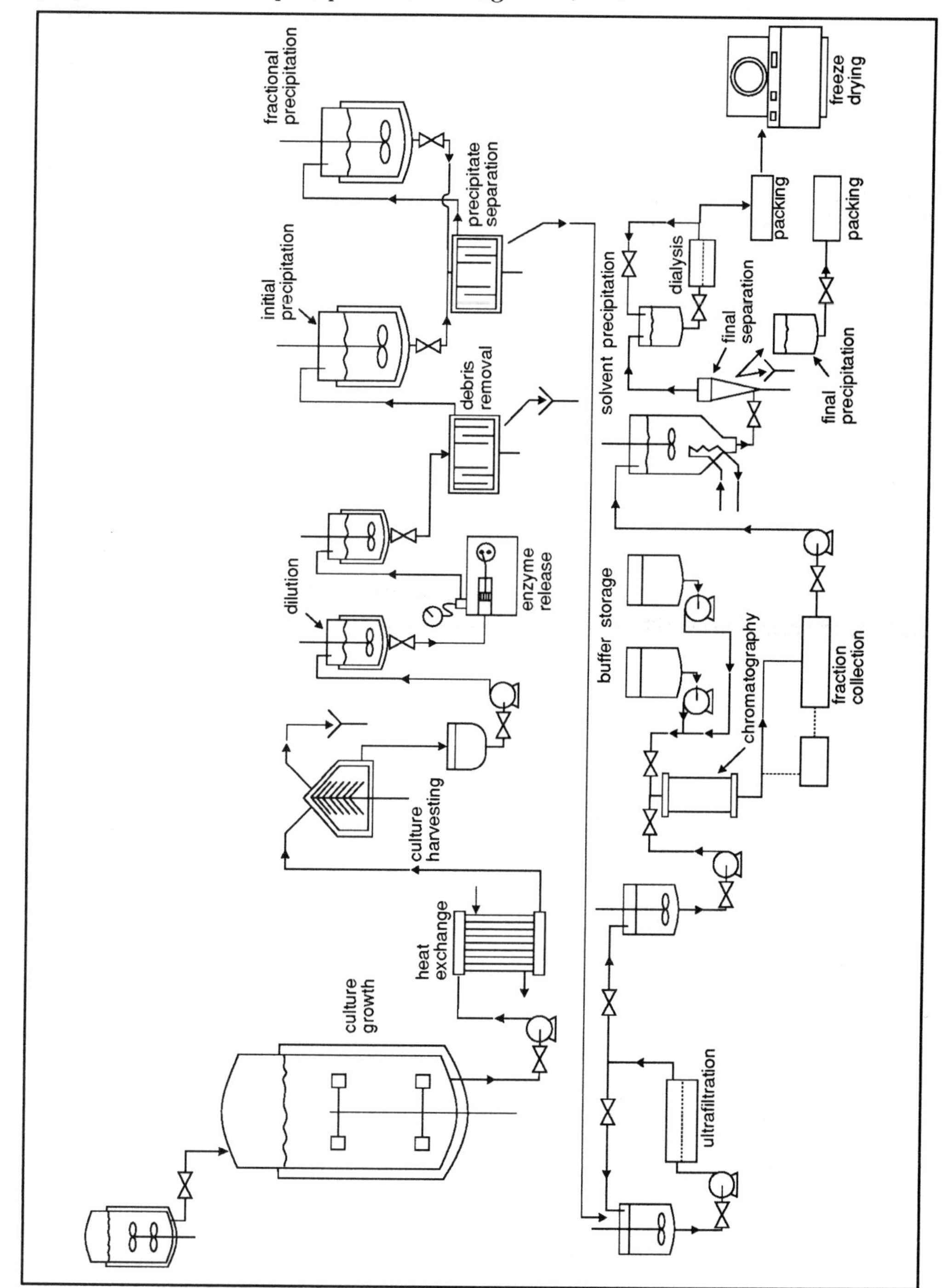

Figure 1.2 Isolation and purification of a food grade intracellular enzyme.

step yield influences overall yield

Associated with the large number of process steps is the overall yield. Having a process with 8 process steps each with a yield of 85% gives an overall yield of 27%. If for this process an overall yield of 94% is required, the step yield needs to be over 99%. This is the cause for one of the major problems in downstream processing.

A process with 5 steps has an average step yield of 75%. 1) Calculate the overall yield for the process. 2) If for this process an overall yield of 50% is required what is the minimum average step yield necessary?

The answer to 1) is:

Step 1: 100 x 0.75 = 75%

Step 2: 75 x 0.75 = 56.25%

Step 3: 56.25 x 0.75 = 42.19%

Step 4: 42.19 x 0.75 = 31.64%

Step 5: 31.64 x 0.75 = 23.73%

The overall yield is therefore 24%.

The answer to 2) is that since there are 5 steps we know that the step yield reduction can be greater than 50/5 = 10% ie step yield smaller than 90%. In fact, a step yield of 88% gives an overall yield of 53% and a step yield of 87% gives an overall yield of 50%. The minimum step yield is therefore 87%.

SAQ 1.2

Which of the following characteristics are more commonly associated with classical biotechnology than with modern biotechnology.

1) Unit recovery operations have low resolution.

2) Production scale < 10 m^3.

3) Purification factor is low.

4) Recovery has large number of unit operations.

5) Process organism is a mammalian cell.

6) Process is well described mathematically.

7) Overall yield is relatively high (90%)

8) Separation based on biospecific interactions.

Figure 1.2 showed a large variety in the unit operations applied, which complicates the design of a recovery plant even further, but forms an outstanding challenge to the bio-recovery engineer. Let us look at this process a little more closely.

recovery of biomolecules often has five stages

In the recovery of biomolecules 5 stages can be distinguished: pretreatment, solid-liquid separation, concentration, purification and formulation. In each stage a wide range of unit operations are available, as indicated in Figure 1.3.

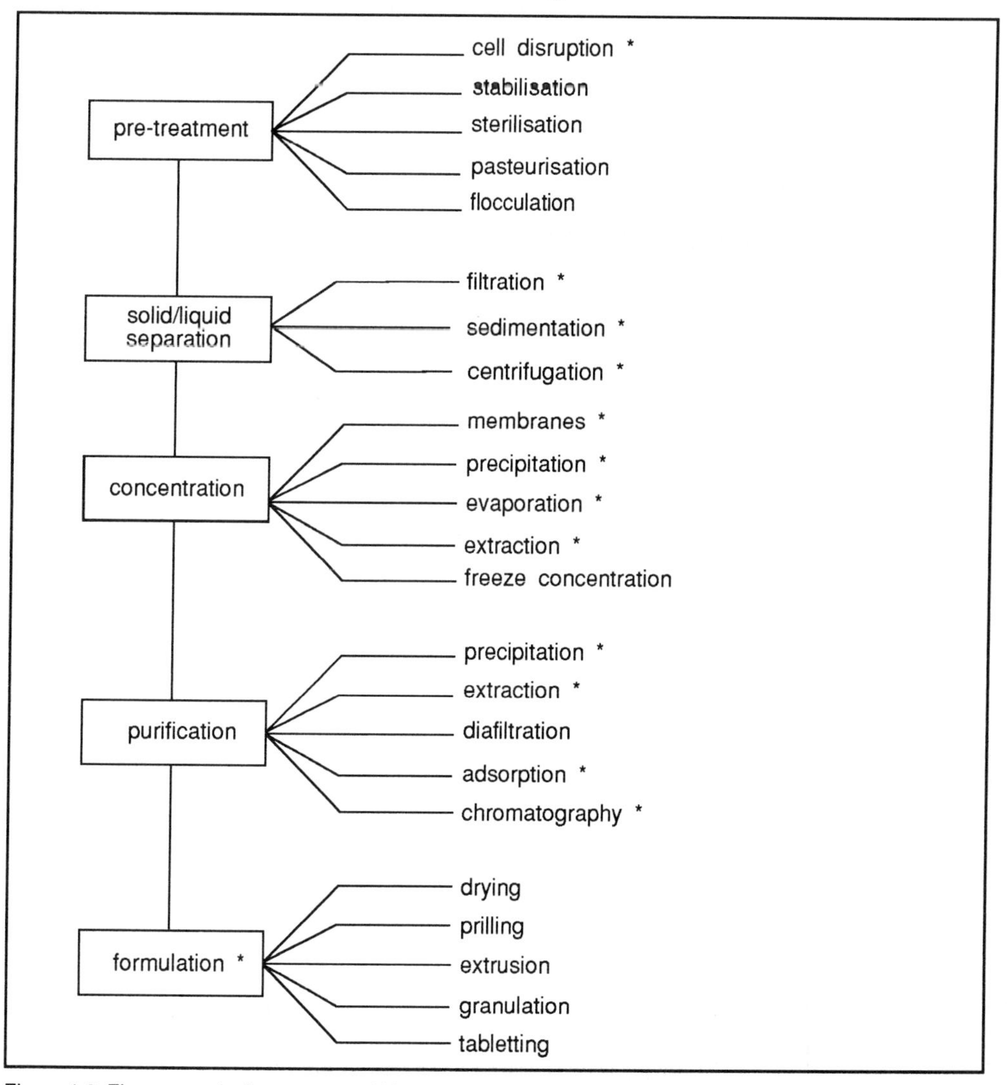

Figure 1.3 Five stages in the recovery of bioproducts (* indicate unit operations described in more detail later in this book).

Figure 1.4 can be helpful in selecting unit operations. In this Figure a listing of unit operations is given in relation to their primary separation factor and the (molecular) size of the material to be recovered.

As can be seen from this Figure, in most cases more than one unit operation is available to achieve separation.

Π Figure 1.4 provides a very useful set of information so it may be worthwile trying to remember it. Read it through carefully and perhaps write out your own version and pin it up where you can see it. It would be a good exercise to arrange the unit operations according to some alternative criteria.

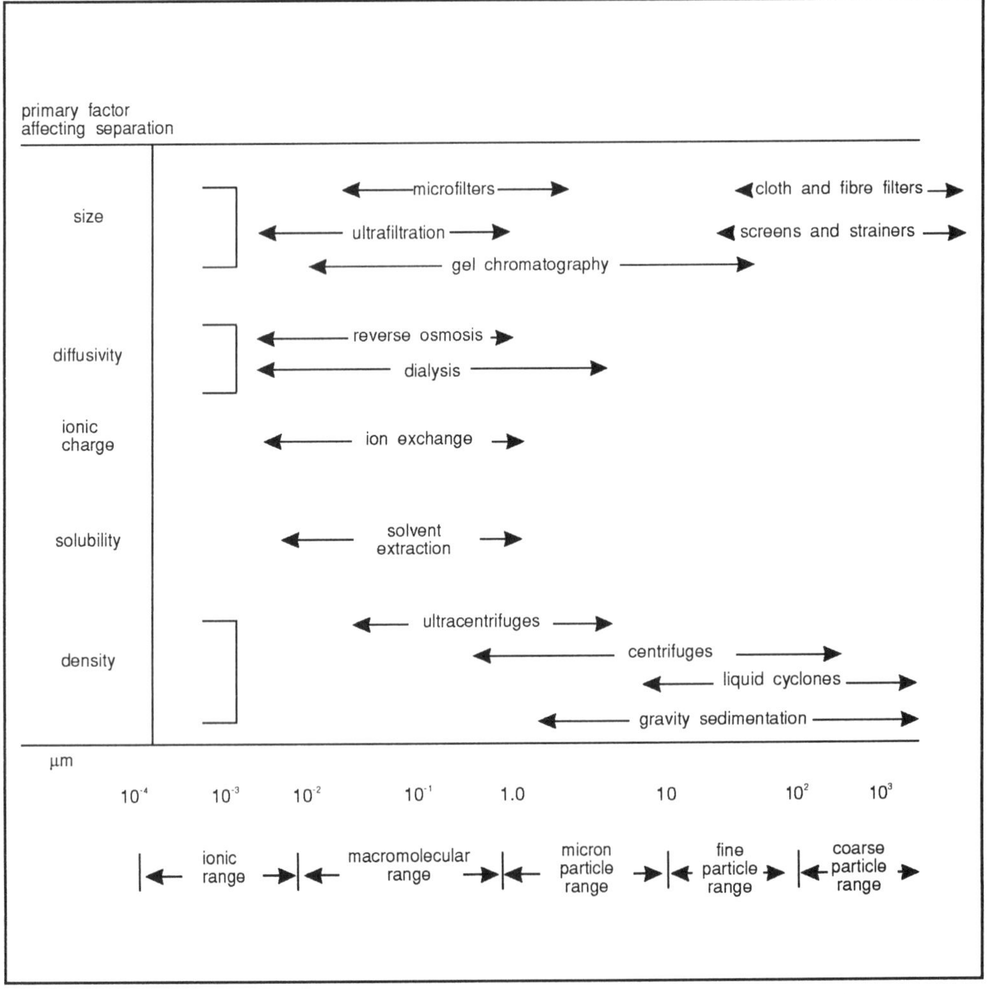

Figure 1.4 Separation range of unit operations ranked according to their primary separation factor.

When one or more unit operations have been chosen, a further selection will be made in the pilot plant.

After the reliability and feasibility of the unit operation chosen have been proven, the next step is scale up.

The rules required for a proper scale-up can be derived from a mathematical formulation of the unit operation involved. Later in the book we will demonstrate this for those unit operations indicated with an asterisk in Figure 1.3.

SAQ 1.3

Match each of the unit operations listed with one or more of the recovery stages.

Unit operation	Recovery stage
Drying	Pre-treatment
Sedimentation	Solid/liquid separation
Cell disruption	Concentration
Precipitation	Purification
Chromatography	Formulation
Membranes	
Sterilization	
Centrifugation	
Tabletting	

SAQ 1.4

For each of the following size ranges, select at least three appropriate unit operations from the list provided.

Size ranges	Unit operations
1) Ionic	a) Microfilters
2) Macromolecular	b) Cloth and fibre filters
3) Micron particle	c) Ultrafiltration
4) Fine particle	d) Screens and strainers
5) Coarse particle	e) Gel chromatography
	f) Reverse osmosis
	g) Dialysis
	h) Ion exchange
	i) Solvent extraction
	j) Ultracentrifuges
	k) Centrifuges
	l) Liquid cyclones
	m) Gravity sedimentation

1.6 Additional information

Appendix 1 provides suggestions for further reading for those who wish to study the material covered in each chapter more extensively. These suggestions have been arranged on a chapter by chapter basis. Throughout the text, a wide variety of symbols are used to represent various quantities. Although these are explained when they are first introduced into the text, a summary of the major symbols used is provided in Appendix 2.

Summary and objectives

We have seen in this chapter that three market sectors can be distinguished and downstream processing may differ substantially from one sector to another. Characteristics of biomolecules influences their stability and, in modern biotechnology, separation is based largely on biospecific interactions. Such interactions often allow high purity to be achieved using a small number of unit operations. In classical biotechnology, however, often a large number of unit operations of low resolution is necessary to achieve the required purity. In all processes the overall yield is determined by the step yields and the number of unit operations.

Five stages in the recovery of bioproducts can be distinguished and for each stage a wide range of unit operations is available. The separation range of unit operations ranked according to the primary separation factor can be helpful in selecting unit operations.

Now that you have completed this chapter you should be able to:

- distinguish three market sectors based on market volume and selling price;
- list common characteristics of modern and of classical biotechnological processes;
- determine overall yield from step yields;
- match unit operations with recovery stages;
- select appropriate unit operations based on size of material to be recovered.

Characterisation of fermentation broth

Characterisation of fermentation broth

2.1 Introduction

In this chapter we will discuss those properties of fermentation broths important to downstream processing.

Fermentation broths can be characterised by the various microorganisms used, their size, shape (morphology), rheological (liquid flow) behaviour and last but not least the concentration of cells, products and byproducts.

2.2 Morphology of cells

The broad range of bioproducts requires a large variety of cells as production host ranging from bacteria of 1 μm to cellular agglomerates of over 4000 μm.

The sizes of a number of solids, representative of those organisms suspended in typical whole broths, are given in Figure 2.1.

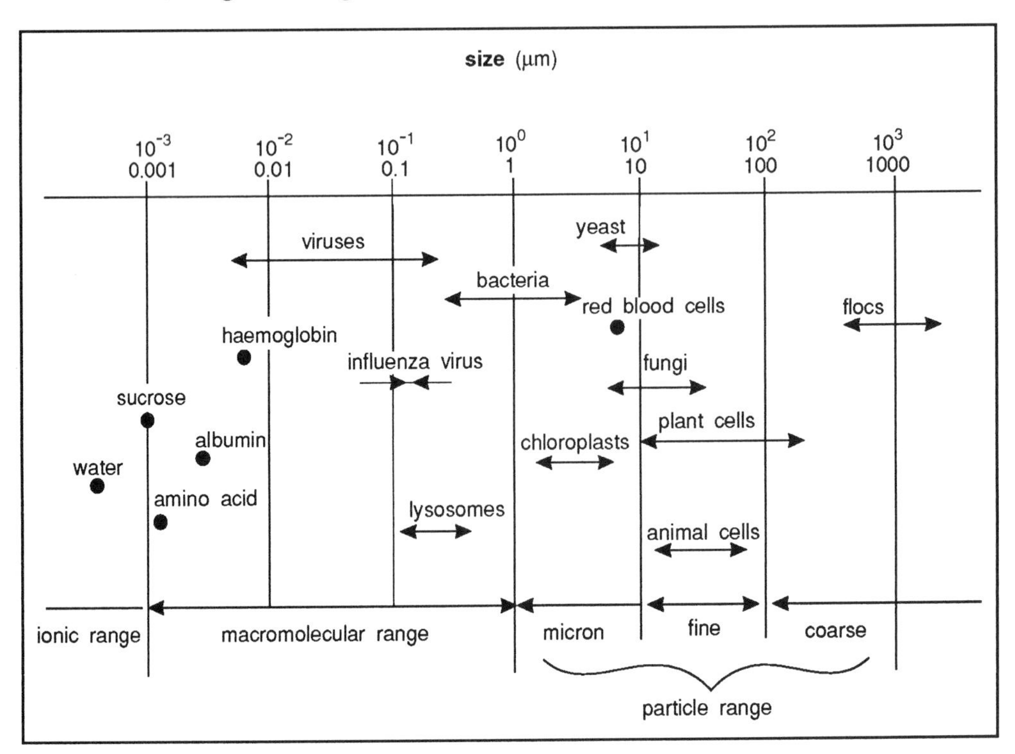

Figure 2.1 Size of representative solids in whole broths.

It generally holds that decreasing cell size gives decreasing separation capacity resulting in higher costs.

Π List the following solids in order of decreasing size (largest to smallest): bacteria; viruses; flocs; albumin; sucrose; plant cells.

Now check your order by referring to Figure 2.1.

The morphology of some representative solid phases is given in Figure 2.2.

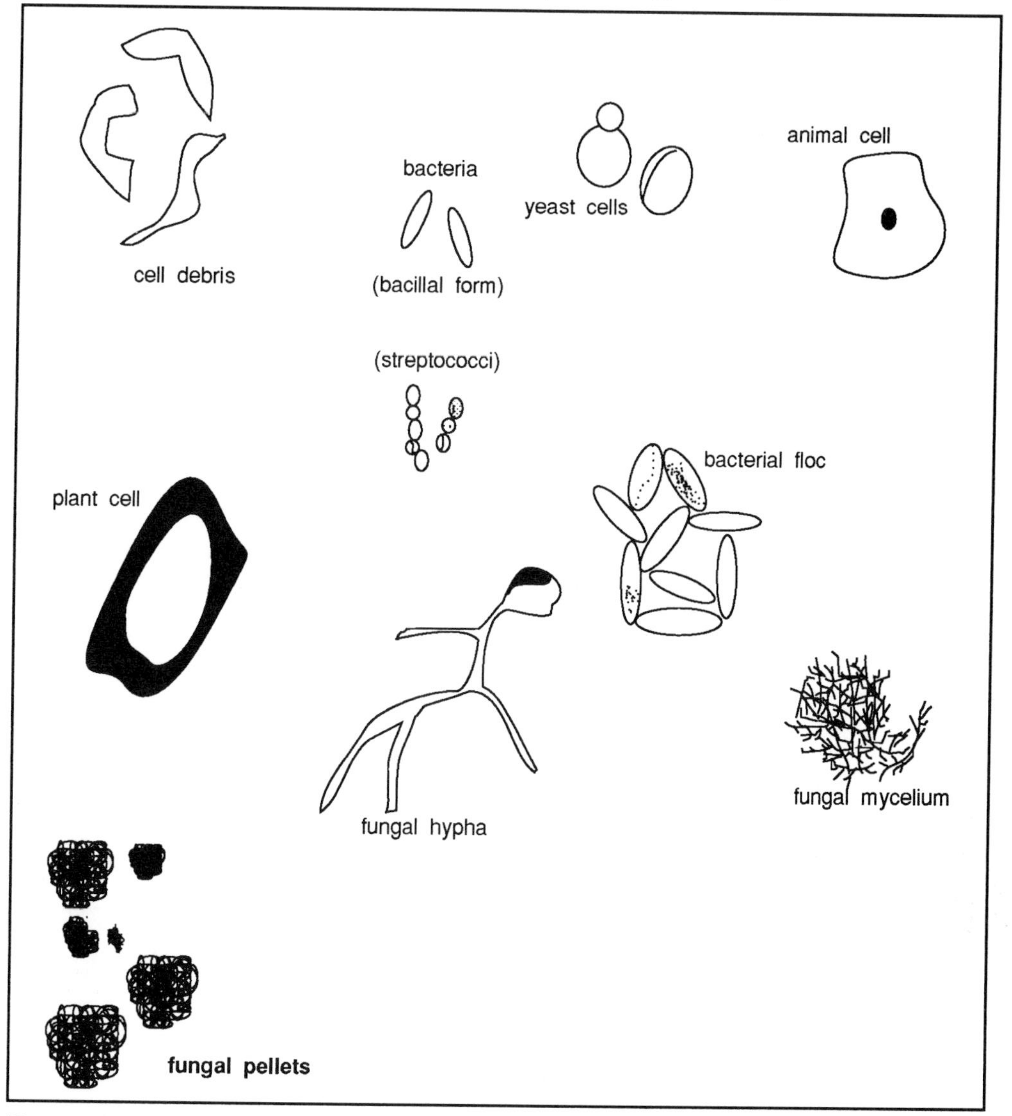

Figure 2.2 Morphology of cells representative of fermentation broths (not to scale).

A large variety of sizes and shapes of solids is characteristic of fermentation broths. Yeast and bacterial cells are usually homogeneously suspended in the liquid. In the case of fungi, frequently a network of hyphi will be formed which increases the viscosity greatly.

Under certain conditions fungi will form agglomerates called 'pellets'. Owing to their large size (100 - 4000 µm) these pellets are relatively easy to recover.

Bacteria may form slime layers depending on the strain and fermentation conditions. Slime forming bacteria are very difficult to separate in recovery processes.

Π Write down how you think that slime forming bacteria would present problems in downstream processing.

The main reasons are:

1) slime tends to retain liquid;

2) slime may block certain unit operations eg membranes;

3) slime increases the viscosity of the broth, which in turn decreases the efficiency of certain unit operations eg filtration and centrifugation.

2.3 Structure of the cell wall

Gram discovered in 1884 that some bacteria show a dark violet colour after a staining process, while others do not. He explained this phenomenon in terms of the cell wall structure. Cells that remain stained are called Gram positive, the others Gram negative.

We will not deal with details of the Gram staining technique here but if you would like to follow this up in more detail, we recommend the BIOTOL text 'Infrastructure and Activities of Cells'.

Gram positive bacteria have a simple cell wall

The cells of Gram positive bacteria (such as *Lactobacillus* and *Staphylococcus*) are surrounded by a cytoplasmic membrane covered by a structural murein network composed of polysaccharide and amino acids. The cytoplasmic membrane is formed by a double layer of phospholipids and should be considered very deformable, like a soap bubble. Conversely, the murein layer is quite rigid and maintains the characteristic shape of the bacterium (see Figure 2.3).

Gram negative bacteria have a complex cell wall

The structure of the cell wall of Gram negative bacteria (such as *Escherichia coli*, *Pseudomonas* sp.) is much more complicated, as shown in Figure 2.4. In this case there is an outer membrane and the murein layer is much thinner. A detailed discussion of the chemical composition of bacterial cell walls is given in the BIOTOL text 'Infrastructure and Activities of Cells'. Outside of the cell wall may be other structures such as capsules and slime layers.

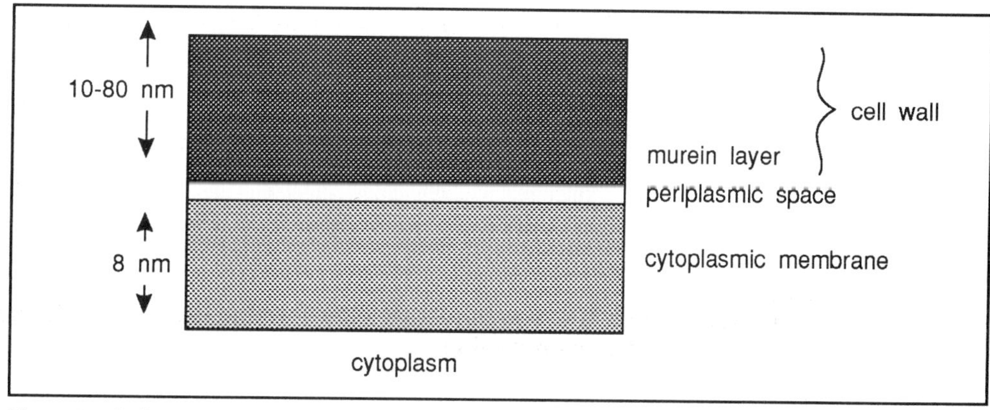

Figure 2.3 Stylised cell wall structure of Gram positive bacteria.

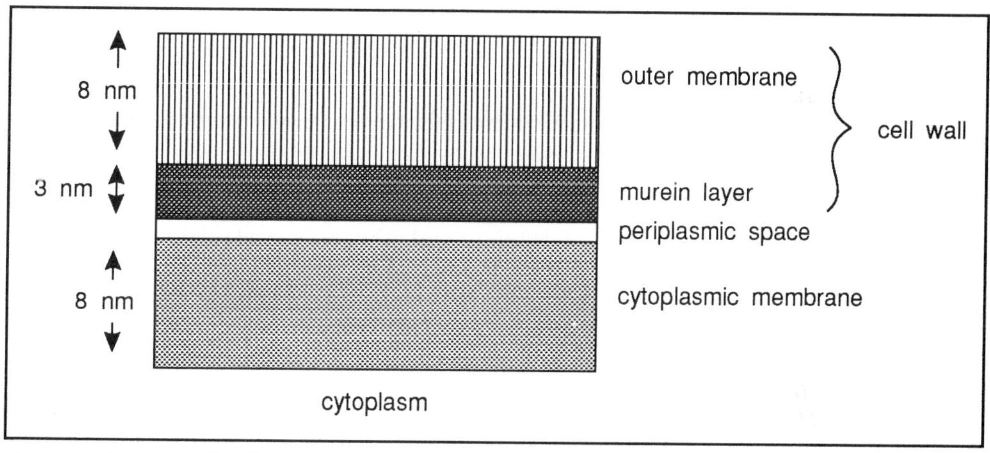

Figure 2.4 Stylised cell wall structure of Gram negative bacteria. Note that there is considerable variation between different species.

Π Would you expect Gram positive cells to be more difficult to disrupt mechanically than Gram negative cells? Give a reason for your response.

Gram positive cells are generally more difficult to disrupt mechanically than Gram negative cells because the structural layer (murein) is much thicker in Gram positives. There is, however, substantial variations between different species and the presence of capsules and slime layers also influence the mechanical disruption of cells.

yeast cell walls contain glucan and mannan

The cell wall structure of yeasts is much more difficult to elucidate than that of bacteria. The cytoplasmic membrane consists of a bilayer of phospholipids and outside of that is a highly crosslinked structural layer of glucan and mannan (types of polysaccharide). The cell wall cross section is given in Figure 2.5. The cell wall thickness of yeasts is approximately 70 nm.

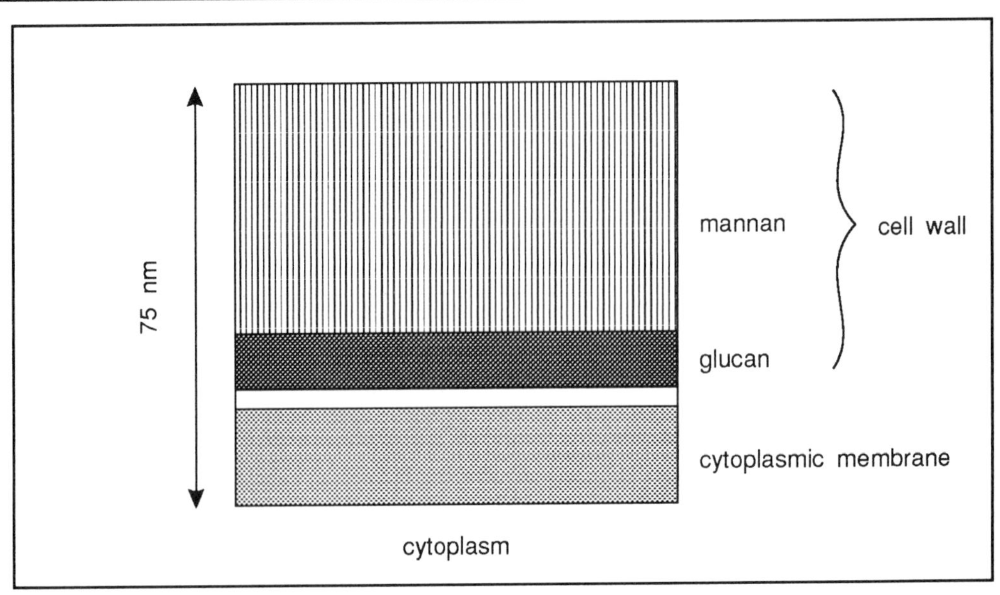

Figure 2.5 Stylised cross section of yeast cell wall.

The cell wall structure of fungi is highly diverse and although it is often composed of cellulose or chitin little is known about the precise structure. Plant cell walls are also generally composed of cellulose and other polysaccharides. Conversely, animal cells do not have a cell wall but they do have a cytoplasmic membrane which is stiffened by sterols.

The importance of the cell wall structure in downstream processing will be explained in Chapter 3, 'Release of intracellular components'.

SAQ 2.1

Identify each of the following statements as True or False. If false give a reason for your response.

1) The murein layer is thicker in Gram positive cell walls than in Gram negative cell walls.

2) Yeast cell walls are composed of protein.

3) Plant cells do not have cell walls.

4) If the cell wall of a rod shaped bacterium is removed, the cell would become spherical if placed in an isotonic solution.

5) Fungal cell walls may be composed of cellulose.

2.4 Concentrations

product concentrations

The choice of the most appropriate recovery method is greatly determined by the product concentration in the whole broth. Table 2.1 shows this concentration for various products. As can be seen the product concentrations may differ more than a factor 1000.

type of cell	product	concentration (w/w %)
Yeast	ethanol	7-12
Fungus	organic acid	5-10
Fungus	penicillin	3-5
	cephalosporin	3
	streptomycin	2-4.5
Bacteria	enzymes	0.05-1
	vitamin B_{12}	0.005

Table 2.1 Product concentration in fermentation broths.

biomass concentration

The biomass concentration is also an important consideration for downstream processing. The reason for this is that it may have a strong effect on the viscosity and consequently on all subsequent recovery steps in which the whole broth is processed. Table 2.2 shows biomass concentrations for various types of cells.

type of cell	product	concentration (% dry weight)
Bacteria/yeast	single cell protein	3-6
Fungus	citric acid	2-3
Fungus	penicillin	2-4
Bacteria	enzymes	3-5
Animal cells	--	0.05-0.02
Plant cells	--	0.1-5

Table 2.2 Biomass concentration for different types of products and cells.

The biomass concentration can also be expressed as the number of cells per ml. The cell concentrations for bacteria at the end of fermentation range from 10^{11} - 10^{12} cells ml^{-1} and for yeast from 10^8 - 10^9 cells ml^{-1}.

In addition to the product and biomass, the broth may also contain various amounts of biopolymers (polysaccharides, proteins) DNA, RNA, non-consumed media components, cell organelles etc, up to 3% dry solids.

Π Examine Tables 2.1 and 2.2 and compare ethanol from yeast with penicillin from fungus. Which of these products has the highest concentration range when expressed as g (g dry weight biomass)$^{-1}$.

For ethanol from yeast the concentration range is:

$$\frac{7}{6} \text{ to } \frac{12}{3} = 1.2 \text{ to } 4.0 \text{ g (g dry weight)}^{-1}$$

For penicillin from fungus the concentration range is:

$$\frac{3}{4} \text{ to } \frac{5}{2} = 0.7 \text{ to } 2.5 \text{ g (g dry weight)}^{-1}$$

We can conclude that ethanol from yeast has the highest concentration range when related to biomass dry weight.

2.5 Biomass density

Dry biomass has a density of about 1400 kg m^{-3}. Because the cells have a water content of 70-80%, the density of the broth is about 1100 kg m^{-3}. Pellets and flocs have lower densities due to the water entrapped between the cells. The density of the medium after fermentation (supernatant/filtrate) is about 1030 kg m^{-3}.

2.6 Rheological behaviour

The rheological behaviour of fermentation broths is of minor importance in downstream processing. It mainly affects the performance of (membrane) filters and centrifuges.

If the cells are separated from the broth the supernatant or filtrate behaves generally like water (except in cases of fermentation of polysaccharides such as xanthene etc.). Therefore we will restrict ourselves to simple Newtonian and non-Newtonian models which cover 90% of the field. We shall see that the viscosity of a cell suspension is strongly influenced by cell shape and by cell concentration.

Π What factors could change the viscosity of a fermentation broth during growth of the process organism?

During growth an increase in biomass concentration and changes in particle (cell or flock) shape can influence viscosity. In addition, substrate is taken up during metabolism and the proportion of undissolved substrate may be reduced. At the same time, metabolites are excreted. These factors may also influence the viscosity of a fermentation broth.

Rheological behaviour of liquids has been considered in the BIOTOL text 'Bioprocess Technology - Modelling and Transport Phenomena'. For the purposes of this chapter, you are reminded of the meaning of some terms:

1) Viscosity is the resistance of a liquid to flow.

2) Shear rate is the relative velocity of parallel adjacent layers in laminar flow of a liquid body under shear force.

3) Shear stress (force) is the force applied which causes liquid flow.

2.7 Newtonian fluids

The flow behaviour of liquids can be described most simply by the Newtonian model:

$$\tau = \eta \mathring{\gamma} \qquad \text{(E - 2.1)}$$

where:

τ = shear stress (Nm^{-2}; Pa)

η = dynamic viscosity ($Ns\ m^{-2}$; Pas)

$\mathring{\gamma}$ = shear rate (s^{-1})

In this model the shear stress is a linear function of the shear rate $\mathring{\gamma}$. The viscosity η is a proportional constant and independent of $\mathring{\gamma}$. Examples of Newtonian liquids are: water and oil.

particles in liquid alter viscosity Einstein equation ($\phi < 0.1$)

Particles such as cells suspended in a liquid (mostly water) alter the viscosity. The phenomenon was first formulated by Einstein for rigid particles which show no interaction.

$$\eta_{relative} = \frac{\eta_{suspension}}{\eta_{liquid}} = 1 + K_1\phi \qquad \text{(E - 2.2)}$$

where:

K_1 = constant

ϕ = volume fraction

K_1 depends on the particle shape and ϕ represents the volume fraction of particles. For spherical particles $K_1 = 2.5$ and for rod shaped particles K_1 varies from 20-1200.

Eilers equation ($\phi > 0.1$)

Equation 2.2 holds only for highly diluted suspensions ($\phi < 0.1$). For more concentrated suspensions Eilers derived a more useful equation:

$$\eta_{relative} = \left\{1 + K_1\phi \;.\; 0.5\left(1 - \frac{\phi}{\phi_{max}}\right)^{-1}\right\}^2 \qquad \text{(E - 2.3)}$$

in which ϕ_{max} is the maximum packing density (0.6 - 0.7) for the type of cells under investigation. Thus a ϕ_{max} of 0.6 means that 60% of the total volume consists of cells when the cells are packed together as closely as can be achieved.

Figure 2.6 shows the relative viscosity as a function of the volume fraction of cells for both Equations 2.2 and 2.3.

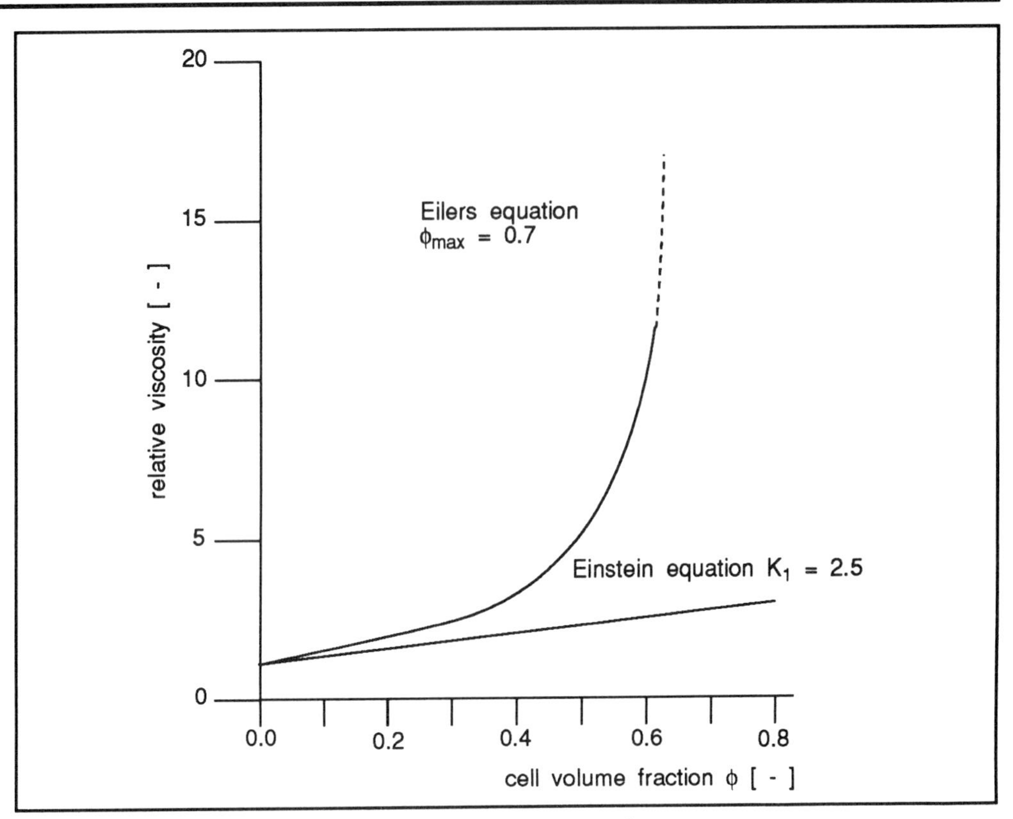

Figure 2.6 Relative viscosity as a function of the volume fraction cells.

Π Determine the shear stress for a suspension of spherical cells in water with a cell volume fraction of 0.2 and a maximum packing density of 0.6 when the shear rate = 60 s^{-1}.

Since the cell volume fraction is 0.2 we use Eilers equation (Equation 2.3) to determine relative viscosity ($\eta_{relative}$); in which, K_1, has a value of 2.5 because the cells are spherical. With ϕ at 0.2 and ϕ_{max} at 0.6, we obtain a value of 1.89 for $\eta_{relative}$. Substituting this value into Equation 2.1, at a shear rate ($\dot{\gamma}$) of 60 s^{-1}, we obtain a value of 113.4 Nm^{-2}, for the shear stress (τ).

SAQ 2.2

Complete the following statements concerning Newtonian fluids:

1) For highly dilute suspensions, relative viscosity is best described by the [ ] equation.

2) Shear stress is a [ ] function of shear [ ] for Newtonian fluids.

3) The value of K_1 is determined by the particle [ ] and equals 2.5 for [ ] particles.

4) For concentrated suspensions of Newtonian fluids, relative viscosity is best described by the [ ] equation.

5) For Newtonian fluids, the viscosity is [ ] of shear rate.

2.8 Non-Newtonian fluids

non linear function of shear rate

In the case of non Newtonian fluids the shear stress (τ) is a non linear function of the shear rate $\mathring{\gamma}$.

$$\tau = K \cdot f(\mathring{\gamma}) \qquad \text{(E - 2.4)}$$

power law model

In many cases the power law model can be applied:

$$\tau = K(\mathring{\gamma})^n \qquad \text{(E - 2.5)}$$

where:

K = consistency index ($Nm^{-2}\, s^n$)

n = power law index

shear stress

Both parameters can be measured with a rheometer.

To give you an example of the values of K and n, for Actinomycetes, K = 2000-3000 and n = 0.4-0.8 depending on cell concentration, type of strain and fermentation conditions.

Bingham model

Some broth viscosities can be characterised by use of the Bingham model:

$$\tau = \tau_o + \eta\, \mathring{\gamma} \qquad \text{(E - 2.6)}$$

The yield stress (τ_o) is the minimum stress required to cause liquid flow. For example mold suspensions (antibiotic production and yoghurt) are often best described by the Bingham model. An alternative expression has been derived relating τ, τ_o, K and γ:

$$\tau = \tau_o - K(\mathring{\gamma})^n \qquad \text{(E - 2.7)}$$

Herschel-Buckley model

This is called the Herschel-Buckley model.

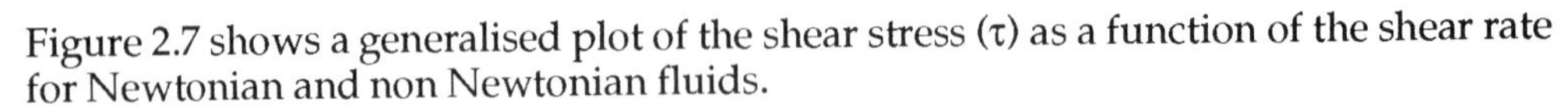

Figure 2.7 shows a generalised plot of the shear stress (τ) as a function of the shear rate for Newtonian and non Newtonian fluids.

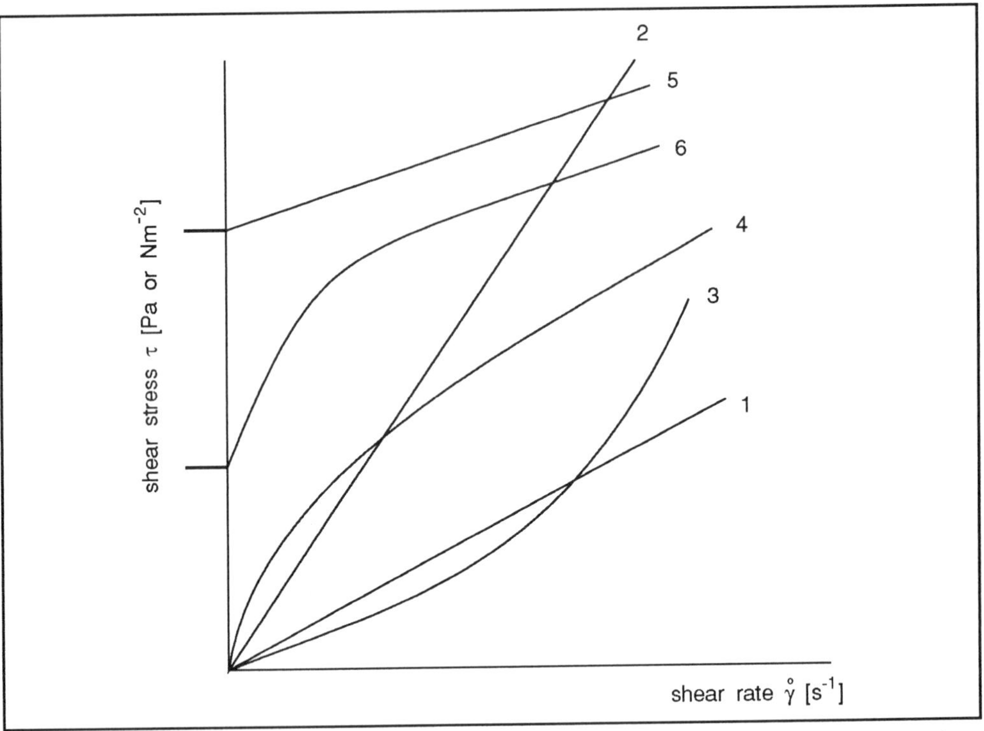

Figure 2.7 Relation between shear stress and shear rate for various rheological models. 1) Newtonian: low viscosity, 2) Newtonian: high viscosity, 3) Dilatant, 4) Pseudoplastic, 5) Bingham, 6) Casson. Many of these models are described in the text.

Π Examine each of the plots in Figure 2.7, which one shows a dramatic increase in shear stress (τ) at high shear rate ($\overset{o}{\gamma}$)?

You should have spotted plot 3 - the fluid is described as dilatant. What this means is the greater the movement, the greater the viscosity.

Some concentrated cell suspensions show dilatancy, (Figure 2.7 curve 3). For example it is found for yeast at high concentrations (15% dry solids). The index n in Equation 2.5 is larger than 1. Another typical example of the phenomenon is quick sand, the more movement the higher the viscosity.

Table 2.3 shows the viscosities for some commonly occurring materials.

The actual rehological behaviour of culture broths can of course only be determined by experiment. Nevertheless you should be able to make some predictions. Try it on the following example.

Π Which of the following is most likely to deviate from Newtonian rheology? A culture of unicellular bacteria; a culture of filamentous fungi; a culture of unicellular algae; a culture of protozoans.

You should have suggested the culture of filamentous fungi. Clearly the filamentous nature of the culture will influence the flow behaviour of the broth. Generally such cultures follow a power law where n is usually less than 1.

Material	Viscosity (mPas)	Temperature (°C)
Air	1.7×10^{-2}	0
Water	1×10^{0}	20
Egg albumin	1.2×10^{1}	20
Glycerol	10^{2}	60
Castor oil	10^{3}	20
Glucose	10^{4}	10

Table 2.3 Viscosities of common materials.

SAQ 2.3

For each of the following conditions/statements, select appropriate generalised plot(s) from those shown in Figure 2.7.

1) n = 1.

2) n less than 1.

3) n greater than 1.

4) A finite yield stress has to be applied to cause liquid flow.

5) τ_o is finite, n is less than 1.

6) The plot is influenced by the viscosity of the fermentation broth.

SAQ 2.4

Identify each of the following statements as True or False. If False give a reason for your response.

1) Broths containing Actinomycetes behave as a dilatant.

2) Increasing cell density always increases viscosity.

3) Spherical cells generally have a lower viscosity than rod shaped cells.

4) $\tau = \eta\dot{\gamma}$ describes a Newtonian model for liquid flow.

5) For solutions that exhibit dilatant behaviour, the apparent viscosity decreases as the shear rate increases.

Summary and objectives

Fermentation broths are highly complex mixtures. Overall they consist of 85-95% water and 5-15% dissolved and undissolved solids. The product is generally a minor constituent of the broth. The structure and the composition of the broth can be affected by heat treatment, pH variation and ageing. The sizes of the organisms used may vary strongly, ranging from 0.4 to 100 μm and over 1000 μm in the case of pellets. The range of shapes include spherical cells, rods and highly filamentous forms. Cell wall composition varies strongly depending on the type of organisms used.

The viscosity depends on the cell types used, cell morphology and concentration and end product accumulation. Storage and pretreatment may affect the viscosity strongly.

Now that you have completed this chapter you should be able to:

- list various cell types in order of size;
- compare morphology and cell wall structure of cell types representative of fermentation broths;
- interpret equations that describe various rheological models;
- distinguish between the various rheological models.

Releases of intracellular components

Releases of intracellular components

3.1 Introduction

In many cases, but in particular in the production of proteins, the product is not excreted into the medium but remains in the cell.

Although liberation of intracellular products has been practised for many years, modern biotechnology gave a new impulse to the development of new or improved liberation techniques.

rDNA products

There is the need for cell breakage to liberate the product from the cell, especially in the case of recombinant DNA (rDNA) technology in which prokaryotic cells are used to produce intracellular proteins.

Table 3.1 lists some important traditional and rDNA intracellular products.

traditional intracellular products	rDNA intracellular products
glucose isomerase	chymosin (yeast / *E.coli)*
β-galactosidase	insulin (*E.coli* / mammalian)
phosphatase	immunoglobulin
ethanol dehydrogenase	plasminogen activator
DNase, RNase	interferons (mammalian)
$NADH/NAD^+$	human growth hormone (*E.coli)*
alkaloids	somatostatin
	human serum albumin
	F VIII (mammalian)
	streptokinase (mammalian)

Table 3.1 Traditional and rDNA intracellular products.

There are many ways to disrupt cells to liberate the product. However, only a few are applied on large scale; these are shown in Figure 3.1.

The mechanical methods will be discussed in more detail. The non-mechanical methods will only be discussed in brief, not because they are less important but because of lack of information: most companies are unwilling to publish their large scale recovery methods.

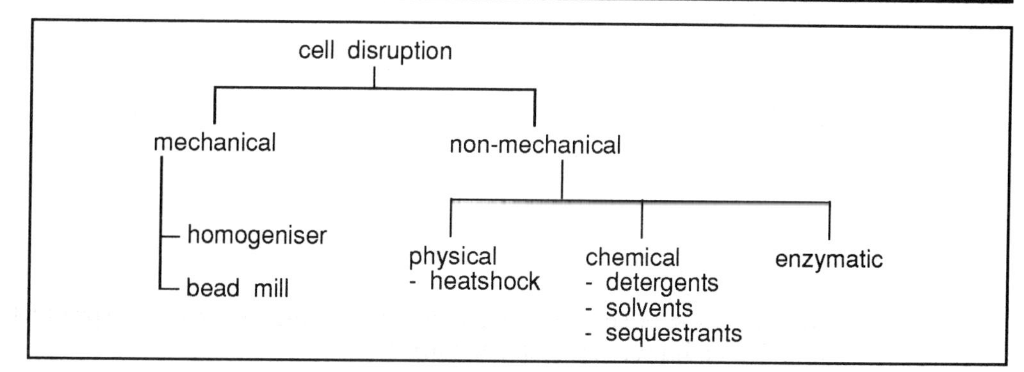

Figure 3.1 Methods for cell disruption on large scale

3.2 Mechanical cell disruption

3.2.1 The bead mill

A bead mill consists of a horizontal or vertical grinding cylinder. The central shaft is fitted with a number of impellers and driven by an electromotor via a belt.

The grinding cylinder is made out of stainless steel or, in case of a laboratory machine, made out of glass.

The cylinder is partly filled with small beads. A typical bead loading value is 80%.

The beads are moulded from wear resistant materials such as:

- zirconium oxide;
- zirconium silicate;
- titanium carbide;
- glass;
- alumina ceramic.

The impeller design has been based on an efficient energy transfer to the beads. A typical value for tip speed is 15 m s^{-1}.

Figure 3.2 shows an example of a bead mill.

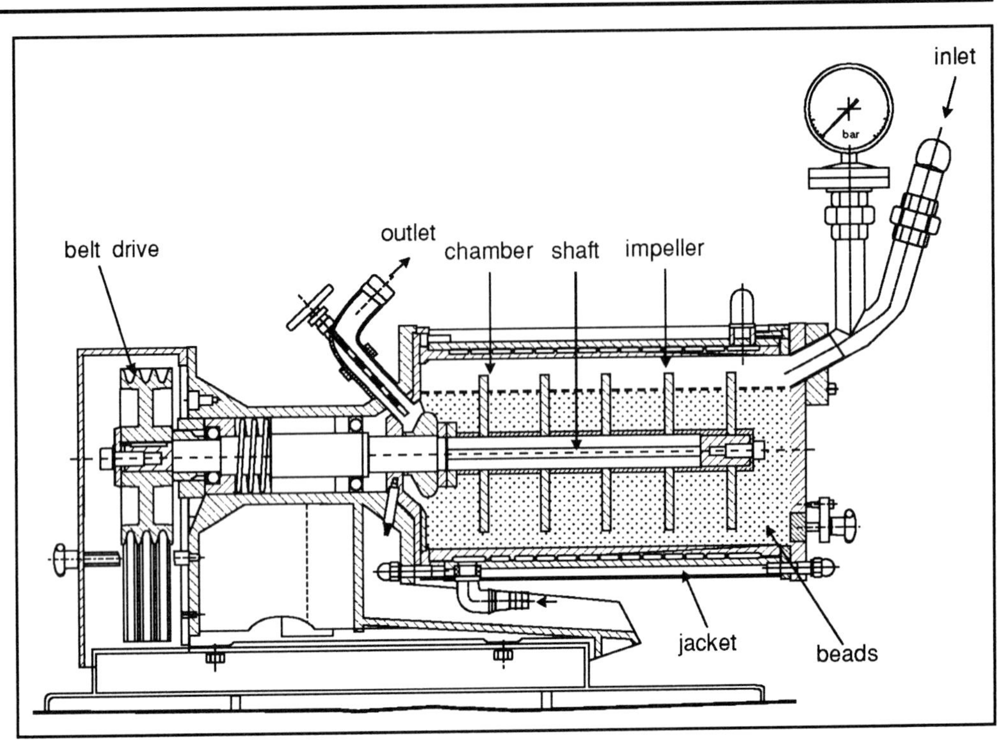

Figure 3.2 Horizontal bead mill.

Π Before reading on, think about breaking cells in a bead mill and see if you can make a list of the factors which will influence the efficiency of releasing intracellular components. When you have made your list compare it with the factors we have cited below.

The optimal conditions for the release of intracellular components depend on:

1) The micro-organism used:

- cell wall thickness and composition;
- size.

2) Location of the product:

- in cytoplasm;
- in cell organelles;
- in periplasmic space.

3) Type of bead mill:

- bead diameter;
- type of impeller;

- bead loading;
- tip speed of impeller.

4) Residence time of cell suspension.

5) Cell concentration.

6) Temperature (rise).

The relationship between these variables and product release is generally unknown and must be determined experimentally.

3.2.2 Cell disruption kinetics for a bead mill

Let us represent the concentration of the product that can be released from the biomass by C_r^{max}.

first order kinetics for release

Under constant conditions the rates of product release for bacteria and yeasts can be described by first order kinetics. The concentration of released product (C_r) will be proportional to the concentration of disrupted cells.

The rate of release dC_r/dt will be proportional to the concentration of unreleased product, $C_r^{max} - C_r$, so:

$$\frac{dC_r}{dt} = K_b\,(C_r^{max} - C_r) \tag{E - 3.1}$$

where K_b is the first order release 'constant' for bead milling.

Integration gives:

$$\ln \frac{C_r^{max}}{C_r^{max} - C_r} = -K_b t \tag{E - 3.2}$$

$$C_r = C_r^{max}\,(1 - \exp(-K_b t)) \tag{E - 3.3}$$

A value for the maximal attainable release (C_r^{max}) has to be determined experimentally. The value of K_b depends strongly on the type of impeller, bead size, bead load, impeller speed and temperature. K_b has to be determined experimentally too.

Equations (3.1) to (3.3) hold only for a batch mode with a defined residence time t. In case of a continuous mode the mean residence time and the residence time distribution or the number of continuous stirred tanks reactors (CSTR's) in series should be taken into account when predicting the release.

Figure 3.3 shows a result of disruption of yeast cells in batch mode for two types of impellers.

SAQ 3.1

From the information presented in Figure 3.3, determine as accurately as possible the values of K_b for both types of impellers.

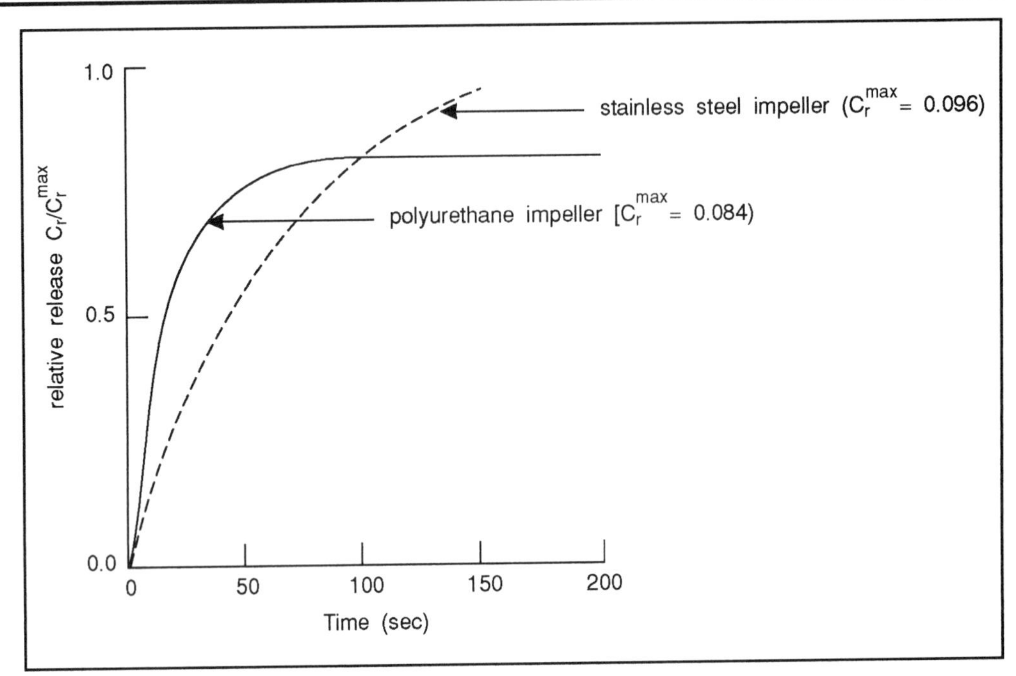

Figure 3.3 Batch disruption of yeast with two different types of impeller.

3.2.3 Scale up of bead mills

The main problem in the scaling up of bead mills is the removal of the energy dissipated in the broth. A bead mill can be considered as a jacketed cylinder with an internal heat source. Almost all of the power input via the impeller is dissipated into heat. This heat has to be removed via the cylinder wall. The ratio of heat transfer area to the mill volume can be expressed as follows:

$$L = \frac{\text{surface area (A)}}{\text{mill volume (V)}} = \frac{\pi T L}{\frac{\pi}{4} T^2 L} = \frac{4}{T} \qquad \text{(E - 3.4)}$$

where T = cylinder diameter (m);

L = length of bead mill (m).

On scaling up, the cylinder diameter will be increased, so this ratio will be diminished.

The power input (P) can be mathematically described as follows:

$$P = c\rho N^3 D^5 \qquad \text{(E - 3.5)}$$

where c = dimensionless constant

ρ = suspension density (kg m^{-3})

N = rotational speed of the impeller (s^{-1})

D = impeller diameter (m)

c depends on the type of flow in the mill (laminar or turbulent) and the type of impeller.

Increasing the impeller diameter will result in a considerable increase in power input (at constant speed). Also if the cylinder diameter is doubled then the heat transmission surface to volume ratio is reduced (Equation 3.4). As nearly all the power input will be dissipated in the broth, scale-up using bead mills will be limited.

Π Determine 1) the fold decrease in the heat transmission surface to volume ratio and 2) the fold-increase in power input required to maintain constant speed, when the diameter of a bead mill and the diameter of the impeller are doubled.

1) We can see from Equation 3.4 that a doubling of the diameter will produce a 2-fold decrease in the heat transmission surface to volume ratio.

2) We can see from Equation 3.5 that a doubling of the diameter will produce a 2^5 = 32-fold increase in power input at constant speed.

3.2.4 The homogeniser

A homogeniser is basically a high pressure positive displacement pump which pumps the cell suspension through an adjustable orifice discharge valve.

The pressure may vary from 200 - 1000 bar depending on the type of micro-organism and concentration.

cell suspension is pumped through an orifice

A disadvantage of plunger pumps is their irregular discharge characteristic due to the plunger kinematics. Therefore homogenisers are equipped with 3-5 cylinders and pulsation dampers.

A typical homogeniser discharge valve assembly is illustrated in Figure 3.4.

During discharge, the suspension passes between the valve and the seat. The backpressure is controlled by a hand wheel. This provides the pressure on the seat via a spring mechanism. The pressure can be read from a pressure gauge at the inlet of the seat.

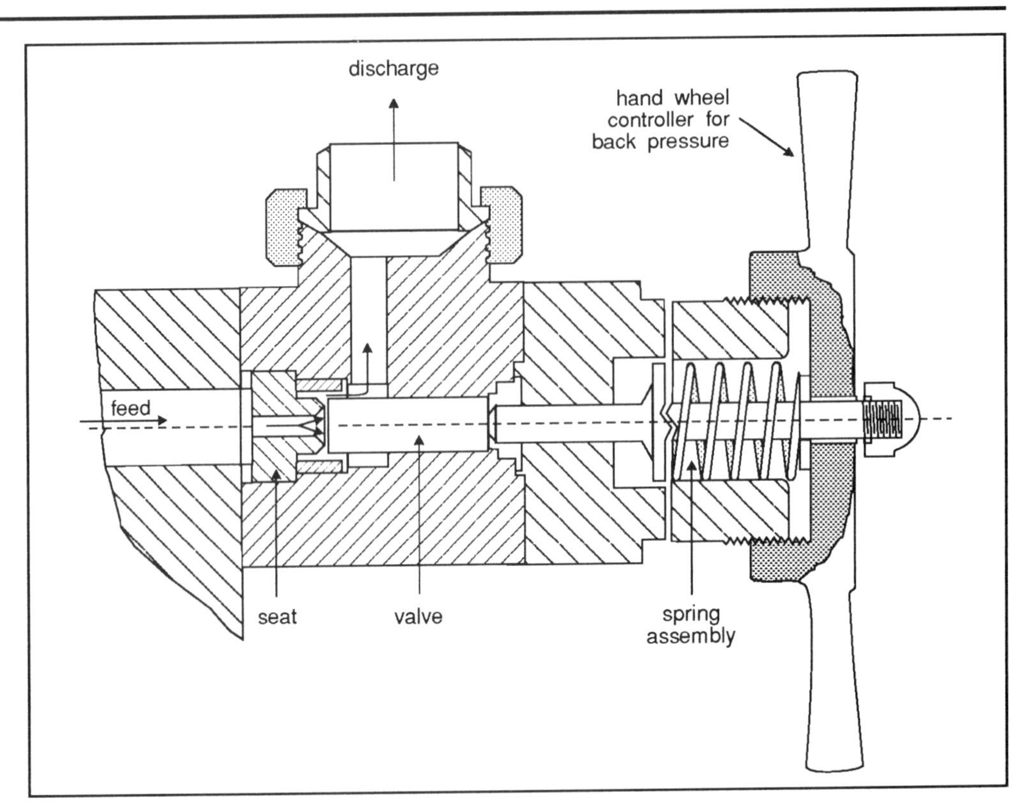

Figure 3.4 Homogeniser discharge valve.

biological and physical factors influence disruption

Different types of high efficiency discharge valves have been developed (see Figure 3.5). Valve and seat are both subject to abrasion. Abrasion resistant materials must be used, such as stellate and tungsten carbide.

What exactly happens during passage through the valve is unknown. It is expected that cavitation, which causes very high pressures locally, plays an important role in the disruption of cells.

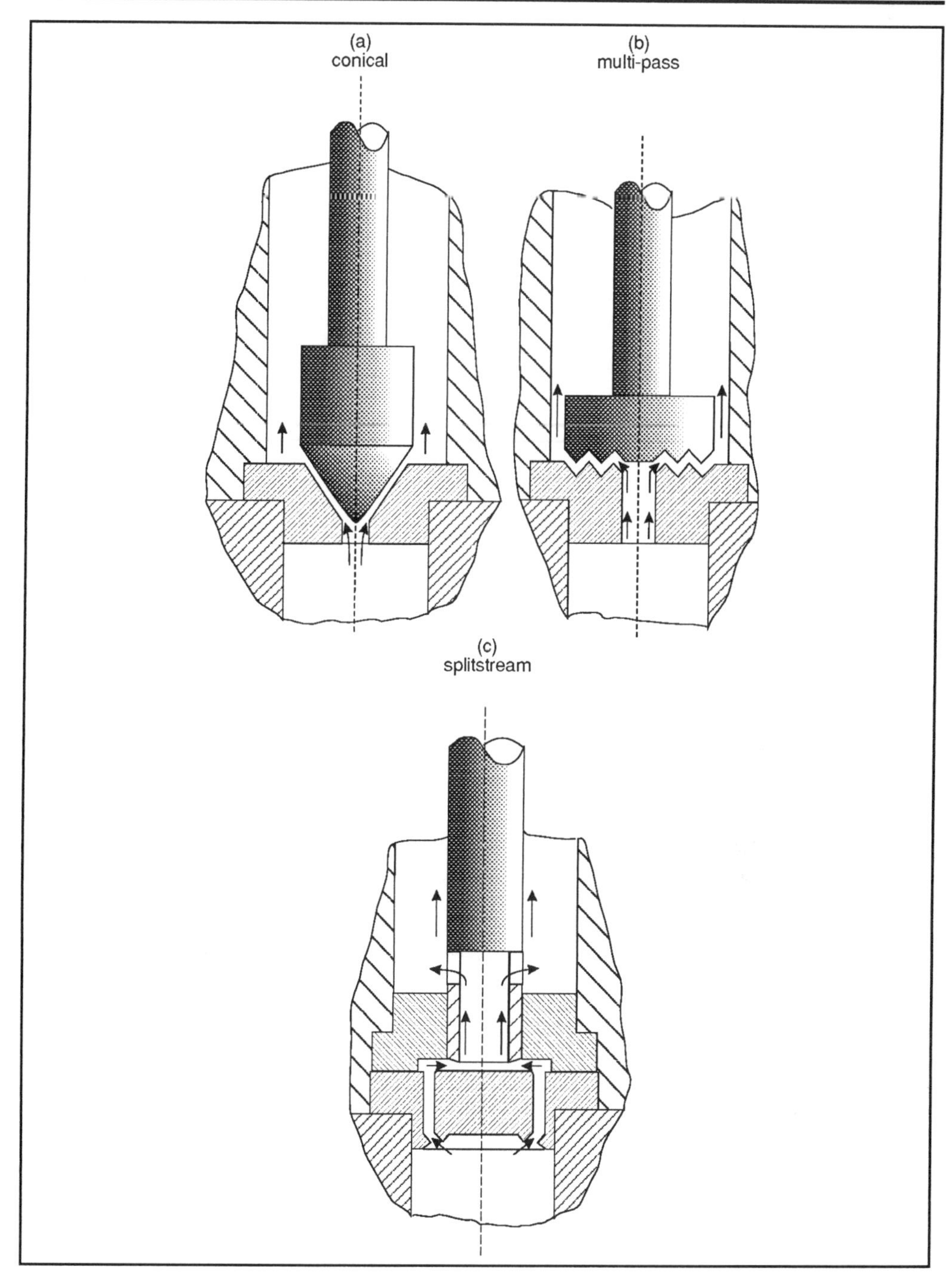

Figure 3.5 Discharge valves and seats.

Π Before reading on make a list of factors which you believe will influence the efficiency of the homogeniser for disrupting cells. Then compare it with our list given below (your list should be similar, but not identical, to the one we generated in discussing milling).

The optimal conditions for cell disruption depend on:

1) micro-organisms used:
   - cell wall thickness and composition;
   - size.
2) location of the product:
   - in cytoplasm;
   - in cell organelle;
   - in periplasmic space.
3) type of homogeniser:
   - pressure;
   - type of valve and seat;
   - temperature (rise);
   - number of passages.

As for the bead mill, a general relation between the optimal conditions and the product release is unknown and must be determined experimentally.

3.2.5 Cell disruption kinetics for a homogeniser

Under constant conditions the rate of cell release can be described by a first order kinetics.

Similar to Equation 3.2 we may write:

$$\ln \frac{C_r^{max}}{C_r^{max} - C_r} = K_h f(\Delta p) N \qquad \text{(E - 3.6)}$$

where: f is a function of the pressure difference (Δp). N = number of passages.

Thus:

$$C_r = C_r^{max} \left(1 - \exp\left(-K_h\ f(\Delta p) N\right)\right) \qquad \text{(E - 3.7)}$$

The exposure time t in Equation 3.2 is replaced by the number of passages through the homogeniser (N). Generally speaking $f(\Delta p)$ is unknown and will have to be determined experimentally, but can in many cases be approximated by:

$$f(\Delta p) = \Delta p^{\beta}$$

For baker's yeast the exponent (β) has a value of 2.9, so for baker's yeast the following equation hold:

$$C_r = C_r^{max} (1 - \exp(-K_h \Delta p^{2.9} N) \quad (E - 3.8)$$

Note Δp is expressed in bar (not in Pa) in this relationship.

Figure 3.6 shows the relationship according to Equation 3.8 for baker's yeast.

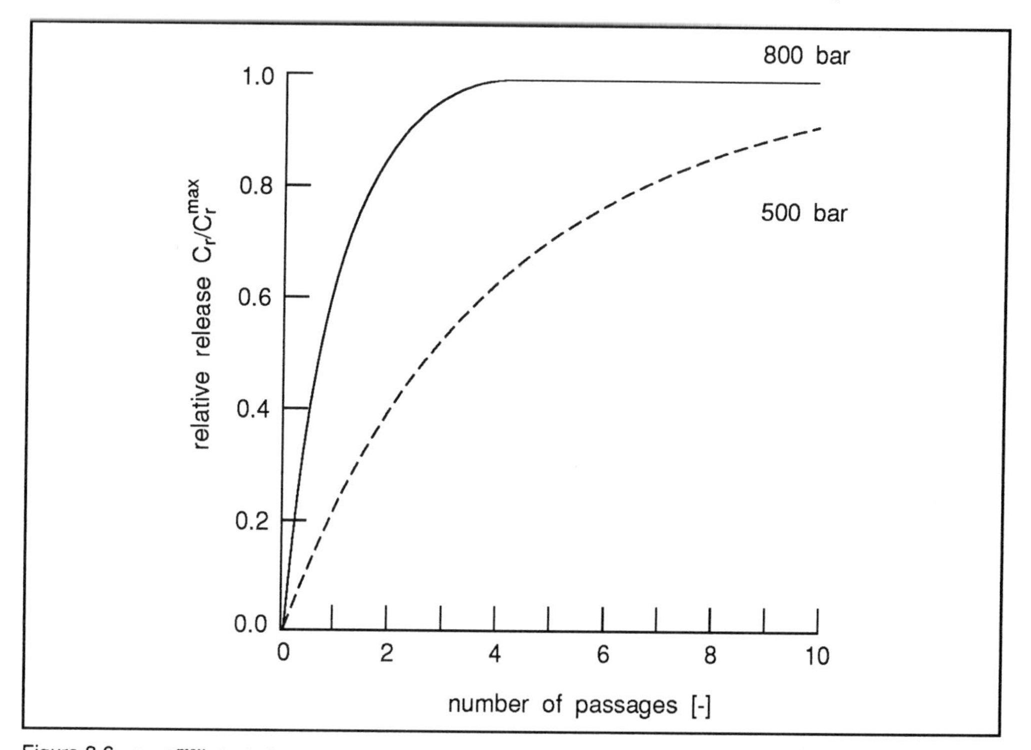

Figure 3.6 C_r/C_r^{max} for baker's yeast as a function of the number of passages at different pressure differences in an homogeniser (the relationship is as described by Equation 3.8). Temperature 5°C.

increasing temperature improves efficiency

From this graph and Equation 3.8 we may conclude that a single passage is not sufficient to liberate the product from the yeast cells. Even at high pressures two or more passages are necessary. Temperature also influences the efficiency of homogenisation since higher temperatures reduce the broth viscosity. Homogenising at a constant pressure at increasing temperatures also has a positive effect on the release. However, temperatures are usually kept low within certain limits because of heat denaturation of the product.

Π Explain, in your own words how an increase in pressure difference influences the performance of a homogeniser.

We can see from Figure 3.6 that an increase in pressure difference increases both the rate of release of product and the final amount of product recovered.

3.2.6 Scale up of homogenisers

Basically the scale-up of a homogeniser is simple. Simply all we need to do is to keep all variables constant and install a bigger plunger pump and discharge valve!

Here again nearly all the power input of the homogeniser will be dissipated into heat. The temperature rise of the broth can be predicted as follows:

The power input P is proportional to the homogenisation pressure Δp and the volumetric flow processed, so:

$$P = \phi_v \Delta p = \phi_v \rho c_p \Delta \theta$$

where ϕ_v = volumetric flow rate ($m^3 s^{-1}$);

ρ = density (kgm^{-3});

c_P = specific heat ($J\ kg^{-1}\ K^{-1}$);

$\Delta\theta$ = temperature difference (K)

$$\text{or } \Delta\theta = \frac{\Delta p}{\rho c_p} \qquad \text{(E - 3.9)}$$

Note that the same relationship holds for the bead mill.

temperature rise not a function of scale

As can be seen the temperature rise is only a function of the applied pressure and the broth properties, not a function of scale!

Π What does each of the following symbols represent?

K_b, N, C_r, C_r^{max}, c_p and β

Now check your response by referring to Equations 3.1, 3.6 and 3.9.

Π Let us do a calculation using the relationship established in Equation 3.9. If the homogenising pressure is 700 bar ($7 . 10^7\ Nm^{-2}$) and the broth density is 1100 kg m^{-3} and its specific heat is 4000 J $Kg^{-1}K^{-1}$, what is the temperature rise?

$$\text{Since } \Delta\theta = \frac{\Delta p}{\rho\ c_p} \text{ then } \Delta\theta = \frac{7 . 10^7}{4000 \times 1100} = 15.9 \text{ K (or °C)}$$

SAQ 3.2

Complete the following statements concerning homogenisers (insert or select as appropriate).

1) Increasing temperature increases/decreases broth viscosity and has a positive/negative effect on cell disruption.

2) $\frac{\Delta p}{\rho c_p}$ equals [ ].

3) Power input is [ ] to the homogenisation pressure.

4) $(C_r^{max} - C_r)$ represents the concentration of [ ] product.

5) The [ ] and cell wall [ ] and composition of a micro-organism influences cell disruption.

3.3 Non-mechanical disruption

cell lysis frequently used in large scale production

The non-mechanical disruption mostly designated as lysis is frequently used in large scale production because it does not require additional or expensive equipment.

3.3.1 Heatshock

thermolysis, easy and cheap

Heatshock or thermolysis is very popular because it is relatively easy and cheap. It can however only be used if the product is stable to the heatshock treatment. It inactivates the organisms while the product is still active. A temperature rise disturbs the cell wall.

The effectiveness of this disturbance is a function of:

- pH;
- ionic strength;
- presence of sequestering agents eg EDTA, a chelating agent which binds certain ions, such as Mg^{2+}, which tend to stabilise the cell wall;
- contaminants (eg proteases).

Product yields of up to 90% can be obtained, but 60-80% yields are more common.

3.3.2 Chemical cell lysis

Addition to the fermentation broth of detergents (such as cationic, anionic and non-ionics) and solvents (such as octanol and acetone) will partly solubilise the cell walls. Frequently a non-ionic detergent (Triton X-100) is used. To enhance the efficiency guanidine-HCI may be added. A typical dosage is guanidine HCI 2 mol l^{-1} - with 2% Triton X-100.

Π Can you think of two disadvantages to chemical cell lysis in product recovery?

We believe the two most important disadvantages are:

- the chemical agent may adversely affect the product;
- the chemical agent may affect downstream processing of product and thus the product yield.

3.3.3 Enzymatic cell lysis

Cell lytic enzymes have the ability to dissolve or degrade microbial cell walls. Lysozyme is an example of an enzyme that has the ability to hydrolyse murein in Gram negative and positive bacteria.

A pretreatment of EDTA will enhance the effectiveness of lysozyme. Glucanase and mannase, often in combination with proteases, are used for the degradation of yeast cell walls. For plant cells, cellulase and pectinase can be used.

Most cell wall components are degradable by enzymes but it is not always feasible to use this method.

efficiency influenced by temperature and pH

The enzyme activity largely depends on the temperature and pH, and in many cases metal ions will be required to enhance their activity or specificity. Enzymes can be inhibited by a number of medium components. The optimal temperature and pH of the cell wall degrading enzymes hardly ever coincide with the optimal broth conditions.

In nearly all cases an overdose of enzyme is required, which pushes up the costs. On the other hand, cell wall degradation has several advantages:

- low energy consumption
- specific reaction
- small risk of product damage
- harmless to the environment

SAQ 3.3

Rearrange Equation 3.6 to obtain an expression for Δp^{β}.

Now use the values presented below to determine the value of β. To do this we suggest you plot a graph of $\log \Delta p^{\beta}$ against β for different values of β.

$C_r = 0.051$

$C_r^{max} = 0.063$

$N = 3$

$\Delta p = 600$ bar

$K_h = 5.5 \ . \ 10^{-9} \ (Nm^{-2})^{-n}$

SAQ 3.4

Identify each of the following statements as TRUE or FALSE. If FALSE give a reason for your response.

1) Scale-up of homogenisers is limited by an increase in temperature.

2) Glucanase is an enzyme that degrades the murein layer in bacterial cell walls.

3) EDTA enhances the effectiveness of lysozyme.

4) Cellulase and pectinase enzymatically digests cell walls of yeast.

5) Scale-up of homogenisers is achieved by increasing pressure by installation of a bigger plunger pump and discharge valve.

Summary and objectives

Mechanical and non-mechanical methods for cell disruption are used on a small as well as a large scale. Non-mechanical methods do not require expensive additional equipment but addition of chemicals may have a negative effect on the performance of the separation equipment. Mechanical methods are relatively expensive and the equipment used is subject to heavy wear which results in high maintenance costs.

The kinetics of cell disruption can adequately be described by a first order approximation.

Scale-up for bead mills is limited, but for homogenisers there are relatively no limits.

Non-mechanic disruption is frequently used in large scale production because it does not require additional or expensive equipment.

Now that you have completed this chapter you should be able to:

- describe the principles of operation for bead mills and homogenisers;
- list factors influencing the performance of bead mills and homogenisers;
- determine parameters using equations describing the kinetics of product release;
- list merits and limitations of various mechanical and non-mechanical methods of disruption;
- explain how scale-up is achieved for mechanical methods of cell disruption.

Solid-liquid separation

Solid-liquid separation

4.1 Introduction

In many cases the first step in biotechnological processes is a solid-liquid separation. However, further downstream solid-liquid separation can also occur, as in the case of separating enzyme precipitates or antibiotic crystals from the mother liquor.

Solid-liquid separations are usually mechanical separation methods and are applied to heterogeneous mixtures, not to homogeneous mixtures. They are based on physical differences between the suspended solids (particles) such as differences in size, shape and density. To separate suspended solids from their liquid two principles are used:

- retaining the solid component by using a sieve, screen or filter medium;
- utilising the difference in rate of sedimentation of particles moving through a liquid. This can be done under gravity (settler) or an induced field, for example by using a centrifugal field (centrifuge).

In this chapter we consider both of these principles, firstly filtration then sedimentation and centrifugation.

4.2 Filtration

cake and filtrate

Filtration is a technique to separate a solid-liquid suspension (eg a fermentation broth) into a concentrated (cake) and a diluted (filtrate) part. The filter medium is the separating agent. The principle of filtration is based on the difference in geometry of the suspended solids and the geometry of the pores in a filter medium, see Figure 4.1.

In fact filtration defined in this way is a clarification process. Particles are retained depending on their size. In bio-filtration however nearly all the particles are retained. In this section we will restrict ourselves to broth filtration. According to Figure 4.1 this type of filtration is called cake filtration, because the cells themselves also act as a filter medium.

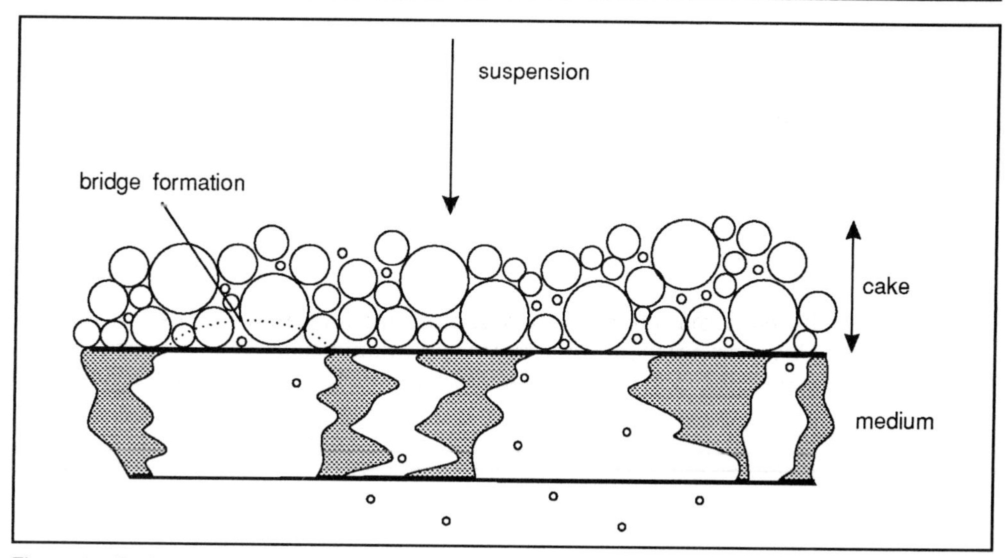

Figure 4.1 Basic principle of cake filtration. Large particles are retained; the very small particles are not, aperture 3-7 times particle size.

4.2.1 Filter media

In cake filtration the operational function of the filter medium is limited to the first few particle layers of the cake. The aperture of the filter medium used is approximately 3 to 7 of the size of the cells. Because of bridge formation, particles will not pass and a cake will be formed. The filter medium itself can be woven (filter cloth) or non woven (filter plates and felts). Important parameters for selection are:

- cleaning procedures/fouling behaviour;
- chemical resistance;
- resistance to flow;
- durability.

woven and non woven media

Because the filter medium is the heart of any filter it is essential to be familiar with the various types of filter medium available. Unfortunately there is no proper filter medium classification method available. So for typical data we will have to rely on experimental data and suppliers information.

Some important media are listed in Table 4.1

Type	Example
Metal sheets	perforated sintered metal, woven wire
Woven fabrics	cloth, natural and synthetic fabrics
Non woven sheets	paper (cellulose), glass wool
Ceramics	silica, alumina
Plastic sheets	synthetic membranes

Table 4.1 Types of filter media.

4.2.2 Types of filters used

Many types of filters are in the market. Because fermentation broths contain 10-40% solids by volume and particles with sizes of 0.5 - 10 μm, only a limited number of filter types can be used. The equipment most frequently used are rotary drum vacuum filters and (membrane) filter presses.

Vacuum filter

vacuum filters commonly used for yeast and fungi

Rotary drum vacuum filters are mostly employed in the filtration of yeast cells and fungi. Figure 4.2 shows a commonly applied rotary drum filter for the filtration of baker's yeast. The drum rotates with its circumference dipping into a tank of broth from which the filtrate is drawn and the cake is deposited on the drum surface. As the drum rotates the cake becomes drier and will be discharged. The drum rotates at low speeds, usually in the range from 0.25 - 5 rpm.

There are several types of cake discharge:

- knife discharge;
- string discharge;
- belt discharge;
- roller discharge.

The choice of one particular type of discharge depends on the behaviour of the filter cake. Dry cakes can be removed from the cloth by a roller or a string. Sticky cakes require a scraper (knife).

Figure 4.2 Rotary drum vacuum filter for the filtration of baker's yeast with knife discharge. By permission of Gist-brocades, Delft, The Netherlands.

If necessary the cake may be washed by spraying water (solvent) onto the surface of the cake.

Mechanical dewatering of the cake as well as cloth washing can be introduced into the basic filtration cycle.

Rotary drum vacuum filters are available ranging from 2 to 80 m^2. A schematic drawing of a rotary drum vacuum filter is shown in Figure 4.3.

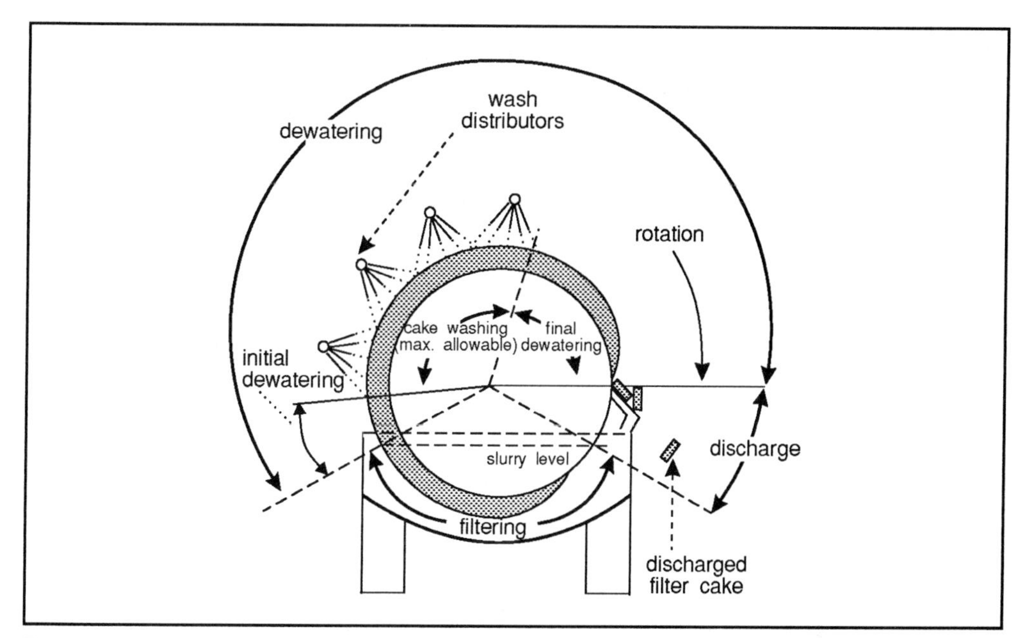

Figure 4.3 Schematic drawing of a rotary drum vacuum filter.

(Membrane) filter press

filter press - a series of cloth walled chambers

The basic unit of a filter press is a sequence of perforated plates alternating with hollow frames mounted on suitable supports. Each face of every plate is covered with a filter medium (cloth) to create a series of cloth walled chambers into which slurry can be forced under pressure. The plates and frames are held together by pressing with hydraulic or screw rams. The solids are retained within the chambers while the filtrate discharges into hollows on the plate surface, and hence to drain points.

Figure 4.4 shows a plate and frame filter assembly. At the end of the filtration cycle, the cake has to be discharged. This is done by releasing the hydraulic pressure. The plates come apart so that the cake can be removed from the cloth (manually).

rubber membranes can improve performance

Washing of the cake can in some cases be facilitated by including washing plates in the filter assembly. In practice the slurry inlet is frequently used as an inlet for the washing liquor (water).

In membrane filter presses a rubber membrane is moulded on the cake chamber. Compressed air or pressurised water is used as a force to inflate the membrane and compact the cake *in situ*.

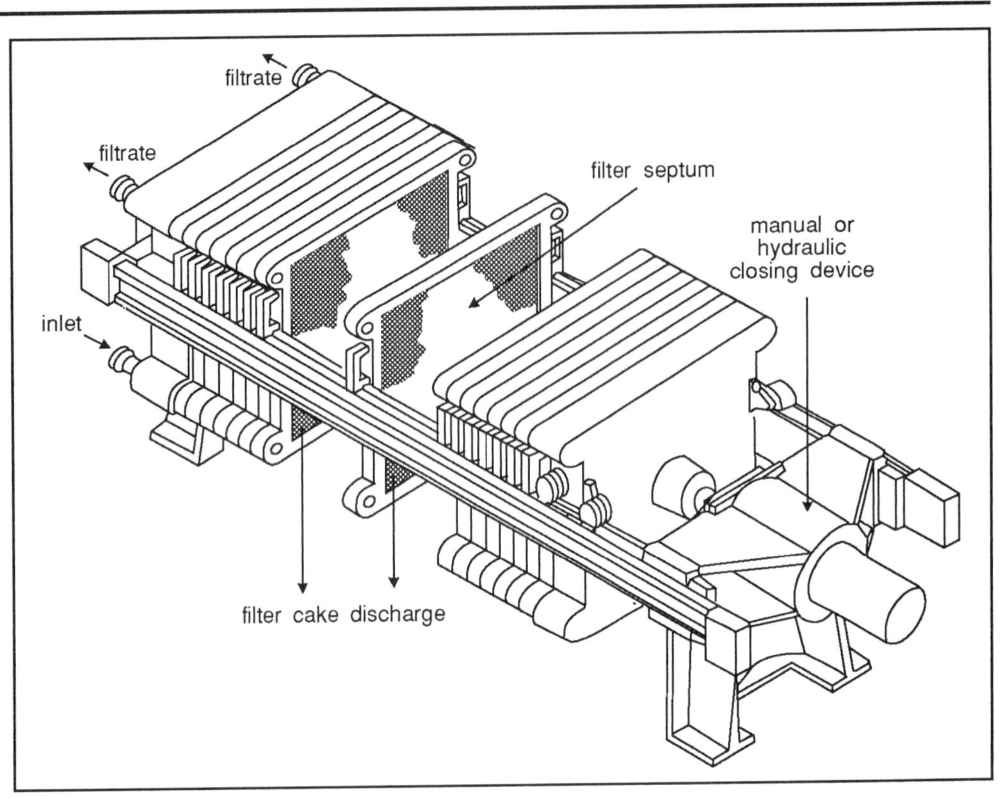

Figure 4.4 Plate and frame filter assembly

The advantages of membrane filter presses are:

- higher yield;
- drier cake;
- easy cake removal.

The disadvantage is higher capital investment.

SAQ 4.1

1) Give two advantages that membrane filter presses have over non-membrane filter presses.

2) List four types of filter media and give an example of each type.

3) List three parameters that influence the choice of a filter medium.

4) List three different mechanisms of cake discharge in vacuum filtration.

5) Name the three main stages of vacuum filtration.

4.2.3 Filtration fundamentals for pressure filtration

Darcy's law

A filter cake can be considered a porous medium. The flow through a porous medium is governed by Darcy's law. Applying it to a system of cake and filter medium, we obtain:

Darcy's law: $\Delta p = \eta\,(r_c + r_m)\,\phi_V''$ (E - 4.1)

where Δp = pressure difference (Nm^{-2});
η = dynamic viscosity (Pas);
r_c = cake resistance (m^{-1});
r_m = medium resistance (m^{-1})
ϕ_V'' = volumetric flow rate per unit area ($m^3\,s^{-1}\,m^{-2} = ms^{-1}$)

The volumetric flow rate per unit area ϕ_V'' is defined as follows:

$$\phi_V'' = \frac{dV}{dt}\frac{1}{A} \qquad \text{(E - 4.2)}$$

where A = filtration area (m^2);
V = volume of filtrate (m^3).

The resistance in the cake is assumed to be directly proportional to the amount of cake deposited. This is only true for incompressible cakes! It follows that:

$$r_c = \alpha w \qquad \text{(E - 4.3)}$$

where w = mass of cake per unit area ($kg\,m^{-2}$).

α is the specific cake resistance ($m\,kg^{-1}$)

Substitution and rearrangement of E - 4.1, 4.2 and 4.3 gives:

$$\frac{dV}{dt} = \frac{\Delta p A}{\alpha \eta w + \eta r_m} \qquad \text{(E - 4.4)}$$

This Equation shows that the mass of cake deposited per unit area (w) is a function of time in batch operation. However, as you might expect it is also related to the concentration of solids (cells) in the broth:

$$wA = CV, \qquad \text{(E - 4.5)}$$

where C is the concentration of solids in the broth ($kg\,m^{-3}$).

Substitution of Equation 4.5 into Equation 4.4 and rearranging gives:

$$\frac{dt}{dV} = \frac{\alpha \eta C\,V}{A^2 \Delta p} + \frac{\eta r_m}{A \Delta p} \qquad \text{(E - 4.6)}$$

w is a function of time and solids concentration

This differential equation can be solved for either constant pressure and constant rate filtration. Because the first mode is much more common we will solve Equation 4.6 for constant pressure. By integration we obtain:

$$\int_o^t dt = \frac{\alpha \eta C}{A^2 \Delta p}\int_o^V V dV + \frac{\eta r_m}{A \Delta p}\int_o^V dV \qquad \text{(E - 4.7)}$$

or:

$$t = \frac{\alpha \eta C}{2A^2 \Delta p} V^2 + \frac{\eta r_m}{A \Delta p} V \quad \text{(E - 4.8)}$$

for simplicity we define two constants:

$$a = \frac{\alpha \eta C}{2A^2 \Delta p} \text{ and } b = \frac{\eta r_m}{A \Delta p} \quad \text{(E - 4.9)}$$

Substitution of Equation 4.9 into Equation 4.8 gives:

$$\frac{t}{V} = aV + b \quad \text{(E - 4.10)}$$

which gives a straight line if $\frac{t}{V}$ is plotted as a function of V (see Figure 4.5). This is a useful relationship to remember.

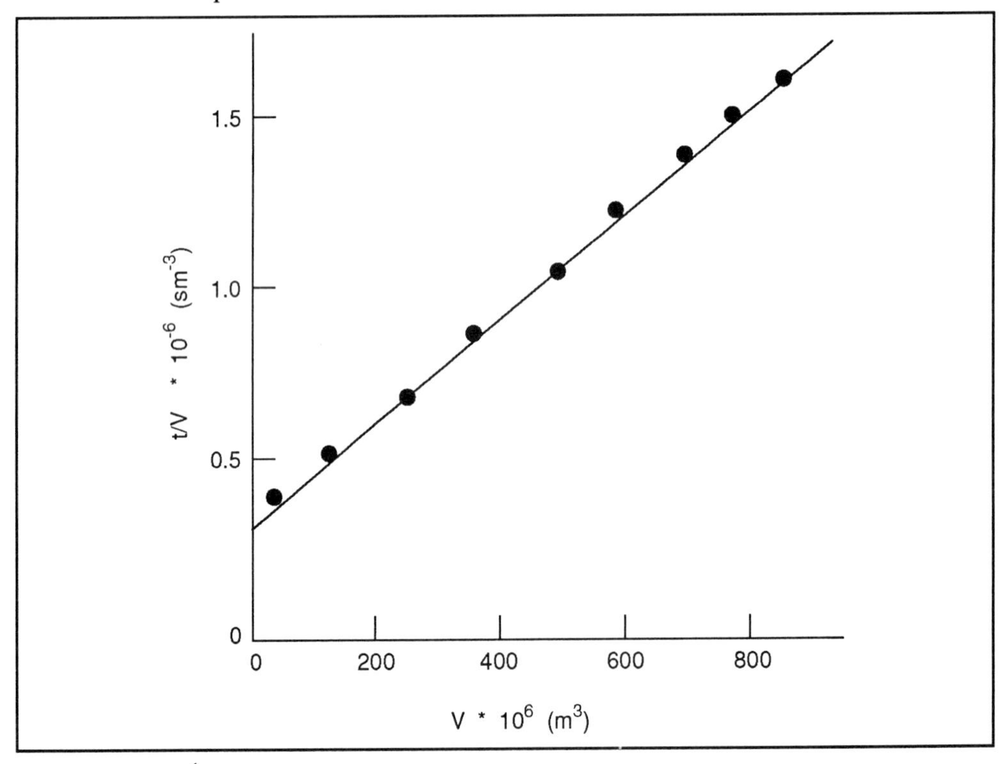

Figure 4.5 Plot of $\frac{t}{V}$ versus V for constant pressure filitration of yeast (*Saccharomyces carlbergensis*).
Dry solid concentration = 0.5%, Δp = 34 kPa.
Note that the slope = $\frac{\alpha \eta C}{2A^2 \Delta p}$ and the intercept = $\frac{\eta r_m}{A \Delta p}$ (see text for details).

The slope of the line in Figure 4.5, a, is a measure for the specific cake resistance. The intercept b is a measure for the medium (cloth) resistance.

determination of a and r_m

The specific cake resistance α and the medium resistance r_m can be evaluated from a simple laboratory scale test.

Π Determine the specific cake resistance (α), cake resistance (r_c) and medium resistance (r_m) from Figure 4.5 (filter diameter 7.5cm (3"), viscosity of the filtrate 1.2 mPas, solid concentration 5 kg/m^3). What can you conclude from these values?

The slope of the curve is:

$$a = \frac{0.90 \,.\, 10^6}{600 \,.\, 10^{-6}} = 1.50\; 10^9 \;sm^{-6}$$

Applying Equations 4.9 and 4.10 and rearranging gives:

$$\alpha = \frac{a \,.\, 2A^2\Delta p}{\eta C}$$

Substitution gives:

$$\alpha = \frac{1.50 \,.\, 10^9 \,.\, 2\left[\frac{\pi}{4}(7.5 \,.\, 10^{-2})^2\right]^2 34 \,.\, 10^3}{1.2 \,.\, 10^{-3} \,.\, 5}$$

$$\alpha = 3.31 \,.\, 10^{11}\;\frac{m}{kg}$$

The medium resistance:

Intercept from Figure 4.5: $b = 0.32 \,.\, 10^6$

from Equation 4.9, it follows that:

$$r_m = \frac{b \,.\, A \,.\, \Delta p}{\eta}$$

Substitution gives:

$$r_m = \frac{0.32 \,.\, 10^6 \,.\, \frac{\pi}{4}(7.5 \,.\, 10^{-2})^2 \,.\, 34 \,.\, 10^3}{1.2 \,.\, 10^{-3}}$$

$$r_m = 4 \,.\, 10^{10}\;m^{-1}$$

To compare this value with the specific cake resistance they should have the same units. From Equation 4.3 it follows that:

$$r_c = \alpha \,.\, w = \frac{m}{kg}\,\frac{kg}{m^2} = m^{-1}$$

ie the same unit as r_m.

From Equation 4.5 it follows that when 800×10^{-6} m^3 of filtrate have been collected:

$$w = \frac{CV}{A} = \frac{5 . 800 . 10^{-6}}{\frac{\pi}{4} (7.5 . 10^{-2})^2} = 0.9 \frac{kg}{m^2}$$

so $r_c = \alpha w = 0.9 * 3.3 * 10^{11} = 3.0 * 10^{11} \ m^{-1}$

We can conclude that compared to the cake resistance, the medium resistance is negligible.

Typical values for α: $10^{12} < \alpha < 10^{15}$ ($m \ kg^{-1}$) and typical values for r_m: $10^8 < r_m < 10^{11}$ (m^{-1})

It is usual that, compared to the cake resistance, the medium resistance r_m is negligible in broth filtration. For crystal filtration (size 50 > 500 μm) the medium resistance is high compared to the cake resistance.

4.2.4 Factors affecting the specific cake resistance

Various factors affect the specific cake resistance in broth filtration:

- type of organism;
- particle size/distribution;
- pH of the broth;
- fermentation time;
- broth temperature.

In practice, the effect of all these factors on specific cake resistance α will have to be evaluated experimentally.

Figure 4.6 shows the effect of fermentation time after inoculation on the specific cake resistance for *Penicillium chrysogenum*. Changes by a factor of 10 or more are possible. The explanation for this phenomenon is that the mycelium will be more fragmented at prolonged fermentation times.

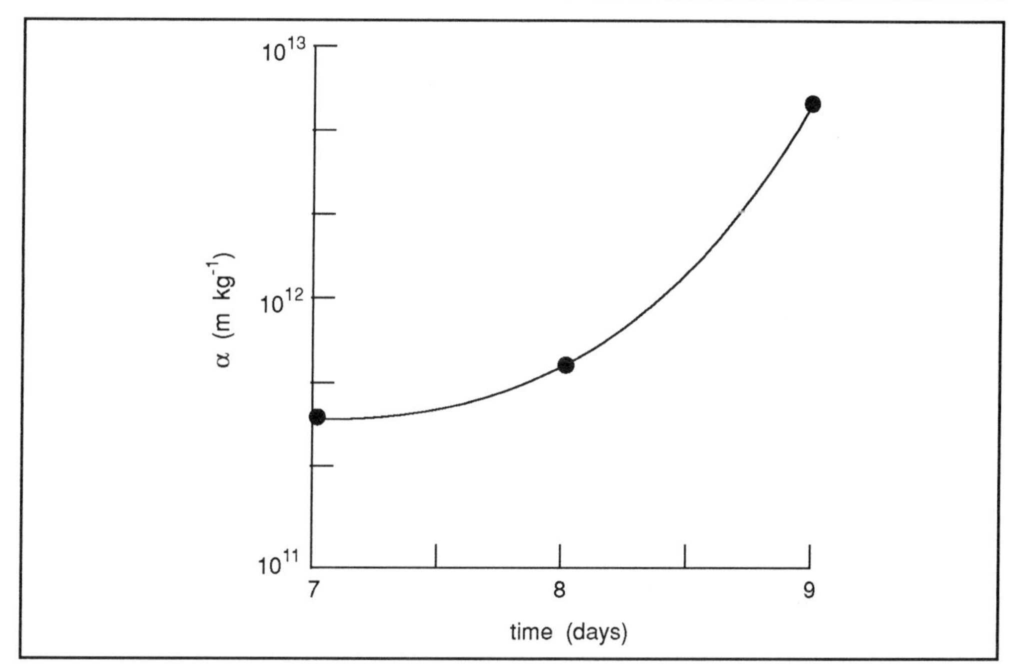

Figure 4.6 Effect of fermentation time on the specific cake resistance of a *Penicillium chrysogenum* fermented broth.

SAQ 4.2

Complete the following statement or circle the correct responses as appropriate:

1) For compressible/incompressible cakes, the resistance in the cake is assumed to be inversally / directly proportional to the amount of cake deposited.

2) The mass of cake deposited per unit area is a function of [ ] in batch operation and concentration of [ ] in the broth.

3) For constant pressure filtration, the slope of a $\frac{t}{V}$ versus V plot is a measure of [ ] and the intercept a measure of the [ ].

4) Compared to the medium resistance, the cake resistance is usually negligible for broth filtration (TRUE/FALSE).

5) For constant pressure filtration, a plot of α versus fermentation time always gives a straight line (TRUE/FALSE).

4.2.5 Broth treatment

Pretreatment

pretreatment - bodyfeed filtration

If the specific cake resistance is too high, $\alpha > 10^{14}$ m kg^{-1}, the broth can be pretreated in several ways in order to reduce the resistance. In biotechnology two types of pretreatment are common:

- addition of filter aid (bodyfeed);
- addition of flocculant.

In bodyfeed filtration the broth is mixed with 0.5-5% (w/w) filter aid to form a permeable, less compressible cake during filtration.

Filter aids are incompressible discrete particles of high permeability. The particle size ranges from 2 to 20 μm for the principal types of filter aid. Filter aids should be inert to the broth being filtered. Diatomite, Perlite and inactive carbon are most frequently used. On rare occasions starch, woodpulp and cellulose are used as well.

Figure 4.7 shows the effect of the addition of filter aid on the specific cake resistance.

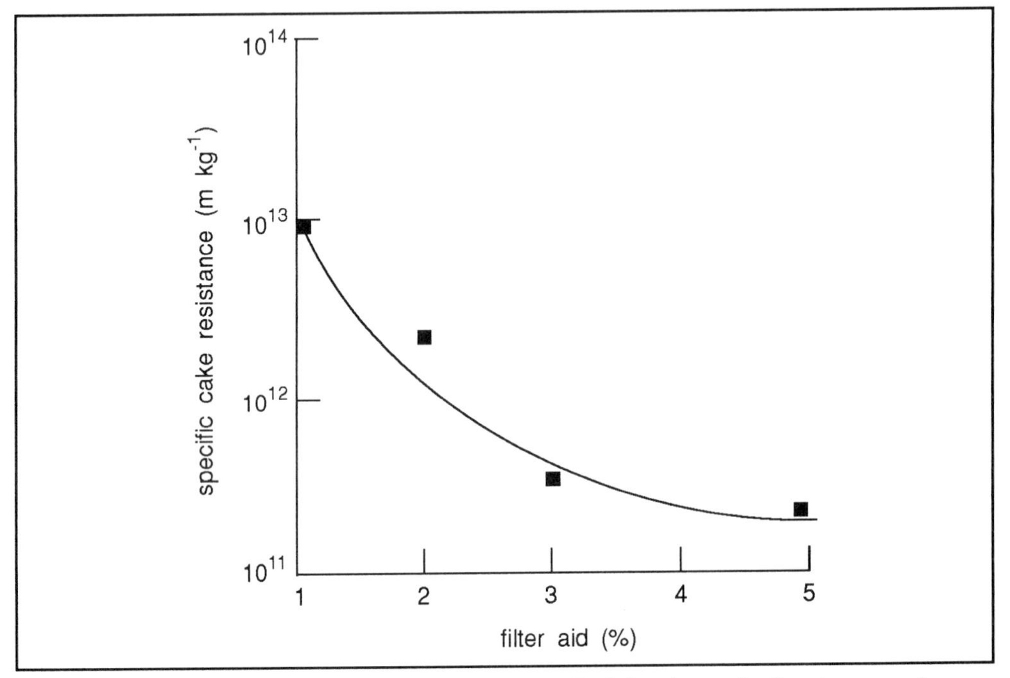

Figure 4.7 Specific cake resistance as a function of filter aid (Radiolite) dosage for *Streptomyces griseus*.

Π Determine the effect of the addition of 2% filter aid on the volumetric flow rate, using the data shown in Figure 4.7. (Make a comparison between the rates with 1% and 2% filter aid). Begin by using Equation 4.8.

If we neglect the medium resistance (ie $r_m = 0$) the filtrate volume can be calculated by rewriting Equation 4.8. So:

$$V = \left(\frac{t \cdot 2A^2 \Delta p}{\alpha \eta C}\right)^{0.5}$$

We assume A, η C and Δp as constant. Adding 2% filter aid instead of 1%, the specific cake resistance decreases from approx 10^{13} m kg^{-1} to 10^{12} m kg^{-1}. In the same time more filtrate passes the filter according to:

$$V_2 = V_1 \left(\frac{\alpha_1}{\alpha_2}\right)^{\frac{1}{2}}$$

Substituting gives:

$$V_2 = V_1 \cdot 10^{\frac{1}{2}} = 3.2$$

Thus, in the same time, 3.2 V, of filtrate will be collected and the volumetric flow rate will have been increased by 3.2 times.

pretreatment - addition of flocculants

Flocculants (electrolytes such as aluminum sulphate) and polyelectrolites (such as polyacrylic and polyamine) are applied in a typical dose of 0.1 - 2%. Because cells have a net negative charge at neutral pH, cationic flocculants (such as polyamine) can be used.

Flocculants are not however very popular in processes in which membranes are used because severe fouling of the membranes usually occurs.

Post treatment

post treatment

Post treatment processes can also be used. These are commonly related to the cake's washing and drying. To enhance the product yield it may be necessary to remove the residual filtrate. Disposal of a product free filtercake can also be a reason for washing.

In practice product removal will be accomplished by displacement with a washing liquid (water). After displacement the remaining washing liquid in the cake can be displaced by air. The cake can then be dried *in situ* with hot air.

4.2.6 Scale up of filter presses

Scale up of filter presses in biotechnology requires substantial experience.

In cake filtration two limitations exist in the scale up of filter presses:

- cake/medium resistance;
- cake volume;

cake/medium resistance.

In the case of cake or medium resistance limited filtration, the constants a and b can be determined based on Equation 4.9.

The next step is the prediction of the filter performance on pilot plant scale using Equation 4.10. Assuming that all the parameters remain constant only the area (A) will be scaled up with a factor of 15-150. Based on these results a further scale up factor of 10-1000 can be achieved. Typical filter sizes are:

Laboratory	120	cm^2
Pilot plant	1,600	cm^2 (per plate)
Production scale	14,400	cm^2 (per plate)

cake volume

The cake volume can also serve as a scale up criterion because all of the cell volume (plus bodyfeed) will have to be collected in a limited number of cake chambers in the filter press.

The cake volume obtained from a laboratory test can serve as an indication for the number of cake chambers and/or the number of filter cycles to be used.

4.3 Sedimentation and centrifugation

Sedimentation and centrifugation are two of the most widely used unit operations in biotechnology. As we have seen, filtration can be described mathematically which can be helpful in scale up. We will use a similar approach for centrifugation.

gravity sedimentation

When a solid particle moves through a viscous medium under the influence of the Earth's gravity it will attain a constant falling velocity after a certain time. This phenomenon is known as sedimentation. Gravity sedimentation plays an important role in biological waste water treatment. Its application in industrial processes for the recovery of bioproducts is however limited, the main reason for this being the limited driving force in a gravity field. Gravity sedimentation requires large-scale equipment and the residence times are unacceptably long.

centrifugation - advantages and disadvantages

In a centrifuge the gravity force field is replaced by a centrifugal force field which allows for smaller equipment and consequently shorter residence times (thus higher capacities per unit volume of equipment).

Centrifuges offer the following advantages over filters and sedimentation tanks:

- continuous processing of large batches;
- large capacity and low hold up of the machine (short residence time);
- sterile product handling (steam sterilisation);
- production under containment conditions.

There are however also disadvantages:

- high capital investment;
- high maintenance costs;
- relatively high power consumption;
- concentrate is a slurry (5 - 20% (w/w) dry solids);
- supernatant is not free of cells (10^3 - 10^5 cells ml^{-1}).

Π It may be helpful for you to draw up a balance sheet of these advantages and disadvantages. Use a table like this.

Centrifugation	
Advantages	Disadvantages

4.3.1 Types of centrifuges

The different types of centrifuges employed in biotechnology are shown in Figure 4.8.

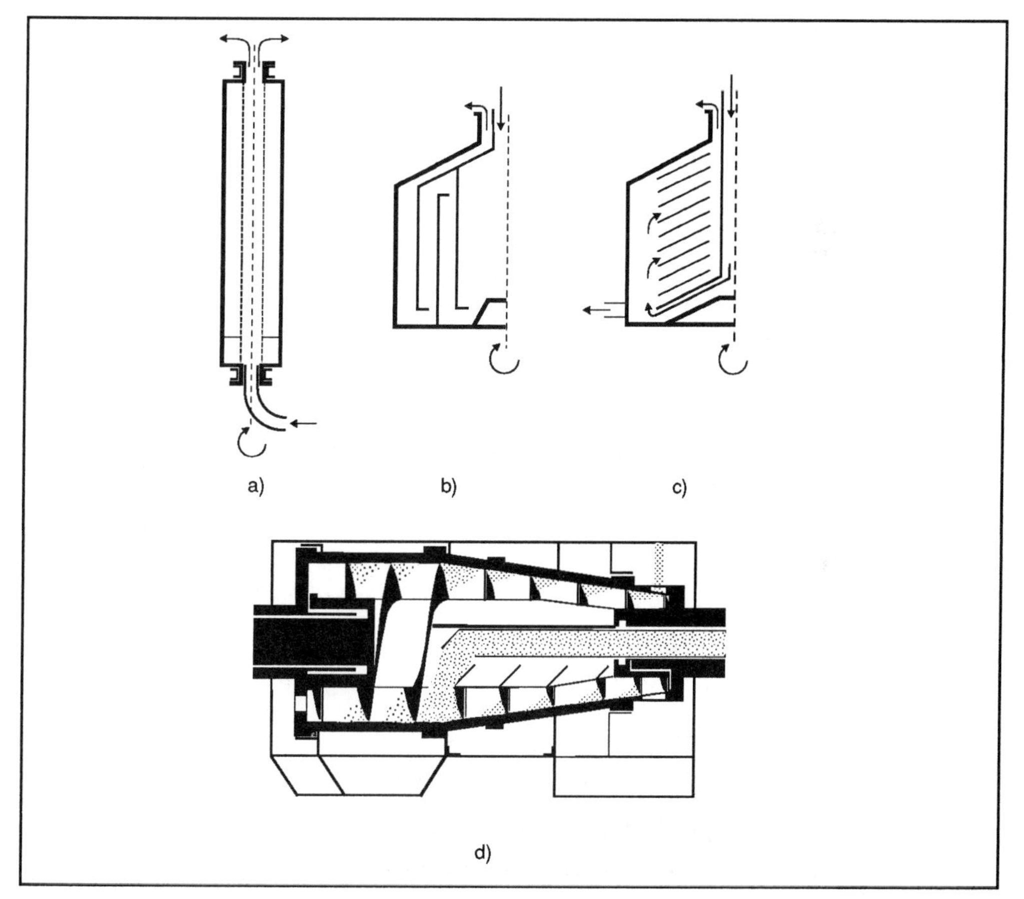

Figure 4.8 Sedimentation centrifuges a) tubular centrifuge, b) multichamber, c) disk stack centrifuge, d) decanter centrifuge (see text for description).

tubular bowl type

The simplest sedimentation centrifuge is the tubular bowl type (Type a in Figure 4.8). Owing to its slender shape and small volume a very high centrifugal force can be applied. This type of centrifuge is commonly used in pilot plants. The solids have to be discharged manually while the feed (broth) and the clarified broth can be batchwise or continuous.

multichamber bowl type

The multichamber bowl centrifuge is derived from the tubular type. It contains a number of concentric tubes connected in such a way that a zigzag flow of the culture broth through the chamber is achieved (Figure 4.8b). The centrifugal force is increasing outwards, so that the smallest particles will be deposited on the outermost chamber wall. Also here the solid discharge has to be done manually and it is a laborious operation. Multichamber bowl centrifuges are frequently employed in the fractionation of human blood plasma.

disc stack type

The disc stack centrifuge is most widely used in biotechnology. Here a stack of hollow truncated cones is used in order to increase the clarification area. Spacer bars keep the discs separated with an in-between distance of 0.4-2 mm. The disc operating angle is about 40-50°, but its value depends on the flow properties of the concentrated cell suspension. The feed enters the centrifuge through a central feed pipe leading to the feed chamber at the bottom of the stack. Solids settle on the lower surface of each cone and clarified liquid moves inwards and upwards to reach the annular overflow channel, situated at the neck of the bowl around the feed pipe. After the solids have settled on the bottom of the cone they tend to slip downwards and outwards. Eventually they come loose from the cone and are collected in the sediment holding space at the bottom of the machine (Figure 4.8c).

Disc stack centrifuges can be discharged continuously by nozzles or discontinuously by opening of the chamber bottom.

decanters or scroll type

A completely different type of centrifuge is the centrifugal decanter or scroll centrifuge. This type is mainly used to concentrate slurries with high dry solid concentrations. A decanter consists of a rotating horizontal bowl, with a length: diameter ratio of 1:4, fitted with a screw conveyer (Figure 4.8d). The screw rotates slightly faster than the bowl. The particles deposited on the wall will then be scraped off and transported to the outlet. The liquid leaves the machine at the other side via an overflow weir. The adjustable feed pipe is located somewhere in the middle of the decanter, depending on the settling behaviour of the slurry (broth).

centrifuge type selection criteria

The selection of the type of centrifuge to be used is based on the physical slurry properties, mainly the volume of solids in the feed and the particle size. Figure 4.9 gives some guidelines.

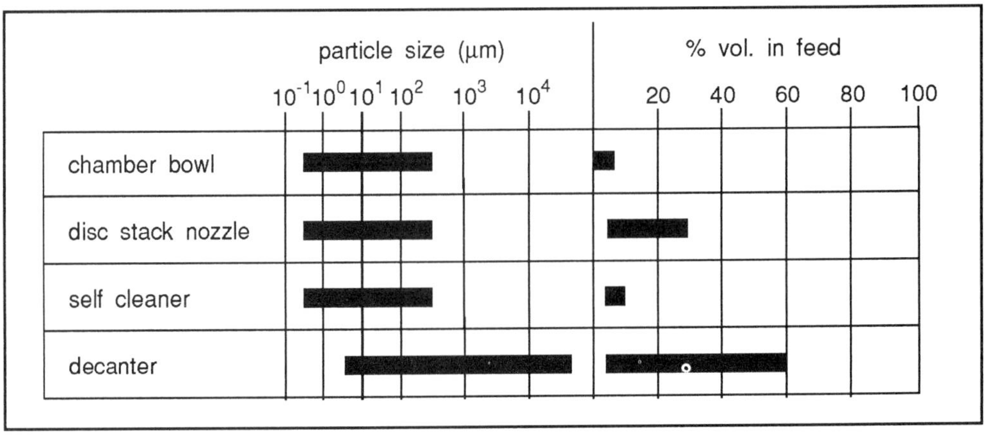

Figure 4.9 Selection criteria for centrifuges and decanters. % vol in feed = % v/v of solids in the feed.

Π From these guidelines, which type device is best suited for collecting larger particles and which type can only handle dilute suspensions?

From the data presented in Figure 4.9 you should have concluded that the decanter type is best suited for larger particles and that the chamber bowl type can only handle dilute suspensions.

RCF

Another important parameter is the relative centrifugal force (RCF) of the rotor which is defined as:

$$\xi = \frac{r\omega^2}{g}$$

where ξ = relative centrifugal force
r = radius (m)
ω = angular speed (rad s^{-1})
g = gravitational constant (m s^{-2})

Table 4.2 shows the RCF factors for commonly used centrifuges.

Laboratory centrifuges	RCF
small table centrifuge	5,000
high speed, chilled, centrifuge	50,000
ultracentrifuge	500,000
Industrial centrifuges	
tubular centrifuge	13,000 - 17,000
disc stack centrifuge	5,000 - 13,000
decanter centrifuge	1,500 - 4,500

Table 4.2 RCF factors for laboratory and industrial centrifuges.

From Table 4.2 we can see that very high RCF values and large volumes cannot be realised simultaneously beyond certain limits.

Π If the angular speed was constant, how would you expect RCF to change on scale up of a centrifuge?

RCF would increase because the radius would be larger ($\xi = \frac{r\omega^2}{g}$). The fact that industrial centrifuges generally have a lower RCF than laboratory centrifuges (Table 4.2) must therefore be due to a far slower speed of rotation caused by limitation of stresses in the bowl.

SAQ 4.3

Match each of the following statements with one or more of the types of centrifuges listed.

Centrifuge type:

a) tubular

b) multichamber

c) disc stack

d) decanter

Statement:

1) A zig zag flow of culture broth through the chambers is achieved.

2) The centrifuge has a high length to width ratio.

3) Solids can be discharged continuously.

4) Frequently employed in the fractionation of blood plasma.

5) Fermentation broth can be discharged continuously.

6) RCF higher than other types of centrifuges listed.

7) RCF lower than other types of centrifuges listed.

8) Can be operated at volumes of solid in feed of up to 60%.

9) Not suited for sedimentation of bacteria around 1 μm diameter.

4.3.2 Sedimentation rates

In this section we will consider the factors influencing sedimentation rates of particles in 1) batch gravity, 2) continuous gravity and 3) centrifugal field types of solid-liquid separations.

Batch gravity sedimentations

Consider a tank containing a highly diluted suspension of spherical particles. Because of the Earth's pull of gravity the particles start to move downwards. See Figure 4.10.

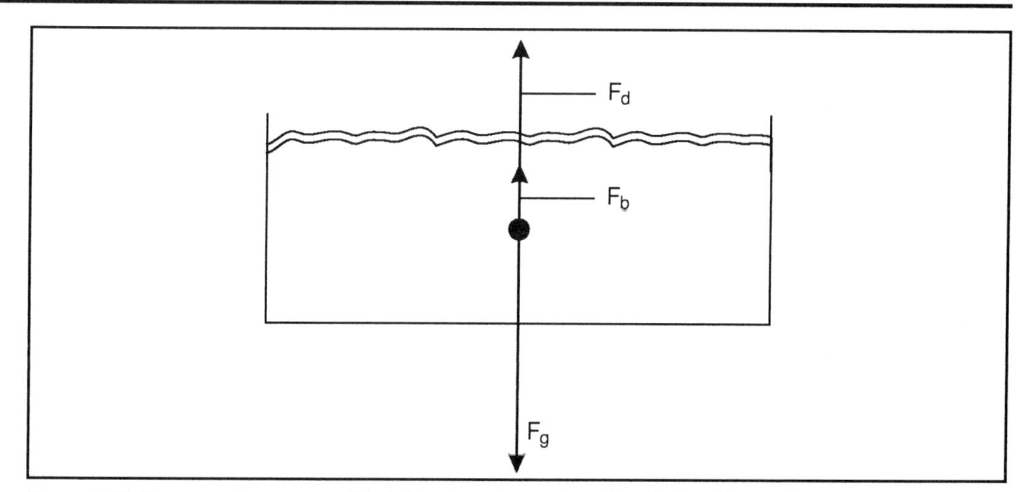

Figure 4.10 Forces acting on a particle in batch sedimentation. Gravitational force (Fg), buoyancy force (Fb), drag force (Fd).

The forces acting on a particle are:

Gravitational force:

$$F_g = \frac{\pi}{6} d_p^3 \rho_s g \quad \text{(E - 4.11)}$$

where: d_p = particle diameter (m), ρ_s = solid density (kg m^{-3}), g = gravitational constant (m s^{-2})

Buoyancy force:

$$F_b = \frac{\pi}{6} d_p^3 \rho_L g \quad \text{(E - 4.12)}$$

where: ρ_L = liquid density (kg m^{-3})

Drag force:

$$F_d = 3\pi d_p \eta v_g \quad \text{(E - 4.13)}$$

where: η = dynamic viscosity (Pas) = Nm^{-2} s, v_g = gravitational sedimentation rate = ms^{-1}

The buoyancy and drag forces will be counterbalanced by the gravitational force, so:

$$F_g = F_d + F_b \text{ or } F_d = F_b - F_g \quad \text{(E - 4.14)}$$

Substitution of Equations 4.11, 4.12 and 4.13 in 4.14 and rearranging gives:

$$v_g = \frac{d_p^2}{18\eta}(\rho_s - \rho_L)g \quad \text{(E - 4.15)}$$

It would be useful if you could try these substitutions and rearrangements to produce E - 4.15.

Equation 4.15 shows that the sedimentation rate (v_g) is determined by the physical characteristics of the particles (d_p, ρ_s) and the medium properties (η, ρ_L).

Π List three factors that would increase the sedimentation rate at constant RCF.

We can see from Equation 4.15 that the sedimentation rate will be large when:

- d_p increases;
- density difference, $\Delta\rho$ increases;
- viscosity (η) decreases.

Continuous gravity sedimentation

Figure 4.11 illustrates a tank for continuous gravity sedimentation. The slurry is fed to one side and the overflow is on the opposite side of the tank.

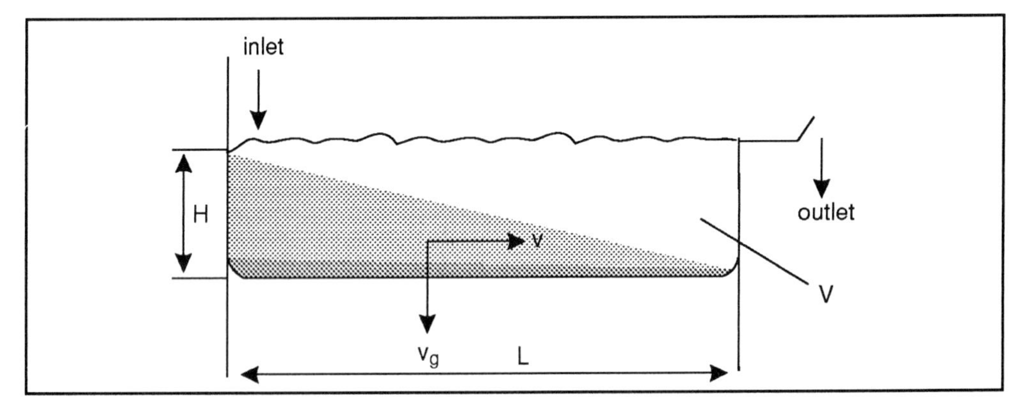

Figure 4.11 Continuous sedimentation tank. Height (H); length (L); width (w); volume (V = H . L . W); gravitational sedimentation rate (v_g); flow velocity (v).

The liquid flows towards the outlet with a velocity v. The particles (cells) are separated out and fall to the bottom of the tank. For this type of separation we can use the sedimentation rate to determine the throughput or the volumetric flow rate (ϕ_V) of the tank, as follows:

The limiting diameter (d_p) of a cell just separated in the tank gives a gravitational sedimentation rate (v_g) as indicated by Equation 4.15. The mean residence time (t) of the liquid in the tank can be obtained as follows:

$$t = \frac{V_{tank}}{\phi_V} = \frac{L \cdot W \cdot H}{\phi_V} \qquad \text{(E - 4.16)}$$

where:

L = length of sedimentation tank (m)

W = width of sedimentation tank (m)

H = height of sedimentation tank (m)

ϕ_V = volumetric flow rate ($m^3\ s^{-1}$)

At the same time the limiting particle must have fallen to the bottom of the tank:

$$t = \frac{H}{v_g} \qquad \text{(E - 4.17)}$$

Equating these expressions and substitution of Equation 4.15 we obtain:

$$\phi_V = v_g \cdot A = \frac{d_P^2\ g(\rho_s - \rho_L)}{18\ \eta} \cdot A \qquad \text{(E - 4.18)}$$

where: A = clarification area (m^2)

tilted plates increase clarification area

So the throughput is proportional to the sedimentation rate v_g and the so-called clarification area A.

The throughput can be increased by enlarging the tank surface. This can be done by installing a large number of tilted plates in the tank, as illustrated in Figure 4.12. The liquid with the suspended solids enters on the left. Some of the particulate matter falls onto the tilted plate and falls to the bottom of the tank. The liquid, now containing less suspended solids moves onto the second section where more of the suspended solids fall onto the second plate and so on.

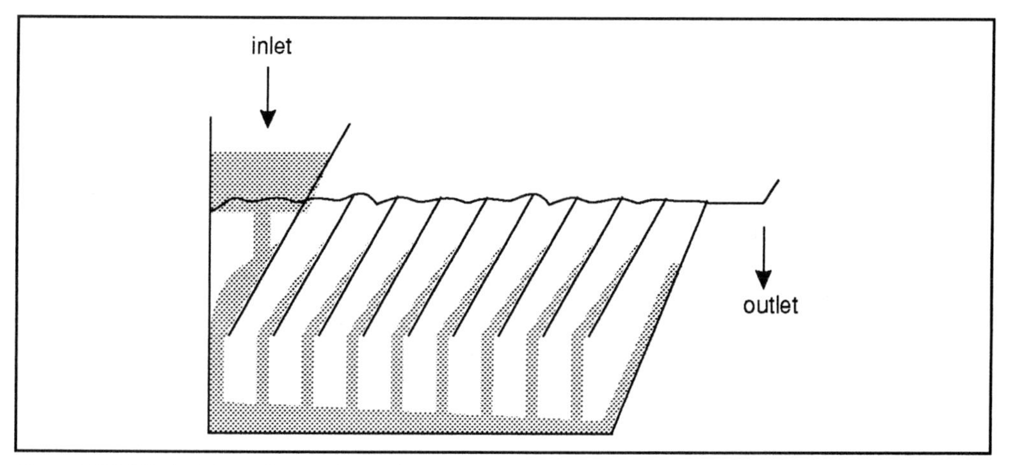

Figure 4.12 Tilted plate sedimentation tank.

Sedimentation in a centrifugal field

As we have already seen, the sedimentation rate in a gravitational field may be calculated by Equation 4.15.

$$v_g = \frac{d_p^2}{18\,\eta}(\rho_s - \rho_L)\,g \qquad \text{(E - 4.15)}$$

$\omega^2 r$ replaces g

Sedimentation in a centrifugal field is similar to Equation 4.15, only the gravitational constant (g) is replaced by $\omega^2 r$. Here ω is the angular speed and r the distance of the particle to the axis of rotation.

Thus:

$$v_c = \frac{d_p^2}{18\,\eta}(\rho_s - \rho_L)\,\omega^{2r} \qquad \text{(E - 4.19)}$$

where: v_c = centrifugal sedimentation rate (m s^{-1}).

Gravity and centrifugal sedimentation of a single particle are illustrated in Figure 4.13.

Based on the sedimentation rate in a centrifugal field (Equation 4.19), the characteristic dimensions of the centrifuge, the flow pattern, and the throughput can be predicted.

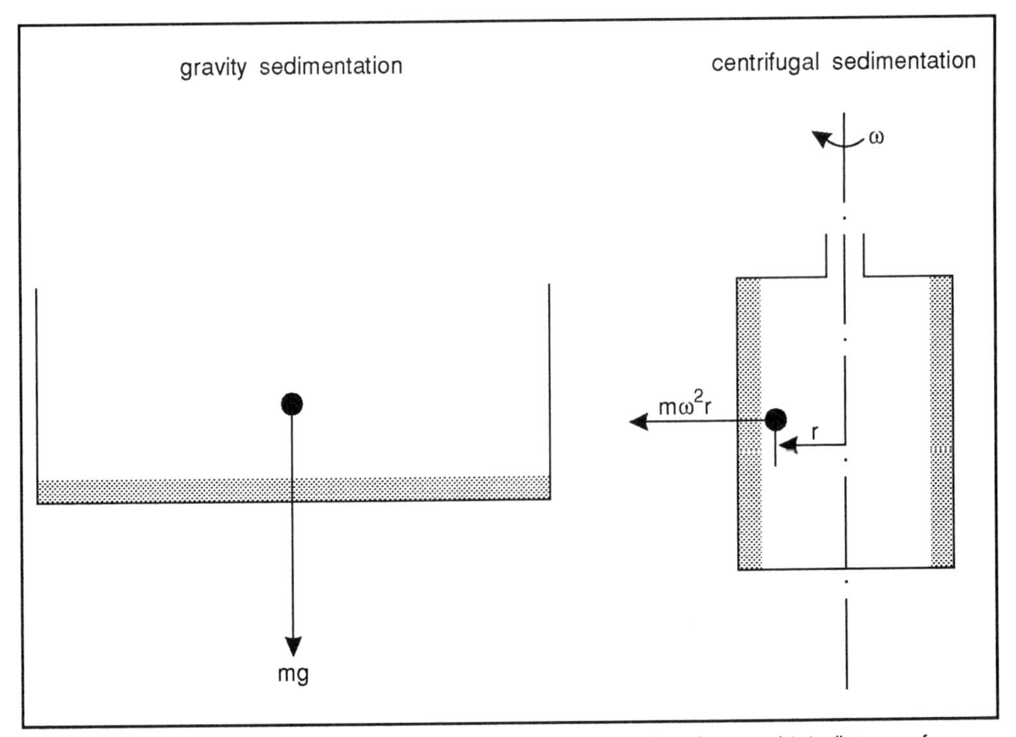

Figure 4.13 Gravity and centrifuge sedimentation of a single particle. Angular speed (ω); distance of particle to axis of rotation (r). m = mass, g = gravity.

SAQ 4.4

You are asked to carry out a number of determinations. Select the parameters required to perform the determinations from the list provided.

Determinations

1) Buoyancy force (F_b).

2) Gravitational sedimentation rate (v_g)

3) Throughput of sedimentation tank (ϕ_v).

4) Centrifugal sedimentation rate (v_c).

5) Gravitational force (F_g).

6) Mean residence time of liquid in a continuous sedimentation tank (t).

Parameters

a) Limiting diameter of particle.

b) Density of particle.

c) Density of liquid.

d) Clarification area.

e) Width of continuous sedimentation tank.

f) Height of continuous sedimentation tank.

g) Length of continuous sedimentation tank.

h) Dynamic viscosity.

i) Angular speed.

j) Gravitational constant.

k) Distance of particle to axis of rotation.

4.3.3 Performance of centrifuges

In this section we will consider the performance of two types of centrifuges: tubular and disc stack centrifuges.

Tubular centrifuge

The tubular (cylindrical) centrifuge is a rotating tube on a vertical shaft. The thickness of the liquid layer on the wall is small compared to the diameter of the tube. The slurry enters at the bottom and the supernatant is withdrawn at the top.

Figure 4.14 shows the characteristic sizes and the flow pattern in the tube.

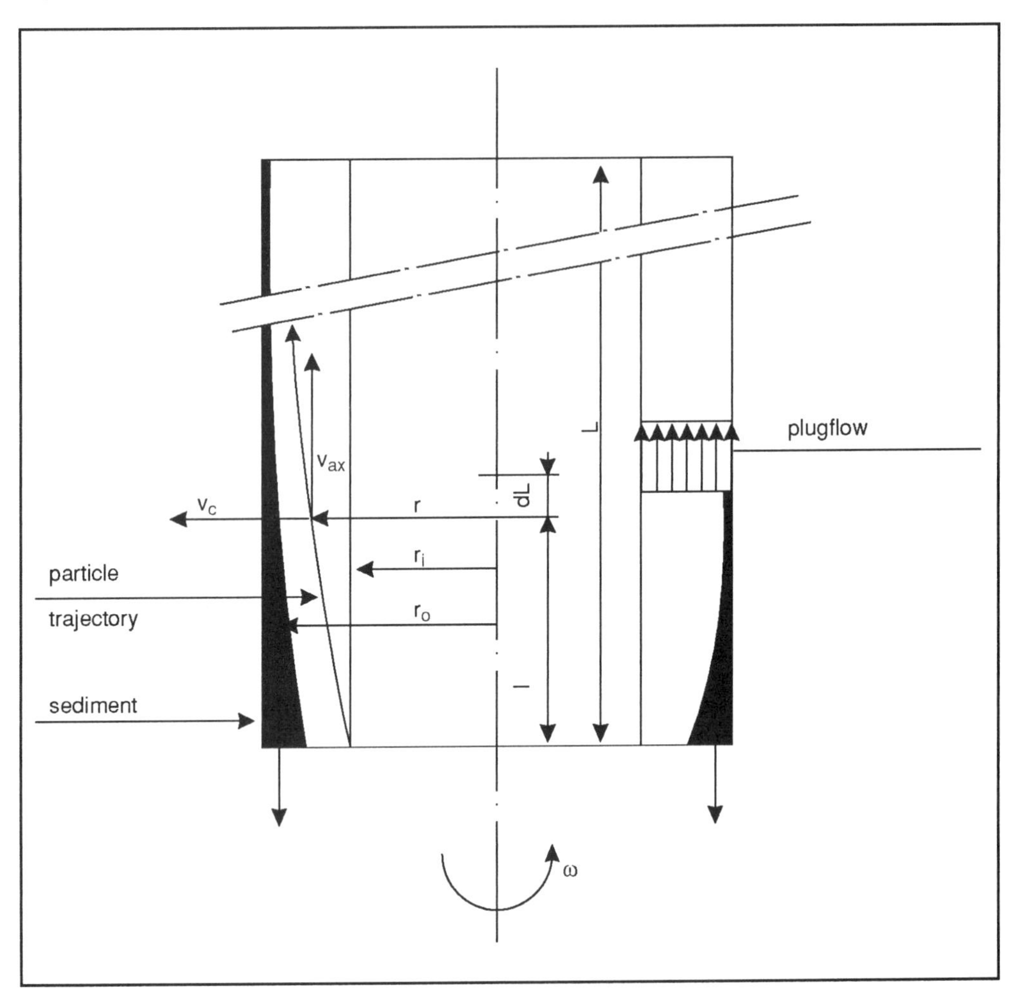

Figure 4.14 Flow pattern in a tube centrifuge. Centrifugal sedimentation rate (v_c); axial velocity (v_{ax}); distance of particle to the axis of rotation (r); inner radius (r_i); outer radius (r_o); length of centrifuge (L); angular speed (ω).

The mathematics of centrifugation may appear a little bit complex, but the conclusions are quite straightforward.

equivalent clarification area

The liquid is considered to flow upwards in plug flow. Separation will occur when the particle is settled before leaving the centrifuge. An equivalent clarification area (Σ) can be defined which is very useful in the scale up of centrifuges (Section 4.3.4). The term indicates the required area of a gravity settling tank with the same clarifying capacity as the centrifuge under the same conditions. Σ will now be derived for a cylindrical centrifuge. To understand the derivation it is essential that you are familiar with the notations given in Figure 4.14.

The motion of a particle in the axial direction (v_{ax}) corresponds with the axial liquid flow (no slip between particle and liquid), so we have:

$$v_{ax} = \frac{\phi_V \; v_c}{\pi (r_o^2 - r_i^2)} = \frac{dL}{dt} \qquad \text{(E - 4.20)}$$

where ϕ_V is the volumetric flow rate.

if $r_o - r_i << r_i$, (thickness of broth layer is small compared to the radius) then:

$$v_c = \frac{dr}{dt} = v_g \frac{\omega^2 r}{g} \qquad \text{(E - 4.21)}$$

Combination of Equations 4.20 and 4.21 gives:

$$\frac{dL}{dt} = \frac{\phi_V}{\pi (r_o^2 - r_i^2) \cdot \frac{g}{\omega^2 r v_g}}$$

or

$$L = \frac{\phi_V \; g}{\pi (r_o^2 - r_i^2) \, \omega^2 v_g} \int_{r_i}^{r_o} \frac{dr}{r}$$

Integration and substitution of Equation 4.15 gives:

$$\phi_V = \left(\frac{\pi (r_o^2 - r_i^2)}{\ln \frac{r_o}{r_i}} \cdot \frac{\omega^2 L}{g} \right) \left(\frac{(\rho_s - \rho_L) \, d_p^2 g}{18 \, \eta} \right) \qquad \text{(E - 4.22a)}$$

or in words:

ϕ_V = c . f (dimensions of equipment) . f (properties of suspension)

Let: $\Sigma_{tube} = \left(\frac{\pi (r_o^2 - r_i^2)}{\ln \frac{r_o}{r_i}} \cdot \frac{\omega^2 L}{g} \right)$ where Σ_{tube} is the equivalent clarification area. (E - 4.22b)

and $\ln \frac{r_o}{r_i}$ is approximately $\frac{r_o - r_i}{r_i}$

then we obtain for the Σ-value:

$$\Sigma_{tube} = (2\pi \, r_i L) \left(\frac{\omega^2}{g} \cdot \frac{r_o + r_i}{2} \right) \qquad \text{(E - 4.22c)}$$

We would not expect you to remember this exactly - but it would be useful to remember that it exists! It tells us that the equivalent clarification area is related to the outer and inner radii (r_o and r_i) of the centrifuge tube and the angular velocity ω.

The flow rate can be calculated as follows:

$$\phi_V = v_g \; \Sigma_{tube} \qquad \text{(E - 4.23)}$$

This again is a useful relationship to remember.

In a similar way the following equation can be derived for a conical tube:

$$\Sigma = \frac{1}{2}(2\pi r_i L)\left(\frac{\omega^2}{g} \cdot \frac{r_o + r_i}{3}\right) \quad \text{(E - 4.24)}$$

equivalent clarification areas may be added

If a centrifuge consists of a cylindrical and a conical part similar to a decanter, the equivalent clarification areas may be added:

$$\Sigma_{tot} = \Sigma_{cone} + \Sigma_{cylinder} \quad \text{(E - 4.25)}$$

SAQ 4.5

Determine the throughput of a tubular centrifuge with a length: inside diameters ratio of 10:1 (ie $L/d_o = 10$).

Thickness of fermentation broth layer in the centrifuge is diameter x 0.1.

Length of centrifuge (L) = 0.75 m

Angular velocity (ω) = 650 s^{-1}

$g = 9.8$ m s^{-2}

$\Delta\rho = 50$ kg m^{-3}

$\eta = 2$ mPas

$d_p = 10$ µm

Disc stack centrifuge

Figure 4.15 shows the flow pattern in a separation space within a conical annular compartment of a disc stack centrifuge.

time, cone angle and number of discs

Once the solid has reached the surface of the disc by moving under centrifugal force it is allowed to slide down to the bottom of the centrifuge.

The equivalent clarification area of a disc stack centrifuge can be calculated. Unlike the tubular centrifuge, the expression includes centrifugation time (t), cone angle (φ) and number of discs (Z).

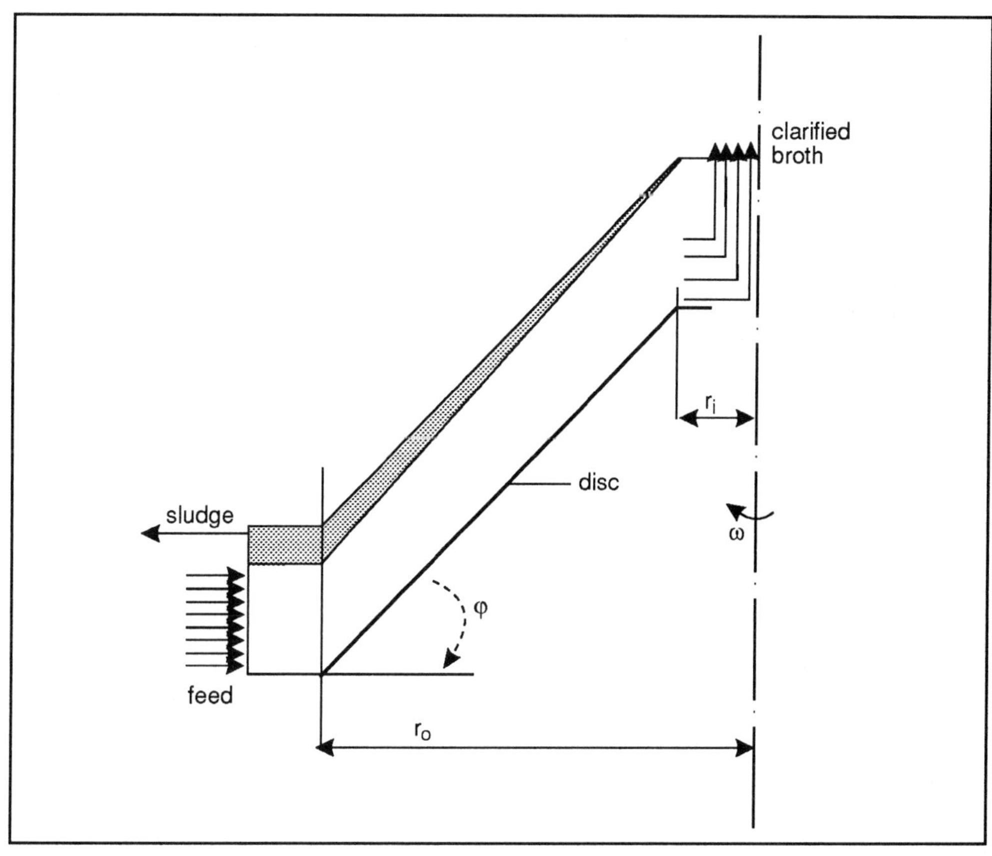

Figure 4.15 Flow in a separating space between two conical discs.

$$\Sigma = \frac{2\pi}{3g} \omega^{2} \operatorname{tg} \phi \, Z \, (r_o^3 - r_i^3) \qquad \text{(E - 4.26)}$$

We will not go into the derivation here and do not anticipate you should have learnt it by rote. Nevertheless it would be helpful to know it exists so that you can calculate Σ and thus ϕ_V.

As we saw for the tubular centrifuge, the flow rate can then be calculated:

$$\phi_V = v_g \; \Sigma_{disc\ stack} \qquad \text{(E - 4.27)}$$

Π Can you explain why, for a tubular centrifuges ϕ_V equals v_g Σ_{tube} and not v_c Σ_{tube}?

The answer is that this is because the angular velocity and the radius are already taken into account in Σ_{tube} (compare Equations 4.15, 4.19 and 4.22b)

Table 4.3 shows some Σ-values for different designs of disc stack centrifuges.

Diameter of discs (mm)	Number of discs (-)	Rotational speed (rpm)	Σ-value ($m^2 . 10^3$)
105	33	10,000	10
240	107	6,500	20
320	98	6,250	40
350	132	4,650	36
500	144	4,250	98

Table 4.3 Numerical Σ-values of disc stack centrifuges.

Π Use the data in Table 4.3 to calculate v_g if the centrifuge has a disc diameter of 500 mm, contains 144 discs with a rotational speed of 4250 and the volumetric flow rate is 0.1 $m^3 s^{-1}$.

From the table a centrifuge with these specifications has an Σ value of $98 . 10^3 m^2$.

Thus using $\phi v = v_g \Sigma$ then $0.1 = v_g . 98 . 10^3$.

$$\text{Thus } v_g = \frac{0.1}{98 . 10^3} = \frac{1}{98 . 10^4} = 1.02 . 10^{-6} ms^{-1}$$

4.3.4 Scale up of centrifuges

scale up with constant v_g

Comparisons of Σ-values have shown that results from experimental work differ from calculated values if different types of centrifuges are considered. It follows that scale up between different types of centrifuges is not reliable. However, scale up between centrifuges of the same type is fairly reliable and is simply based on Equation 4.23, $\phi v = v_g \Sigma$, where v_g remains constant. So:

$$\frac{\phi_{V1}}{\Sigma_1} = \frac{\phi_{V2}}{\Sigma_2} \qquad \text{(E - 4.28)}$$

Π Can you think of a factor that would limit the scale up of all types of centrifuges?

The point we hoped you would consider is that as the size of centrifuge increases, more power is required to maintain angular speed. So, the angular speed may have to be lowered as the size of the centrifuge increases. This, of course influences performance.

In addition to this scale up rule there are two more important scale up parameters in the case of self cleaning separators with nozzle discharge or periodical discharge.

- nozzle capacity (number and diameter of nozzles);
- capacity of sediment holding space (desludge cycle time).

Generally, scale up of centrifuges requires substantial experience and pilot plant trials and is done in cooperation with centrifuge suppliers.

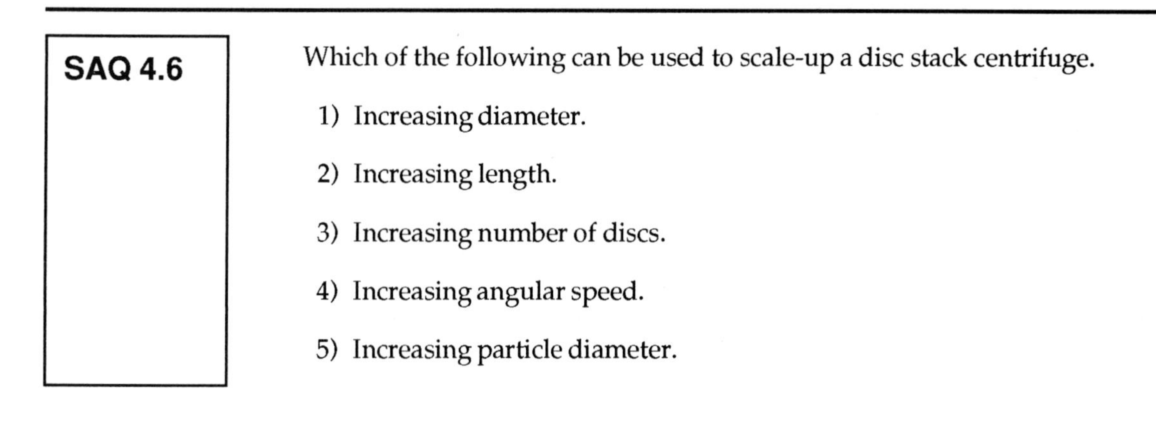

SAQ 4.6

Which of the following can be used to scale-up a disc stack centrifuge.

1) Increasing diameter.
2) Increasing length.
3) Increasing number of discs.
4) Increasing angular speed.
5) Increasing particle diameter.

Summary and objectives

In many cases solid-liquid separations are the first step in biotechnological recovery processes.

Techniques used to achieve solid-liquid separation are based on filtration or centrifugation principles. Which technique will eventually be decided upon largely depends on the broth properties (viscosity, cell dry solids, cell density, cell size, etc.) and requirements such as safety and sterility. If a solid cake (20-35% dry solids) is required, filtration is preferred.

Filter presses and rotary vacuum drum filters are most commonly used in broth filtration. The theoretical background of filter presses is well established. Simple laboratory tests indicate that either the cake and medium resistance or the cake volume is the criterion for scale-up. A proper selection of the filter media is required because they form the heart of the filtration. However, no general guidelines can be given for a proper selection. Laboratory test and subsequent pilot plant tests are essential for scale up.

In the case of concentration prior to cell disruption, centrifugation is preferred. The product losses in the case of extracellular products are generally too high when only one centrifuge is applied. To minimise the losses, the concentrate leaving the first centrifuge should be washed counter-currently in the following two or three stages.

The basic prediction methods of both separation principles are quite well established and show good agreement with experimental results.

Now that you have completed this chapter you should be able to:

- broadly describe the principles of operation of different types of filters and centrifuges;
- determine specific cake resistance and medium resistance for constant pressure filtration;
- use mathematical relationships to predict separation performance, for both filters and centrifuges;
- list factors that could be used to improve separation performance and explain their effect;
- explain how scale up of filters and centrifuges is carried out.

Concentration of products - I

Concentration of products - I

5.1 Introduction

After separating the cells from the whole broth, the filtrate contains 85-98% of water. So the product is only a minor constituent of the broth. Handling large amounts of water is very costly because it requires large equipment. For that reason the next step in the recovery process is almost always aimed at the removal of water. This can be done in different ways:

- evaporation;
- membrane processes;
- liquid-liquid extraction;
- precipitation.

Evaporation and membrane processes will be considered in this chapter, liquid-liquid extraction and precipitation are the subjects of the next chapter.

Evaporation is a simple, but, in many cases, a high energy consuming way of water removal. However, because of its reliability and simplicity it is frequently applied on a large scale.

Membrane processes are more frequently being applied nowadays, they are less energy consuming and can also be used as a first purification step.

5.2 Thermodynamics of water removal

Before we move on to examine specific processes used for water removal, we need to briefly consider the energy costs of water removal.

For concentrating a product by the removal of water (the solvent) an auxiliary phase is used that allows for easy transport of the water rather than the solute. For concentrating non-volatile components such as sugars or proteins, water vapour can be used as the auxiliary phase. Here, the solvent (water) is converted into vapour by heat. After removal of the vapour, we are left with a concentrated solution of the product. This procedure can be carried out continuously. The same can be done by freezing out the solvent and subsequent removal of the (ice) crystals.

Semi-permeable membranes can also be applied for water removal. In this case, the membrane acts as an auxiliary phase. By applying a high pressure on the solution side of the membrane, water can be forced to flow through the membrane.

Irrespective of the processes involved, the removal of water from a solution requires energy. This energy can be expressed as an excess thermodynamic potential of the water in the solution.

For the thermodynamic potential of water in the solution, the following holds:

$$\mu_w(P,T) = \mu_w^o(P,T) + RT \ln \gamma_w x_w \qquad \text{(E - 5.1)}$$

This equation will need some further explanation:

$\mu_w(P,T)$ is the thermodynamic potential of water in the solution at a particular pressure (P) and temperature (T).

$\mu_w^o(P,T)$ is the thermodynamics potential of pure water at the same pressure (P) and temperature (T).

R is the gas constant (kJ mol^{-1} K^{-1}).

T is the temperature (K).

γ_w is the activity of water (-).

x_w is the molar fraction of water (-).

We will explain a little more fully what is meant by molar fraction of water. If we have a solution containing two components (eg water and solute), then the total number of moles of material in the solution = (moles of water) + (moles of solute).

The molar fraction of water (x_w) is then given by:

$$x_w = \frac{\text{moles of water}}{\text{moles of solute} + \text{moles of water}} \qquad \text{(E - 5.2)}$$

Π What is x_w for pure water?

You should have concluded that $x_w = 1$ since x_w in this case will simply be given by moles of water / moles of water.

The molar fraction of solute (x_s) can be written as:

$$x_s = \frac{\text{moles of solute}}{\text{moles of solute} + \text{moles of water}} \qquad \text{(E - 5.3)}$$

By simply adding Equation 5.2 and 5.3 together we can show that:

$x_w + x_s = 1$ or $x_w = 1 - x_s$

Now let us return to Equation 5.1, relating to the thermodynamics potential of water in a solution to the thermodynamic potential of pure water and to the molar fraction of water.

If we re-write this equation to express this in the form of the difference between the thermodynamic potential of pure water and water in the solution. Thus:

$$\Delta \mu_w \; (= \mu_w^o - \mu_w) = - RT \ln \gamma_w x_w \qquad \text{(E - 5.4)}$$

This difference is what we might call the excess thermodynamic potential of pure water over that of water in the solution.

For an ideal binary solution, the activity of water (γ_w) can be taken to = 1 and we have already shown that $x_w = 1 - x_s$. Thus Equation 5.4 can be simplified to:

$$\Delta \mu_w = - RT \ln \; (1 - x_s) \qquad \text{(E - 5.5)}$$

We can simplify Equation 5.5 further, by assuming that x_s is small. If this is done, then:

$$\Delta \mu_w = RT \ln x_s \qquad \text{(E - 5.6)}$$

We will illustrate the significance of Equation 5.6 by the use of an example.

Example:

Suppose we have a sugar (saccharose) solution with a concentration C_s of 0.15 kg sugar/kg water. The temperature is assumed to be 20°C. We would like to predict the energy required to remove 1 kg of water from this solution.

In principle, we can use Equation 5.6. First we need to convert the concentration of the sugar into molar fraction (x_s). To do this we need the molecular weight of water and saccharose. These are 18 and 342 respectively. Thus each mole of water weighs 18g or $18 . 10^{-3}$ kg and lmole of saccharose weighs 342g or $342 . 10^{-3}$ kg.

Remember that x_s is given by $\dfrac{\text{moles of solute}}{\text{moles of solute and moles of water}}$ (see Equation 5.3).

$$\text{Thus: } x_s = \frac{\dfrac{\text{weight of solute}}{\text{mol wt of solute}}}{\dfrac{\text{weight of solute}}{\text{mol wt of solute}} + \dfrac{\text{weight of water}}{\text{mol wt of water}}}$$

$$\frac{\dfrac{0.15}{342 * 10^{-3}}}{\dfrac{0.15}{342 * 10^{-3}} + \dfrac{1}{18 * 10^{-3}}} \quad \text{(Note we have used kg throughout)}$$

$$\frac{0.438}{0.438 + 55.56} = 0.00782$$

The excess thermodynamic potential is:

$$\Delta \mu_w = RT \ln x_s = 8.3 . 293 . 0.00782 = 19.02 \text{ J mol}^{-1}$$

or expressed per kg of water:

$$= \frac{19.02}{18 \cdot 10^{-3}} = 1057 \text{ J kg}^{-1}$$

major energy requirement is heat of ovaporation

Thus it would appear that to remove water from a solution we would only require 1057 $J\ kg^{-1}$ (= 1.057 $kJ\ kg^{-1}$). But remember that in the case of removing water as a vapour, we not only require energy to remove the water from solution, we need to supply the heat of evaporation. The heat of evaporation of water is about 2257 $kJ\ kg^{-1}$. Thus to remove 1 kg of water as vapour we need to supply 2257 kJ for evaporation and 1.057 kJ to remove the water from the solution.

We can represent this in the following way:

$$\text{water in solution} \xrightarrow[= 1.057 \text{ kJ kg}^{-1}]{\text{energy required}} \text{pure water} \xrightarrow[= 2257 \text{ kJ kg}^{-1}]{\text{energy required}} \text{vapour}$$

In other words, in removing water as a vapour, over 99% of the energy demand is to evaporate the water. Evaporation requires a phase transition (water to vapour) which is very energy-demanding. It is the challenge of the designers of evaporators to find ways to reduce this amount of energy.

Let us compare this with membrane processes. For this we will use a different approach to calculate the energy demand.

Van't Hoff's law

Osmotic pressure can be approximated by Van't Hoff's law in which:

$$\pi = RTC_s \qquad \text{(E - 5.7)}$$

where:

π = osmotic pressure and C_s = concentration of the solute (mol m^{-3})

C_s can be expressed as $C_s = C_W x_s$ (E - 5.8)

where C_W is the molar density of the solution and can be approximated by that of water for dilute solution. Thus for dilute solution we can take $C_W = 5.5 * 10^4$ mol m^{-3}.

Substitution into Equation 5.7 gives:

$$\pi = RT C_W x_s \qquad \text{(E - 5.9)}$$

Returning to our example of a saccharose solution, then the osmotic pressure will be:

$$\pi = RT C_W x_s = 8.3 * 293 * 5.5 \ 10^4 \times 0.00782 = 1.05 * 10^6 \text{ Pa}$$

If we remove 1 kg of water from the sugar solution using a membrane (or another selective barrier) the amount of energy required will be:

$$E = \pi V_W$$

where V_W = volume of solution containing 1 kg of water (= 10^{-3} m^3)

Thus: $E = 1.05 * 10^6 * 10^{-3}$ $J\ kg^{-1}$ = 1050 $J\ kg^{-1}$

This is of course the same amount of energy as calculated before. We can represent this situation as:

$$\text{water in solution} \xrightarrow[\text{1050 J kg}^{-1}]{\text{energy required}} \text{pure water}$$

Thus in contrast to evaporative methods, separation of water through a semi-permeable barrier would appear to have cost very little energy. From energetic considerations, it would appear that the separation of water using membranes is very much cheaper. In practice however matters are not that simple. We need to consider the cost of equipment and the working life of equipment. Furthermore, in many situations the use of the membranes present a number of operational difficulties such as clogging and biodeterioration.

Let us now examine the operation of evaporation and membrane concentrators in more detail.

5.3 Evaporation

Evaporation is a process that requires heat as an auxiliary phase to remove water from a solution.

Heat, carried as steam, partially converts the solution into vapour. The vapour is then removed from the solution and condensed.

Evaporators generally consist of four parts:

- a heating section to which the steam (heat carrier) is fed;
- a concentration and separation section where concentrate and vapour are separated;
- a condenser in which the vapour condenses, and;
- the required vacuum and product pumps, control equipment, etc.

Based on these four parts a wide variety of equipment has been developed, ranging from lab scale (0.5 - 1.0 l h^{-1} water evaporation capacity) to very large industrial scale (such as tap water from sea water; 150 m^3h^{-1} water evaporation capacity).

5.3.1 Types of evaporator

There are many different types of evaporators. In this section we will consider the principle of operation of five types.

Natural circulation evaporator

rising bubbles cause circulation

The natural circulation evaporator is based upon natural circulation of the product, which is caused by the density difference generated by heating. The liquid to be evaporated boils in vertical tubes. The rising bubbles cause circulation of the liquid, thus facilitating the separation of liquid and vapour at the top of the heating tubes. The remaining liquid recirculates to the evaporation section so that only part of the total evaporation occurs in one pass (Figure 5.1).

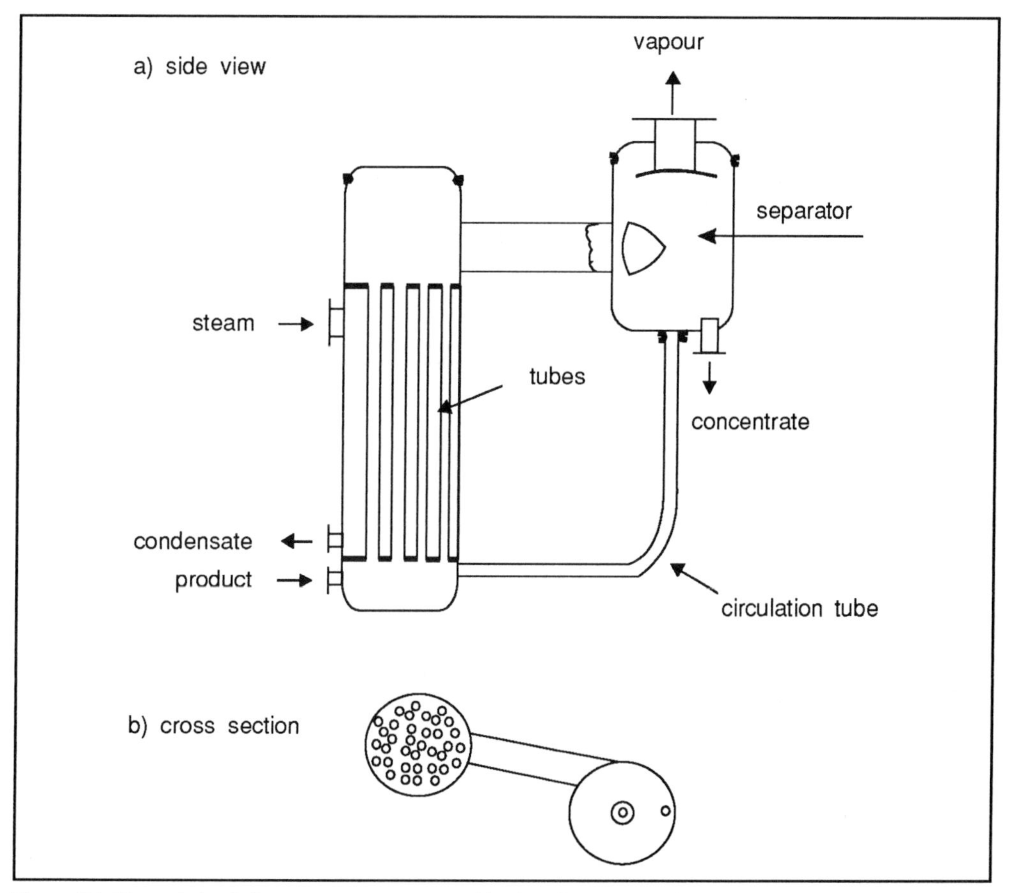

Figure 5.1 Natural circulation type evaporator. a) side view, b) cross section.

dry-out reduces efficiency

The degree of evaporation strongly depends on the temperature difference between the boiling liquid and the steam. The liquid circulation in its turn depends on the degree of evaporation.

If the tubes are insufficiently immersed, a dry-out may occur, reducing evaporation. This, in turn, will reduce the circulation, thus increasing the chance of dry-out. Due to this 'instability' the efficiency of the evaporator will drop considerably. For this reason, a pump may be installed in the circulation channel in order to attain sufficient circulation.

Another reason for installing a circulation pump is that the boiling of a liquid on the tube surface may cause heavy fouling. By running the evaporator under pressure, bubble formation can be suppressed.

A circulation evaporator has a poorly defined residence time, a high steam consumption, but a good heat transfer at high temperature differences. It is relatively cheap, easy to operate and therefore frequently applied.

Falling film evaporator

The falling film evaporator consists of a large number of long tubes (4-8 m length) surrounded by a steam jacket. The liquid enters the tubes at the top and distributes itself over the heating surfaces as a thin film (Figure 5.2). Uniform distribution of the liquid is essential for proper evaporation.

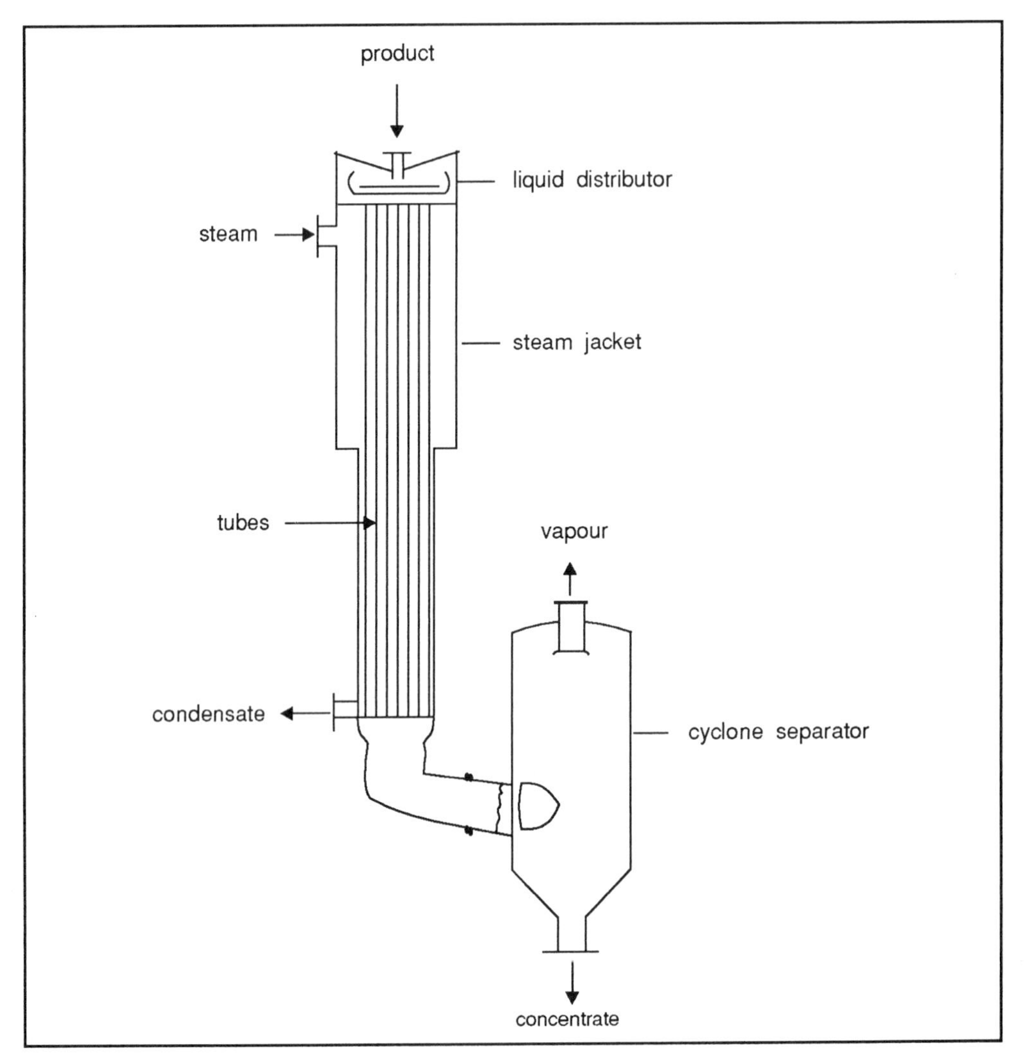

Figure 5.2 Falling film evaporator.

As the liquid flows downwards, its linear velocity increases considerably. This is caused by the vapour, evolved from the liquid, flowing in the same direction. Due to this mechanism, falling film evaporators are well suited for concentrating viscous products. They are frequently applied in the chemical, food and fermentation industry.

concentration of viscous products

Heat sensitive (enzyme) solutions, as well as clear, foaming, or corrosive products can be treated, even at low pressures.

Plate evaporator

Plate evaporators differ from other evaporator types in having a relatively large evaporation surface in a small volume. Metal plates, usually with corrugated faces, are supported by a frame.

large evaporation surface, small volume

During the evaporation process, the heat carrier (steam) flows through the channels formed by the free space between two plates. It alternately climbs and falls parallel to the concentrate in co-current, counter-current mode. The concentrate and vapour are fed to a liquid/vapour separator. Plate evaporators are frequently applied in the dairy and fermentation industry, because of their small construction volume and flexibility. However, the possibility of treating viscous and solids containing products is limited.

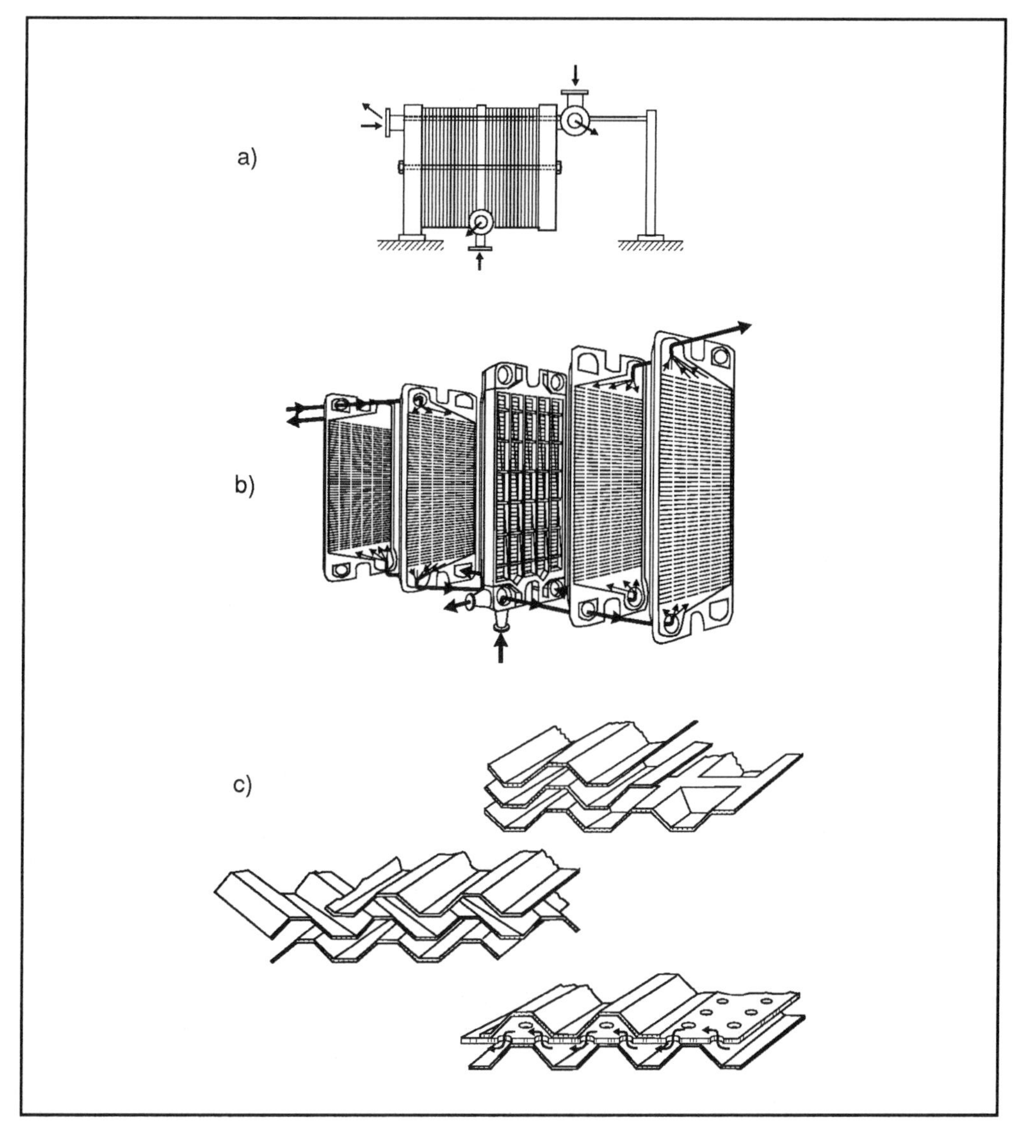

Figure 5.3 Plate evaporator. a) frame assembly, b) exploded view of plate evaporator, c) details of corrugated plates.

Multiple-effect evaporator

multiple effect reduces heat consumption

In single-effect evaporators, such as those already considered, the energy (steam) consumption is rather high. Theoretically, we need 1 kg of steam for the removal of 1 kg of water. To reduce this high steam consumption multiple effect evaporators of up to seven effects may be used. By 'effect' we mean stage.

Adding a second effect to a single-effect evaporator reduces the heat consumption by 50%, a third effect to 33%. The savings from additional effects can be represented by:

$$\text{saving } \% = \left(1 - \frac{N}{N + M}\right) 100\% \qquad \text{(E - 5.10)}$$

N is the original number of effects and M the additional number.

One additional effect to a five-effect evaporator saves 17%. So the larger N, the lower the savings for any additional effect. Because the capital investment is more than proportional to the number of effects, in practice the number of effects is restricted to seven.

Π How many additional effects are required to reduce heat consumption of a three effect evaporator by 50%. (Try to do this before reading on).

Since:

$$\text{Saving } \% = \left(1 - \frac{N}{N + M}\right) 100\%$$

then:

$$\left(1 - \frac{3}{3 + M}\right) 100 = 50\%$$

Number of additional effects = 3.

forward feed systems - suitable for heat sensitive products

In the food industry and in biotechnology, frequently a forward feed system is used (Figure 5.4). Here, the product enters the first effect at the highest temperature. The product, partially concentrated in the first effect, is fed to the second effect, which has a lower temperature. In this way, the medium flows through the evaporator with increasing concentration and decreasing temperature, using the vapour of each effect as a heating medium for the next effect. The combination of lowest temperature and highest viscosity in the subsequent effects is beneficial to heat sensitive products such as enzymes and proteins. However, increasing the heating surface of subsequent effects is usually a requirement.

backward feed evaporator

Another possibility is the backward feed evaporator. Here the diluted product is fed to the last effect, at the lowest evaporation temperature, and then transferred through the successive effects to the first.

The concentrate is collected from the first effect, which is at the highest temperature.

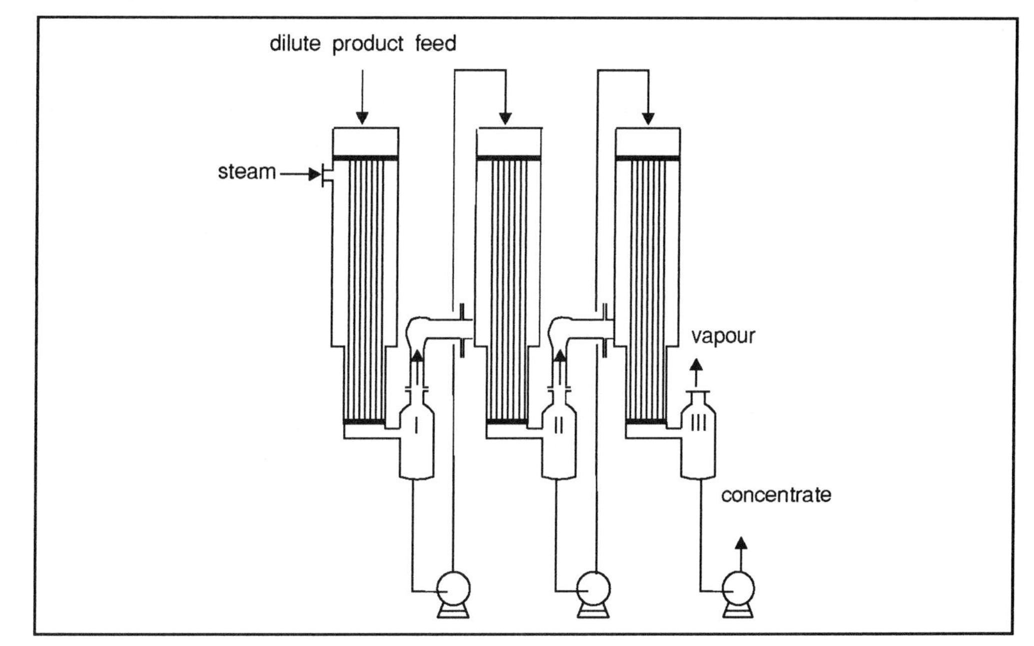

Figure 5.4 Forward feed flow evaporator (see text for a description).

This backward feed system has the advantage that the concentrated product evaporates in the first effect at the highest temperature but the lowest viscosity (because of the high temperature) which improves the heat transfer considerably.

SAQ 5.1

Match each of the following statements with types/modes of operation of evaporators listed below.

Statements

1) The evaporator has a relatively large surface to volume ratio.

2) Liquid circulation depends on the degree of evaporation.

3) Vapour evolved from the liquid increases linear velocity.

4) Concentrated product evaporates at the highest temperature and lowest viscosity.

5) The product enters the first effect at the highest temperature.

Evaporator types

a) Plate evaporator; b) Falling film evaporator; c) Natural circulation evaporator; d) Forward feed evaporator; e) Backward feed evaporator.

SAQ 5.2

For each of the following types/modes of operation of evaporators give an advantage offered by its use.

1) Plate evaporator.

2) Falling film evaporator.

3) Natural circulation evaporator.

4) Multiple-effect evaporator.

5) Multiple-effect evaporator using forward feed.

6) Multiple-effect evaporator using backward feed.

5.3.2 Energy consumption of a single-effect evaporator

The energy consumption of single and multiple-effect evaporators can be calculated from mass and enthalpy balances.

We will illustrate the set up and the use of these equations on a single-effect evaporator.

A single effect evaporator is represented schematically in Figure 5.5.

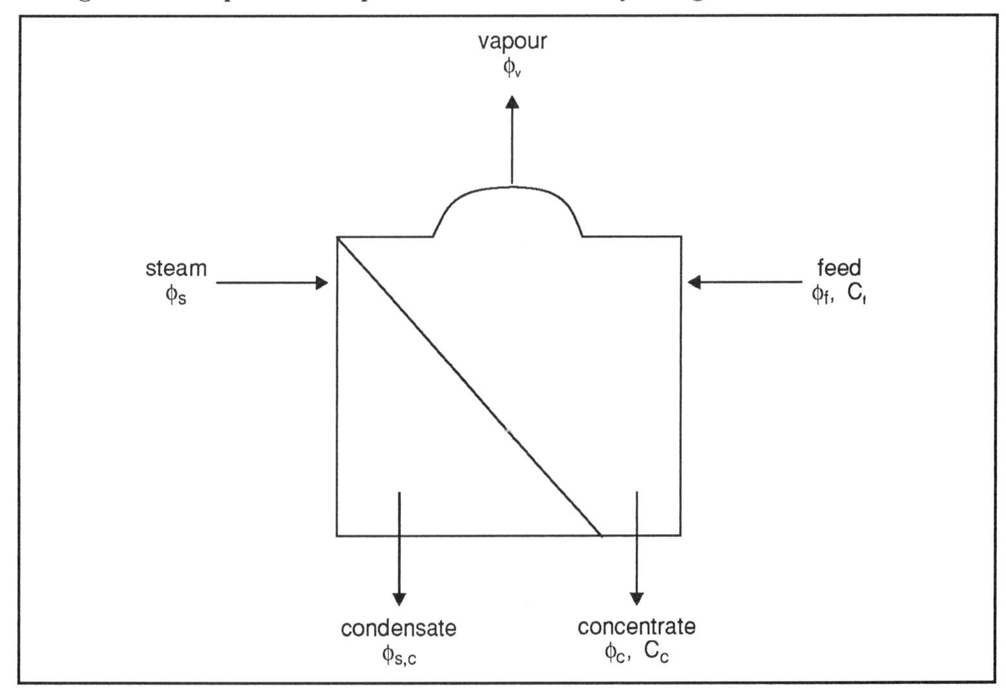

Figure 5.5 Schematic representation of a single effect evaporator. ϕ_v = vapour flow rate, ϕ_s = steam flow rate, ϕ_f = feed flow rate, $\phi_{s,c}$ = condensate flow rate, ϕ_c = concentrate flow rate, C_f = concentration of solute in the feed, C_c = concentration of the solute in the concentrate.

The left-hand side of Figure 5.5 represents the steam jacket, the right-hand side the evaporation room.

The overall mass balance for the solution fed to the evaporator is:

$$\phi_f = \phi_c + \phi_v \quad \text{(E - 5.11)}$$

where:

ϕ_f = mass flow rate of feed (kg s^{-1});

ϕ_c = mass flow rate of concentrate (kg s^{-1});

ϕ_v = mass flow rate of vapour (kg s^{-1}).

mass balance for feed

Since the vapour does not contain solute, the solute balance can be simplified to:

$$\phi_f C_f = \phi_c C_c \quad \text{(E - 5.12)}$$

where:

C_f = product concentration in feed (kg m^{-3});

C_c = product concentration in concentrate (kg m^{-3}).

Combination of Equations 5.11 and 5.12 gives:

$$\phi_v = \phi_c \frac{C_c - C_f}{C_f} \quad \text{(E - 5.13)}$$

If the mass flow of the concentrate (ϕ_c), the initial product concentration (C_f) and the final concentration (C_c) are known, the water evaporation capacity can be calculated according to Equation 5.13.

Π Suppose a feed with 10% solids is concentrated to 50% solids. Now complete the following statement:

The mass flow of the vapour is [ ] times the mass flow of the concentrate and [ ] of the feed.

The completed statement should read:

The mass flow of the vapour is four times the mass flow of the concentrate (using Equation 5.13) and four fifths of the feed (using Equation 5.11).

From a steam mass balance it follows:

$$\phi_s = \phi_{s,c} \quad \text{(E - 5.14)}$$

where:

ϕ_s = mass flow rate of steam (kg s^{-1})

$\phi_{s,c}$ = mass flow rate of steam as condensate (kg s^{-1}).

mass balance for steam

The amount of condensate as well as the amount of vapour (after condensation), can easily be measured in a pilot plant experiment. The ratio of these is an important parameter in evaluating the evaporation process and is called the specific steam consumption (SSC).

$$SSC = \frac{\phi_s}{\phi_v} = \frac{\text{quantity of steam to evaporator}}{\text{quantity of evaporated solution}} = \psi_s$$

ψ_s is, of course, a useful term so it is important to remember what it means. Below we will show how it may be determined.

Dividing Equation 5.14 by 5.13 gives a more quantitative expression for SSC:

$$\psi_s = \frac{\phi_s}{\phi_v} = \frac{\phi_s}{\phi_c}\,\frac{C_f}{C_c - C_f} \qquad \text{(E - 5.15)}$$

enthalpy balance

ϕ_s can be calculated from an overall enthalpy balance.

Feed

We will first consider the enthalpy (or heat content) of the feed.

$$h_f = \phi_f\, c_{p,f}\, \theta_f \qquad \text{(E - 5.16)}$$

where:

h_f = total enthalpy of feed (J);

$c_{p,f}$ = specific heat of feed (J $kg^{-1}K^{-1}$ or J kg^{-1} C^{-1});

θ_f = temperature of feed (°C).

Concentrate

The enthalpy of the concentrate amounts to:

$$h_c\ \phi_c\, c_{p,c}\ \theta_c \qquad \text{(E 5.17)}$$

where:

h_c = total enthalpy of concentrate (J);

$c_{p,c}$ = specific heat of concentrate (J kg^{-1} K^{-1}).

You should note that the specific heats of the feed and concentrate are usually very similar but not necessarily identical because of the differences in solid content.

Vapour

The enthalpy of the vapour can be calculated as follows:

$$h_v = \phi_v \Delta h_v \quad \text{(E - 5.18)}$$

where Δh_v = enthalpy per kg steam at a fixed temperature and pressure. This is also called the heat of evaporation and can be read from steam tables.

Steam

Finally the enthalpy of the steam:

$$h_s = \phi_s \Delta h_s \quad \text{(E - 5.19)}$$

We can thus write an overall energy balance for the evaporator. In a steady state:

enthalpy input = enthalpy output

Thus: $\phi_f c_{p,f} \theta_f + \phi_s \Delta h_s = \phi_c c_{p,c} \theta_c + \phi_v \Delta h_v$ (E - 5.20)

Rearranging the overall energy balance to obtain an expression for steam consumption (ϕ_s), we have:

$$\phi_s = \frac{\phi_c c_{p,c} \theta_c - \phi_f c_{p,f} \theta_f + \phi_v \Delta h_v}{\Delta h_s} \quad \text{(E - 5.21)}$$

The values of Δh_s and Δh_v can be obtained from steam tables.

But since $\phi_v = \phi_c \dfrac{C_c - C_f}{C_f}$ (see Equation 5.13)

then:

$$\phi_s = \frac{\phi_c c_{p,c} \theta_c - \phi_f c_{p,f} \theta_f + \phi_c \left(\dfrac{C_c - C_f}{C_f}\right) \Delta h_v}{\Delta h_s} \quad \text{(E - 5.22)}$$

Note that $c_{p,c}$ and $c_{p,f}$ are usually very similar in value, but not identical. Let us write Equation 5.22 in a different form.

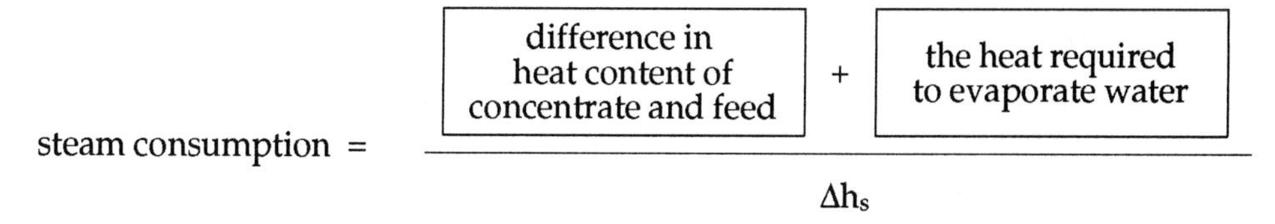

steam consumption = $\dfrac{\text{difference in heat content of concentrate and feed} + \text{the heat required to evaporate water}}{\Delta h_s}$

where the difference in the heat content of the concentrate and feed:

$$= \phi_c c_{p,c} \theta_c - \phi_f c_{p,f} \theta_f$$

and the heat required to evaporate water:

$$= \phi_v \Delta h_v = \phi_c \left(\frac{C_c - C_f}{C_f} \right) \Delta h_v$$

We showed earlier that the difference in heat content of the concentrate and feed is very small compared with the heat of evaporation. So we can reduce Equation 5.22 to:

$$\phi_s = \phi_c \left(\frac{C_c - C_f}{C_f} \right) \Delta h_v$$

Remember however that the specific steam consumption (ψ_s):

$$\psi_s = \frac{\phi_s}{\phi_v} = \frac{\phi_s}{\phi_c} \frac{C_f}{(C_c - C_f)} \quad \text{see Equation 5.15}$$

Thus: $$\psi_s = \frac{\Delta h_v}{\Delta h_s} \qquad \text{(E - 5.23)}$$

which can be simplified further because $\Delta h_v \approx \Delta h_s$. ($\Delta h$ does not differ too much for different pressures, moreover the steam pressures of the steam side and the concentrate side are roughly the same).

SSC = 1 for single-effect

We may conclude that the SSC for a single-effect evaporator is 1. In practice it will be in the order of 1.1 because of 10% heat losses.

SAQ 5.3

For each of the following symbols, name the parameter that it represents and give suitable units:

Δh_c

θ_f

ψ_s

ϕ_v

$c_{p,f}$

$\phi_{s,c}$

For multiple effect evaporators the steam consumption can be calculated by combining the mass and heat balance for each effect.

SAQ 5.4

Identify which of the following statements concerning evaporators are False.

1) $\frac{\phi_s}{\phi_v} = \frac{\Delta h_v}{\Delta h_s}$

2) Steam consumption can be determined if the concentrations, C_c and C_f and the SSC and the flow rate of the concentrate are known.

3) The overall energy balance is given by:

   $h_f + h_v = h_c + h_s$, where h_f = heat content of feed, h_v = heat content of vapour, h_c = heat content of condensate, h_s = heat content of steam.

4) Heat of evaporation is small compared to the difference in heat content of the feed and concentrate ($\phi_c\ c_{p,f}\ \theta_c - \phi_f\ c_{p,f}\ \theta_f$).

5) Heat losses in the evaporator decreases the value of ψ.

6) $c_{p,f}$ and $c_{p,c}$ are generally equal.

5.3.3 Heat transfer

general equation for heat transfer

Prediction of the heat transfer rate in film evaporators requires a substantial background of (heat) transfer phenomena. Here we will restrict ourselves to a simplified approach.

The general equation for the rate of heat transfer is given by:

$$\phi_q = U.\ A.\ \Delta\ \theta_{eff} \qquad \text{(E - 5.24)}$$

where:

ϕ_q = heat transfer rate (kW);

U = overall heat transfer coefficient ($kWm^{-2}\ K^{-1}$);

A = heat transfer surface (m^2)

$\Delta\ \theta_{eff}$ = effective temperature difference (°C) or (K).

Here, the effective temperature difference $\Delta\ \theta_{eff}$ is the difference between the condensing temperature of the steam outside the tube and the boiling temperature of the solution inside the tube. Except for effects such as boiling point elevation caused by an increasing dry solid content of the feed this temperature difference will remain constant along the tubes.

tube material influences thermal conductivity

The overall heat transfer coefficient (U) is influenced by flow conditions and physical properties of both the boiling solution in the tube and the condensed steam film outside the tube. As you might expect, the thickness and thermal conductivity of the tube also influences the value of U. The thermal conductivity (λ) depends on the tube material. (Heat transfer is described in greater detail in the BIOTOL text 'Bioprocess Technology: Modelling and Transport Phenomena).

Π Before examining Table 5.1 see if you can list the following in order of their thermal conductivities. Copper, iron, steel, stainless steel, aluminium, silumin. Then compare your order with the values for λ given in Table 5.1.

solid elements	λ($Wm^{-1} K^{-1}$)	Alloys	λ ($Wm^{-1} K^{-1}$)
Aluminium	220	Silumin	160
Copper	400	Steel	45-60
Iron	60	Stainless Steel	10-20

Table 5.1 Thermal conductivity of solid elements and alloys.

The overall heat transfer coefficient U of evaporators varies from 0.5 - 13 $KWm^{-2} K^{-1}$ depending on the type of evaporator and the physical properties of the feed. Table 5.2 lists some values.

Type of evaporator	U ($KWm^{-2} K^{-1}$)
Long tube vertical evaporator	
Natural circulation	1 - 3
Forced circulation	2 - 13
Short tube evaporator	0.5 - 2.5
Agitated film evaporator	2 - 5
Centrifugal evaporator	3 - 10

Table 5.2 Typical overall heat transfer coefficients in evaporators.

We usually have to evaluate the overall heat transfer coefficient from a pilot plant test with a single tube evaporator, because the physical properties of biosolutions are unknown. By simply collecting the condensate and measuring the $\Delta\theta_{eff}$, U can be calculated from Equation 5.24, as follows:

$$U = \frac{\phi_q}{A\Delta\theta_{eff}} = \frac{\phi_{s,c}\Delta h_v}{A\Delta\theta_{eff}} \quad \text{(E - 5.25)}$$

where:

$\phi_{s,c}$ = mass flow of steam condensate; Δh_v can be read: from a steam table. A is the surface area of the tube. Another method is to measure the amount of (condensed) vapour.

Π If Δh_v is obtained from a steam table, what else needs to be known in order to calculate the heat transfer rate.

We can see from Equation 5.25 that $\phi_q = \phi_{s,c}\ \Delta h_v$ ie provided the mass flow of steam (as condensate) is known, the heat transfer rate (ϕ_q) can be determined.

5.3.4 Selection of evaporation equipment

The selection of an evaporator is influenced by many factors. In biotechnology more often than not, evaporators will be considered as multi-purpose equipment, and should therefore meet the following requirements:

- preferably capable of handling a broad range of product viscosity (1 10 000 mPas);
- handling of heat sensitive products;
- minimum of scale formation, fouling and foaming.

Π Can you think of two desirable characteristics for an evaporation process designed to concentrate heat sensitive products.

The most important desirable characteristics are a small $\Delta \theta_{eff}$ and a short residence time.

agitated film evaporators

Falling film and plate evaporators are restricted in handling viscosities up to 200 mPas. For higher viscosities up to 10 000 mPas, an agitated film evaporator will be necessary.

The more evaporation effects, the longer the residence time and the broader the residence time distribution. Single stage evaporators show short residence times, but have a high steam consumption.

Scale formation and fouling depends strongly on the product to be handled and the way of operation. General guidelines cannot be given.

SAQ 5.5

How would you expect (select appropriate response):

1) the heat transfer rate (ϕ_q) to change if the size of the evaporator was increased - increase/decrease;

2) the heat transfer rate to change if copper replaced iron as tube material in the evaporator - increase/decrease;

3) forced circulation to effect the heat transfer rate in a vertical evaporator - increase/decrease?

5.4 Membrane processes

Concentration techniques using membranes have gained an increasing interest over the past ten years. Membranes are mainly applied in the dairy industry and in biotechnology to achieve separation and concentration of aqueous solutions. Application of membranes is not restricted to cell or molecular separation. By using a membrane as an integral part of a bioreactor, enzymes or even whole cells can be immobilised and thus the membrane module acts as a bioreactor.

Of importance to the installation of membrane systems are:

- average flux;
- degree of fouling;
- cleaning possibilities;
- membrane lifetime and costs;
- ease with which membranes can be replaced;
- containment and/or sterility;
- automation;
- capital investment.

5.4.1 Basic principle of membrane separation

The basic principle of membrane separation is illustrated in Figure 5.6.

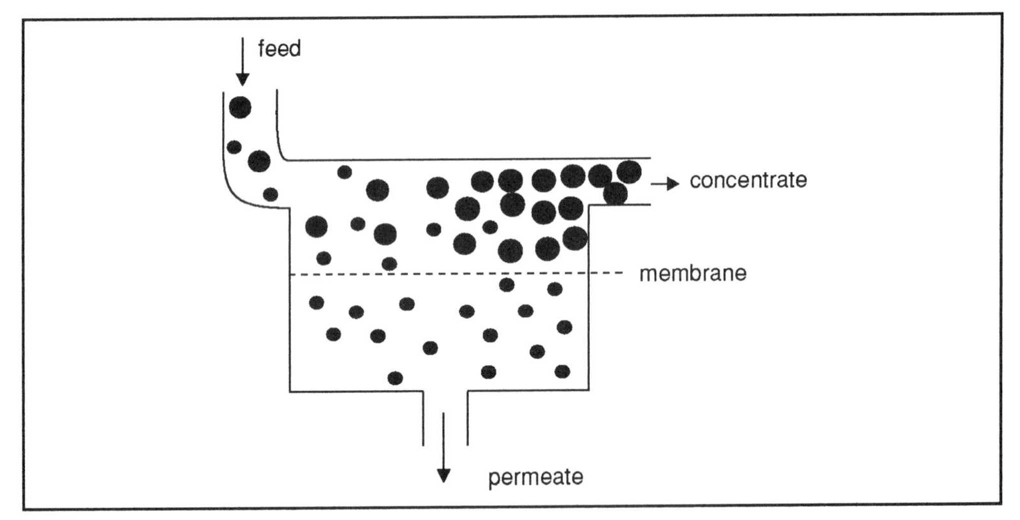

Figure 5.6 Basic principle of a membrane separator.

The feed is pumped along a membrane which acts as a selective barrier to the different components. Some components can freely permeate through the membrane while others will be retained. In this way the feed is separated into two streams: the retained enriched phase, and the permeated stream containing small components.

driving forces for separation

Membrane processes can be distinguished according to the type of driving force ensuring the transport through the membrane (see Table 5.3). You are probably familiar with many of these processes at least by name. The filtration processes are named according to the size of the pores in the filter. Thus microfiltration separates larger particles from a suspension, ultrafiltration is the term used when the pores of the filter are sufficiently small to remove large macromolecules. Hyperfiltration is sometimes referred to as reverse osmosis. We will examine these again a little later.

Driving force	Type of membrane process
Hydrostatic pressure Δp	microfiltration
	ultrafiltration
	hyperfiltration
Electric field $\Delta\psi$	electrodialysis
Concentration ΔC	dialysis

Table 5.3 Membrane processes grouped according to their driving force.

In most bioseparation processes a hydrostatic pressure difference is the driving force. For this reason only these processes will be discussed here. Membrane processes have the following advantages over other concentration techniques:

- no auxiliary phase required;
- process operation at ambient temperatures;
- possibilities for batch and continuous processing;
- relatively simple scale up.

There are also disadvantages, such as:

- little resistance to high/low pH values and temperatures;
- cleaning and sterilisation difficult;
- fouling, causing severe problems.

The hydrostatic pressure driven membrane processes can be classified according to the particle size retained. See Figure 5.7.

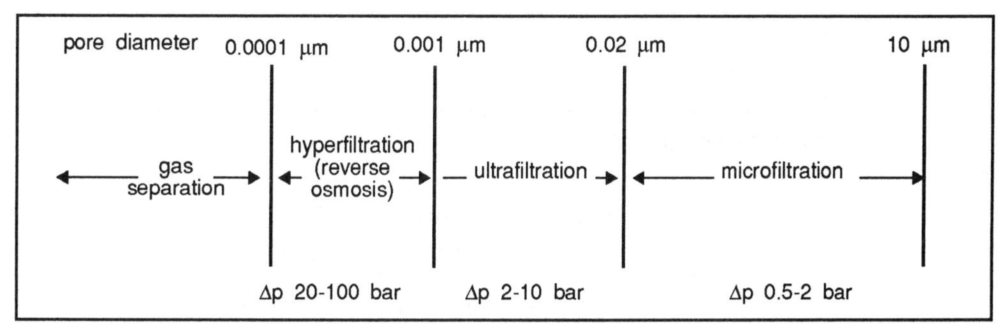

Figure 5.7 Classification of pressure driven membrane processes.

Π Examine Figure 5.7 and see if you can work out which of the processes described are rarely used in biotechnology.

Gas separation and hyperfiltration or reverse osmosis are rarely used in biotechnology. Ultrafiltration (UF) and microfiltration are identical processes, which differ only in the particle size to be retained.

symmetric and asymmetric membranes

Membranes are manufactured from a large variety of materials using several techniques, but they can all be generally classified into two categories: symmetric (or homogenous) and asymmetric membranes. In homogeneous membranes, the diameter of pores are almost constant over the entire cross section of the membrane. Consequently, the entire membrane thickness acts as a selective barrier. This differs from asymmetric membranes, where only a thin top layer determines the selective barrier.

These differences are clearly shown in Figure 5.8.

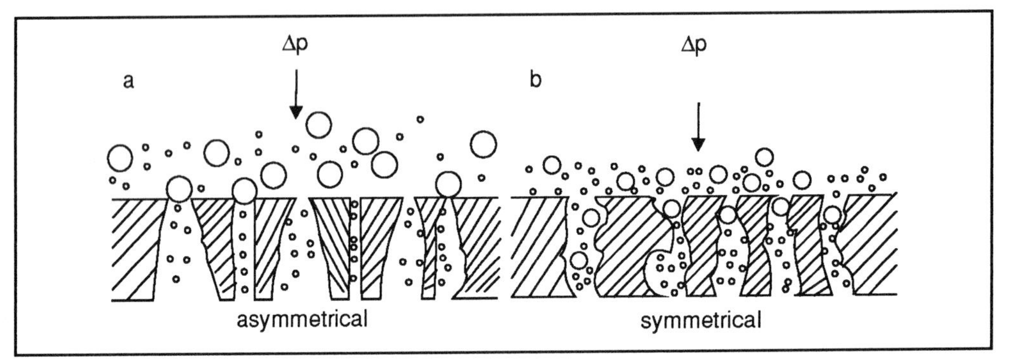

Figure 5.8 Comparison between symmetric and asymmetric membranes: a) asymmetrical, b) symmetrical.

The membrane properties are, to a great extent, determined by the production method and the materials used.

Generally speaking, membranes should have the following properties:

- high selectivity to provide a proper rejection;
- high mechanical strength;
- resistance to: fouling, solvents, high temperatures (sterilisable), low and high pH.

selection of membranes

Some of these properties are contradictory, hence one has to settle for compromises. As selectivity is the most important membrane property, we will briefly focus our attention on this aspect.

Membrane selectivity is mainly the result of the sieve action of the pores, but to some extent it is also caused by hydrophillic/hydrophobic interactions and membrane charge. Since the pores of a membrane are not uniform in size the selectivity shows a certain variation. The smaller the pore size distribution, the better the selectivity.

MWCO

The membrane selectivity is often expressed in terms of molecular weight cut off (MWCO).

pore size influences selectivity

In an ideal UF membrane, all the molecules below the MWCO-value will be permeated, while all the other molecules will be retained. See Figure 5.9, curve a.

Π Examine Figure 5.9 carefully, without examining the legend. Which curves represent membranes with a broad pore size distribution, and which represents a narrow pore distribution?

Curves b and c show membranes with broad and narrow pore size distributions respectively.

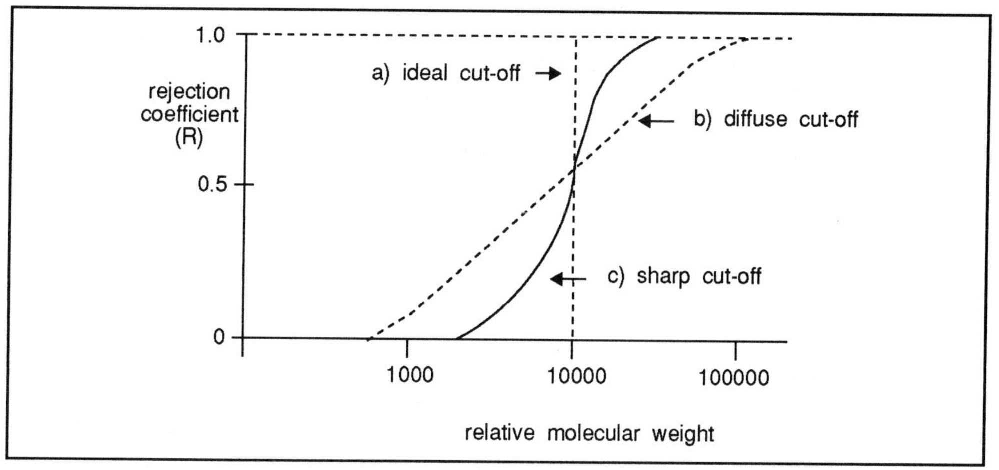

Figure 5.9 Molecular weight cut off for a) ideal membrane, b) broad pore size distribution, c) narrow pore size distribution.

SAQ 5.6

Without looking at Figure 5.7 match each type of filtration with an appropriate particle size and pressure difference.

Type of filtration	Particle size	Pressure difference (bar)
Ultrafiltration	0.01 μm	5
Microfiltration	0.01 nm	500
Hyperfiltration	0.8 nm	0.7
	0.1 mm	40
	5 μm	0.01

SAQ 5.7

Complete the following statements using the words listed below:

1) In membrane separations the [ ] stream contains relatively small components.

2) Membrane selectivity is often expressed in terms of [ ] cut off.

3) For [ ] membranes only a thin top layer determines the selective barrier.

4) A membrane with a sharp molecular weight cut off will have a [ ] pore size distribution.

Word list: molecular weight; asymmetric; small; permeated; narrow; symmetric; size.

In practice because of their limited selectivity, membranes are usually applied to concentrating material rather than for purification.

5.4.2 Membrane configurations

For concentrating materials using membranes, not only is a suitable membrane required, but also an appropriate housing for the membrane.

Most of the membranes for industrial applications are marketed as modules. A modular configuration offers flexibility in process design. Figures 5.10 and 5.11 show the most important configurations.

Selection for industrial application is based on:

- module performance;
- capital investment;
- operational costs.

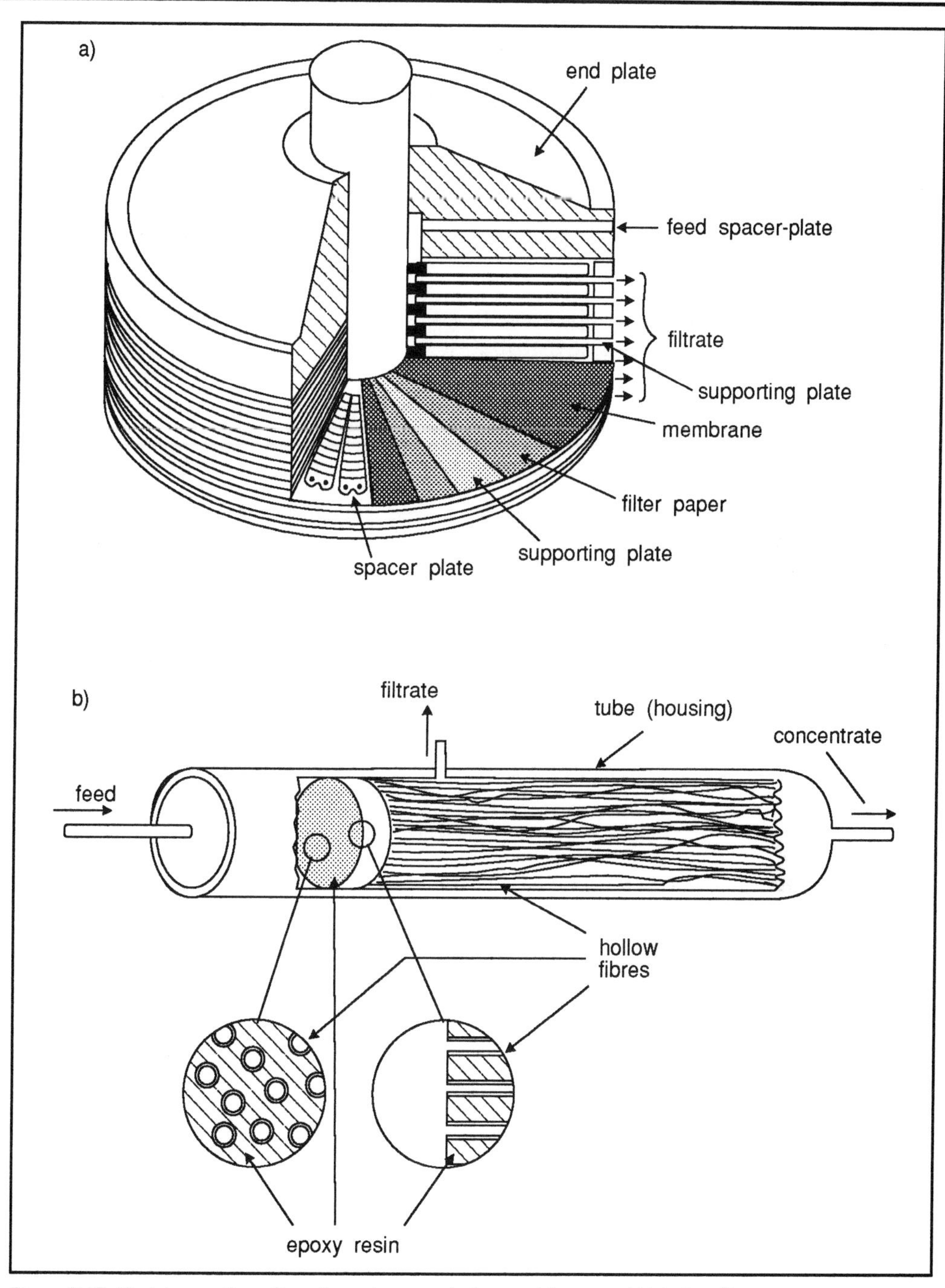

Figure 5.10 Modular configuration of membranes a) plate and frame and b) hollow fibre.

The modes of operation may vary for different applications. Batch, single-stage and multi-stage are the most widely applied configurations.

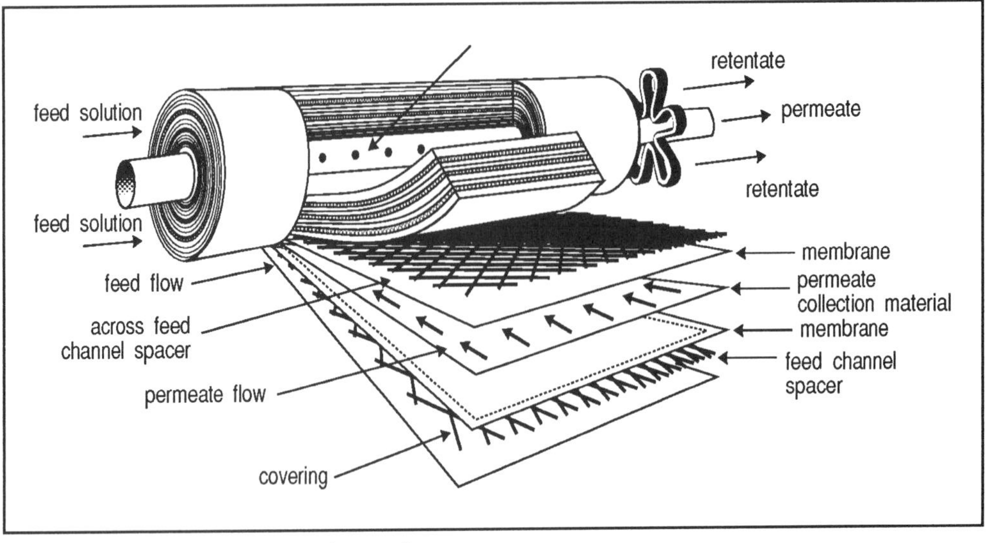

Figure 5.11 Modular configuration of a membrane.

5.4.3 Modes of operation of filters

batch mode retentate returned to feed tank

In a batch mode of operation the retentate is returned to the feed tank to be recycled to the module again. Depending on the required capacity the membrane system may consist of one module or a number of parallel modules. See Figure 5.12.

The concentration in the feed tank will increase during processing. When the final concentration has been reached the circulation is stopped.

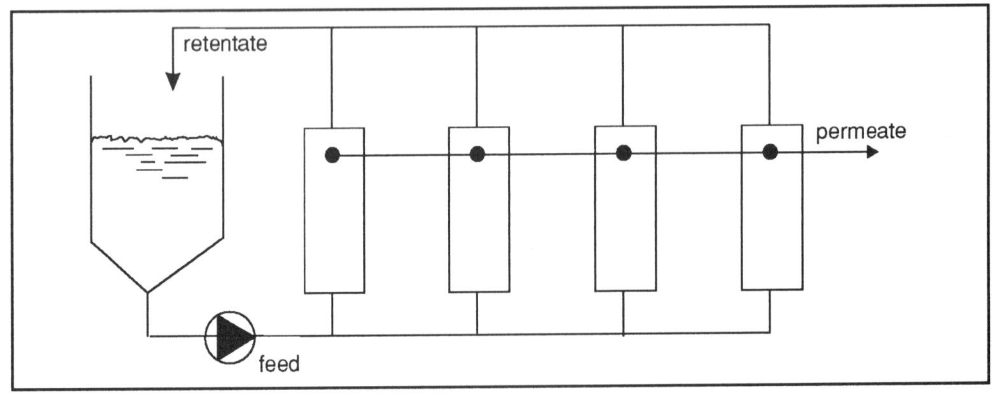

Figure 5.12 Batch operation.

single-stage continuous mode

In the single-stage continuous mode of operation the feed is brought to the required concentration in a single pass and then removed as retentate (concentrate). See Figure 5.13.

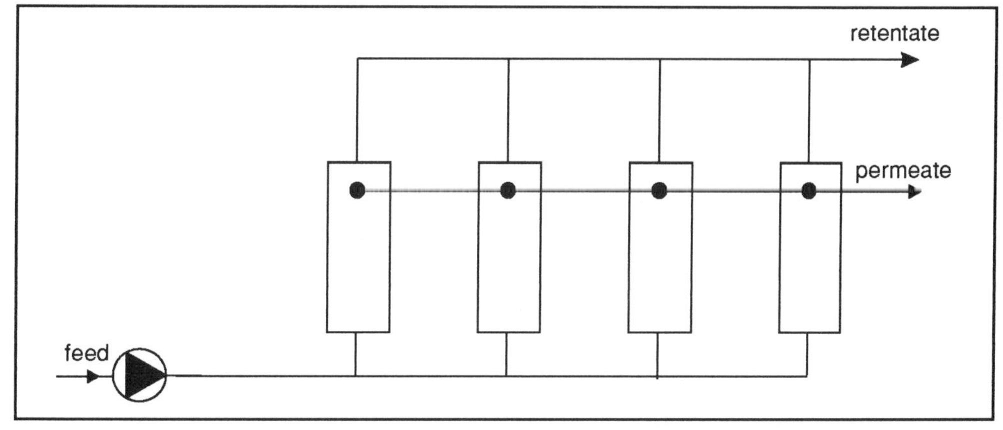

Figure 5.13 Single-stage continuous operation.

multi-stage continuous mode

A more efficient way of concentrating is to increase the concentration stagewise. In each subsequent stage the concentration is brought to a higher level. In the final stage the required concentration will have been reached. See Figure 5.14.

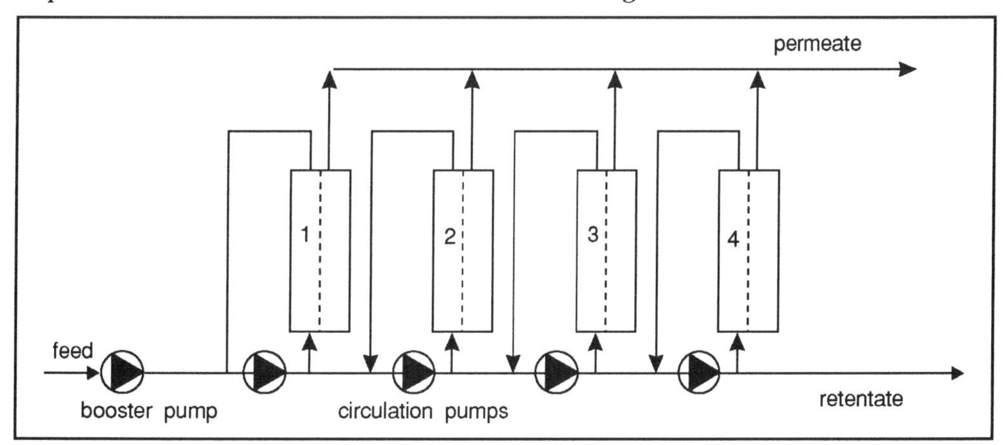

Figure 5.14 Multi-stage continuous operation.

Π Figures 5.12, 5.13 and 5.14 represent three different modes of operation of membrane filter. Which do you think will be the best system? The characteristics of these different modes will be discussed in the next section, so you will have the opportunity to decide which is the best on more through knowledge after you have completed that section.

5.4.4 Characterisation of different modes of operation

As already mentioned, in a membrane process the feed is separated into two flows, a retentate (or concentrate) and a permeate flow, differing in solute concentration. The separation efficiency and concentration factor can be derived from mass balances and membrane properties.

The basic principles can be formulated by using Figure 5.15.

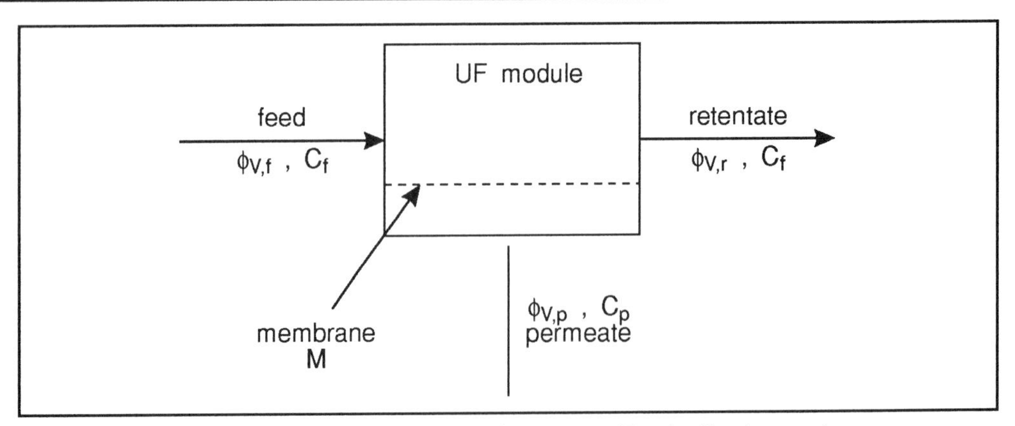

Figure 5.15 Main parameters in a membrane separation process, M = ultrafiltration membrane, UF module = ultra filtration module.

rejection

First we will define the main membrane property: rejection.

$$R = 1 - \frac{C_P}{C_r} \qquad \text{(E - 5.26)}$$

where:

R = membrane rejection coefficient;

C_P = concentration in permeate (mol m^{-3}or kg m^{-3});

C_r = concentration in retentate (mol m^{-3} or kg m^{-3}).

The concentration factor (α) can be defined as follows for batch and continuous modes of processing.

concentration factor

Batch process:

$$\frac{\text{initial feed volume}}{\text{retentate volume}} = \frac{V_f}{V_r} = \alpha \qquad \text{(E - 5.27)}$$

where:

V_f = initial feed volume (m^3)

V_r = retentate volume (m^3)

Continuous process:

$$\frac{\text{feed flow}}{\text{retentate flow}} = \frac{\phi_{V,f}}{\phi_{V,r}} = \alpha \qquad \text{(E - 5.28)}$$

where:

$\phi_{V,r}$ = volumetric flow of feed (m^3 s^{-1})

$\phi_{V,r}$= volumetric flow of retentate ($m^3\ s^{-1}$)

permeate yield, recovery factor

Permeate yield (also called the recovery factor, Δ) is given by:

$$\frac{\text{permeate volume}}{\text{feed volume}} = \frac{V_P}{V_f} = \Delta \qquad \text{(E - 5.29a)}$$

or expressed in flows (for continuous operation):

$$\frac{\phi_{V,p}}{\phi_{V,f}} = \Delta \qquad \text{(E - 5.29b)}$$

solid yield

And finally the non-permeated fraction, also called solid yield (ψ), is defined as:

$$\frac{\text{amount of retained component}}{\text{amount of component in the feed}} = \frac{\phi_{V,r}C_r}{\phi_{V,f}C_f} = \psi \qquad \text{(E - 5.30)}$$

(Do not confuse the solid yield with the specific steam consumption described earlier. We have used the same symbol to represent both. Clearly they are unrelated. To help you remember, for steam consumption we have used the subscript s).

where:

$\phi_{V,r}$ = volumetric flow rate of the retentate ($m^3 s^{-1}$)

$\phi_{V,f}$ = volumetric flow rate of the feed ($m^3\ s^{-1}$)

C_r = concentration of the retentate ($kg\ m^{-3}$)

C_f = concentration of the feed ($kg\ m^{-3}$)

Using these definitions, the various separation modes can be characterised.

Π We have just met with quite a few terms. It would be useful to produce yourself a 'reminder' summary sheet. Thus write out a list like this:

Rejection R =

Concentration factor α = (batch)

Concentration factor α = (continuous)

Let us now practice using these equations.

Π Determine the membrane rejection coefficient if the concentration of product in the retentate is four times higher than that in the permeate?

Your calculation should have been:

$$R = 1 - \frac{C_p}{C_r}$$

$$R = 1 - \frac{1}{4} = 0.75 \quad \text{(see Equation 5.26)}$$

Π If the permeate yield $\Delta = 1$ and the volume of permeate = $1m^3$, what is the volume of the feed?

You should have used Equation E - 5.29a.

$\Delta = \frac{V_p}{V_f}$ thus, since $\Delta = 1$ and $V_p = 1$, V_f also = $1m^3$.

Π If the initial feed volume of a batch process was $4m^3$ and the retentate volume was $1m^3$, what was the concentration factor?

You should use relationship E - 5.27.

$$\alpha = \frac{V_f}{V_r} = \frac{4}{1} = 4.$$

Π In a continuous process, the volumetric flow rate of the feed is 10 times that of the retentate flow. What is the concentration factor for this process?

Since $\alpha = \frac{\phi_{V,f}}{\phi_{V,r}}$ then $\alpha = \frac{10}{1} = 10$ (see Equation 5.28)

Single-stage separation with ideal mixing

In a single-stage separation unit, the feed is considered to be ideally mixed on the high pressure side of the membrane. As a consequence, the concentration along the membrane is constant at C_r (Figure 5.16).

ψ as a function of Δ

Based on the definitions we have just derived, we can find expressions for the performance of membrane separations used in different modes of operation. A useful relationship is the dependence of the non-permeated fraction (ψ) as a function of the permeate yield Δ.

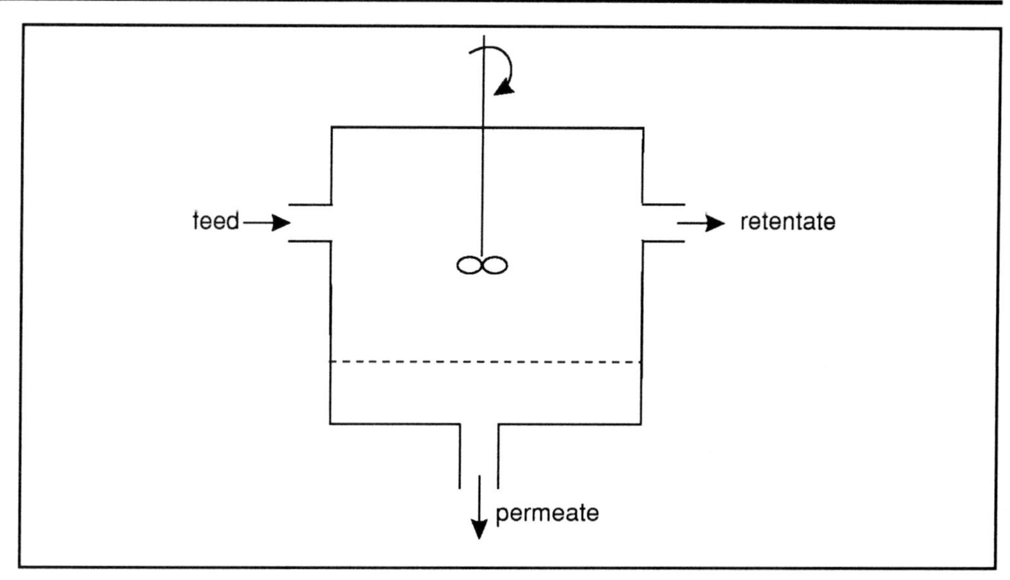

Figure 5.16 Single-stage with internal ideal mixing.

We remind you that:

$$\psi = \frac{\phi_{v,r}\, C_r}{\phi_{v,f}\, C_f} \qquad \text{(E - 5.30)}$$

A mass balance for the solute gives:

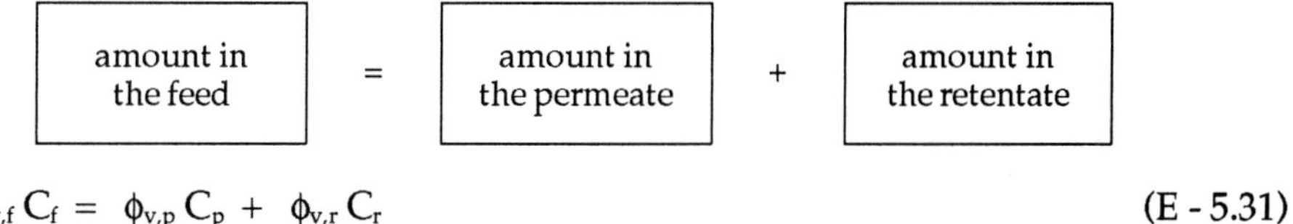

Thus: $\phi_{v,f}\, C_f = \phi_{v,p}\, C_p + \phi_{v,r}\, C_r$ (E - 5.31)

Substitution of Equation 5.31 into 5.30 and dividing by $\phi_{v,r}$ Cr gives:

$$\psi = \frac{1}{\dfrac{\phi_{v,p}\, C_p}{\phi_{v,r}\, C_r} + 1} \qquad \text{(E - 5.32)}$$

Now let us see if we can replace $\phi_{v,p}$, $\phi_{v,r}$, C_p and Cr with R and Δ.

Firstly the ratio $\frac{C_p}{C_r}$ is given by Equation 5.26.

Thus since $R = 1 - \frac{C_p}{C_r}$ then $\frac{C_p}{C_r} = 1 - R$ (E - 5.33)

The overall balance for the liquid flow gives:

$$\phi_{v,f} = \phi_{v,p} + \phi_{v,r}$$

(ie the volumetric flow of the feed = the combined volumetric flows of the permeate and retentate).

Thus $\frac{\phi_{V,f}}{\phi_{V,p}} = 1 + \frac{\phi_{V,r}}{\phi_{V,p}}$

But $\frac{\phi_{V,p}}{\phi_{V,f}} = \Delta$ (see Equation 5.29b)

Thus $\frac{1}{\Delta} = 1 + \frac{\phi_{V,r}}{\phi_{V,p}}$

Therefore $\frac{\phi_{V,r}}{\phi_{V,p}} = \frac{1-\Delta}{\Delta}$ or $\frac{\phi_{V,p}}{\phi_{V,r}} = \left(\frac{\Delta}{1-\Delta}\right)$ (E - 5.34)

We can now substitute Equations 5.33 and 5.34 into Equation 5.32.

Thus $\psi = \dfrac{1}{\left[\dfrac{\Delta}{(1-\Delta)}\right](1-R)+1}$

$$\psi = \frac{1}{\dfrac{\Delta - \Delta R + 1 - \Delta}{(1-\Delta)}} = \frac{1-\Delta}{1-\Delta R} \qquad \text{(E - 5.35)}$$

Thus we have established a relationship between the solid yield ψ, permeate yield (Δ) and the membrane rejection coefficient (R).

Equation 5.35 is quite important.

We can plot the non-permeated fraction (ψ) against permeate yield for different membrane rejection coefficients (R). This has been done in Figure 5.17.

separation achievable governed by rejection coefficient

From this figure it is evident that in an ideally mixed single stage separation unit, any desired separation can be achieved, provided that the membrane has a sufficiently high rejection coefficient.

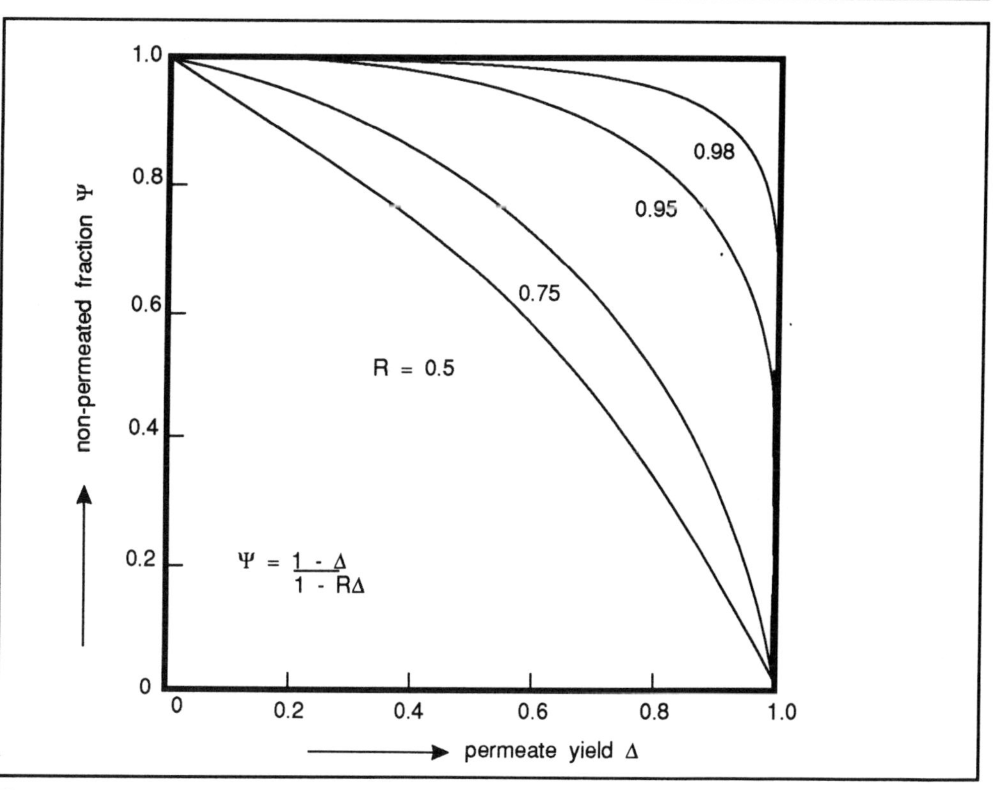

Figure 5.17 Non-permeated fraction as function of the permeate recovery for ideally mixed flow.

Single-stage separation with plug flow

If we have a very long membrane, the concentration will vary gradually from C_f to C_r along the membrane ie a type of plug flow. For the non-permeated fraction the following expression has been derived:

$$\psi = (1 - \Delta)^{1-R} \qquad \text{(E - 5.36)}$$

In Figure 5.18 the non-permeated fraction is plotted as function of the permeate yield.

plug flow more effective than ideally mixed

From this figure, it is evident again that in principle any desired separation can be achieved.

Comparing both types of flow (Figures 5.17 and 5.18) it may be concluded that separation under plug flow conditions is more effective than for an ideally mixed separation. The main reason for this is that the permeate formed out of the retentate in an ideally mixed systems stems from the final concentration C_r. In the case of plug flow however the retentate concentration rises gradually until C_r has been reached (this is represented in Figure 5.19). The average concentration of the retentate is much lower than C_r, this increases the solid yield (see Equation 5.30).

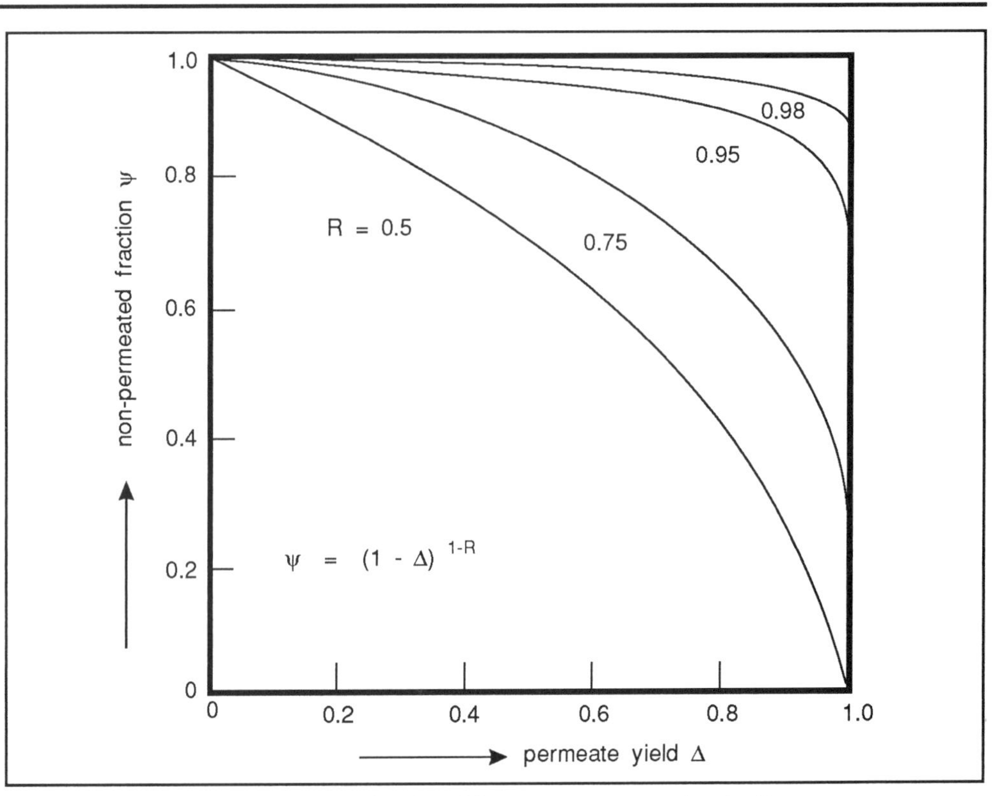

Figure 5.18 Non- permeated fraction as function of the permeate recovery for plug flow.

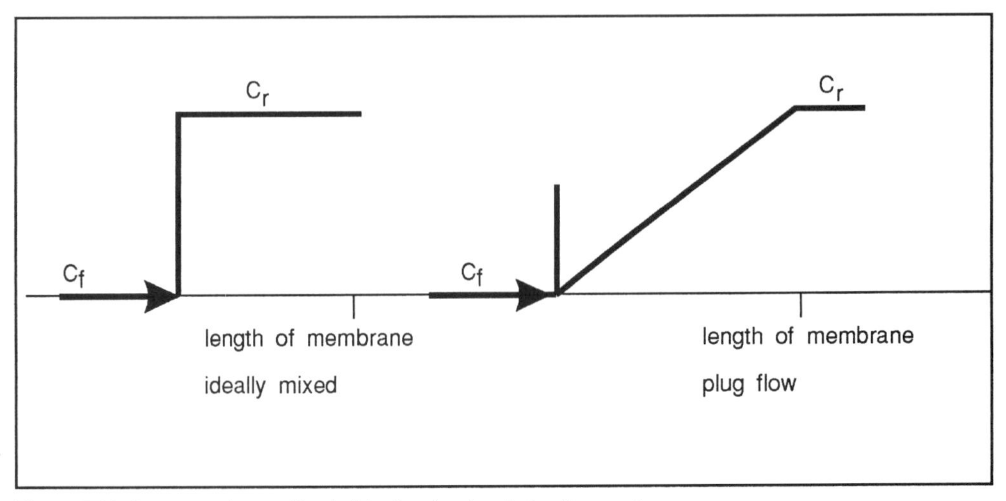

Figure 5.19 Concentration profiles in ideally mixed and plug flow systems.

single-stage plug flow = batch mode

It can be derived mathematically that the non-permeated fraction ψ, permeate yield (Δ) relation of a single-stage separation with plug flow is the same as for separation in batch mode. Equations 5.35 and 5.36 will be applied in Section 5.3.4, here it is used only as an illustration that any further separation with membranes can, in theory, be achieved.

SAQ 5.8

Determine the concentration factor for a single-stage membrane separation with ideal mixing, when the rejection coefficient is 0.5 and the permeate yield is 0.6.

The concentration in permeate = 10 kg m^{-3} and the concentration in feed = 15 kg m^{-3}.

Hint: you will need to use the data in Figure 5.17.

5.4.5 Water flux through an ideal pore membrane (ultrafiltration)

In this section we will consider the theory of water flux through an ideal membrane. In practice this can be used to explain certain influences in membrane concentration and to set up experiments in a systematical way.

An ideal pore membrane consists of a rigid matrix with a network of cylindrical, straight and parallel pores of identical diameter. See Figure 5.20.

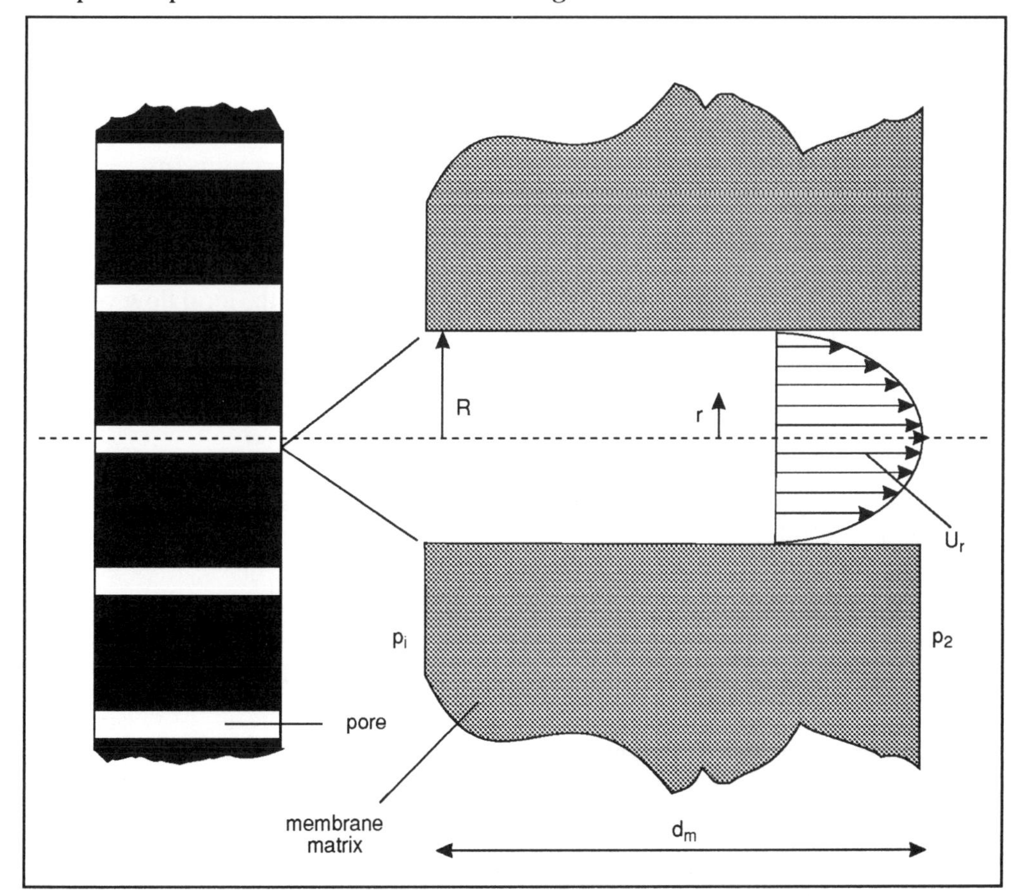

Figure 5.20 Ideal pore membrane. Pore radius (R), membrane thickness (d_m). See text for further details.

The permeability of the solid phase is negligible compared to that in the pores.

Actual flux (j) through the pores is influenced by:

- type of flow in pores;
- membrane properties;
- the driving force;
- fluid properties.

The type of flow has been described mathematically by Poisseulle.

The flow velocity in the pores can be described using Poisseulle's law. In this:

$$U_r = -\frac{R^2}{4\eta}\frac{\Delta p}{\Delta x}\left(1-\left(\frac{r}{R}\right)^2\right) \quad \text{(E - 5.37)}$$

where:

U_r = local flow velocity (see Figure 5.20), R = pore radius (m), $\frac{\Delta P}{\Delta x}$ = pressure gradient where Δp = pressure drop across the membrane and Δx is the pore length, r = distance from pore centre line, η = viscosity

The flow rate ϕ_v can be obtained by integration of the local flow velocity over the entire pore diameter ($r = O \rightarrow R$). We will not carry out this mathematical derivation here, but it yields:

$$\phi_{v,pore} = \frac{\pi R^4}{8\eta}\frac{\Delta p}{\Delta x} \quad \text{(E - 5.38)}$$

flux through a membrane (j)

Assuming that there are n pores in the membrane, the flux j (flow per unit area) will be as follows:

$$j = \frac{\phi_{v,pore}\, n}{A} \quad \text{(E - 5.39)}$$

where A = the area of the membrane

If we define membrane porosity ε as:

$$\varepsilon = \frac{\pi R^2 n}{A} \quad \text{(E - 5.40)}$$

by substituting of Equation 5.39 and 5.40 into Equation 5.38 we produce:

$$j = \frac{\phi_v}{A} \frac{R^2\varepsilon}{8\eta}\frac{\Delta p}{\Delta x} \quad \text{(E - 5.41)}$$

The pore length can be approximated by τd_m in which d_m is the membrane thickness and τ the tortuosity, a factor correcting for not straight but tortuous pores. A reasonable value for τ is 2-3. So Equation 5.41 becomes:

$$j = \frac{R^2\varepsilon}{8\eta}\frac{\Delta p}{\tau d_m} \quad \text{(E - 5.42)}$$

By defining the permeability coefficient K as follows:

$$K = \frac{\varepsilon R^2}{8 d_m \tau} \quad \text{(E - 5.43)}$$

Equation 5.42 can be simplified further:

$$j = K\frac{\Delta p}{\eta} \quad \text{(E - 5.44)}$$

The membrane resistance (R_m) is defined as :

$$R_m = \frac{1}{K} \quad \text{(E - 5.45)}$$

Values are usually of the order 10^{12} - 10^{16} m^{-1}. From Equation 5.44, it can be seen that the membrane flux is proportional to the transmembrane pressure and inversely proportional to the fluid viscosity.

Let us now test your understanding of these relationships.

SAQ 5.9

We have a membrane with a pore radius (R) of 10^{-7} m and a tortuosity $\tau = 3$. If the membrane is 10^{-4} m thick, and has a porosity (ε) = 0.2, calculate its permeability coefficient (K) and its resistance (R_m).

SAQ 5.10

If a membrane has a resistance (R_m) = 10^{13} m^{-1} and a pressure difference Δp = 200 kPa is applied across the membrane, what will the membrane flux (j) be if the viscosity of the medium (η) = 1 mPas. Attempt to convert the value of j you have calculated into litres of medium per square meter of membrane per hour.

If the pores are very small or the solution to be concentrated is far from ideal the osmotic pressure has to be taken into account. The osmotic pressure affects the applied transmembrane pressure:

$$j = \frac{K}{\eta}(\Delta p - \Delta\pi) \quad \text{(E - 5.46)}$$

where:

$\Delta\pi$ = osmotic pressure difference across the membrane (Pa)

Π List four membrane properties that influence flux.

We can see from Equation 5.43 that 1) porosity, 2) membrane thickness, 3) pore radius and 4) tortuosity influence flux through a membrane.

Although these equations show that the flux through the membrane is proportional to the transmembrane pressure, this holds only to a certain extent. If we for example concentrate an aqueous protein solution in an UF module the water will be permeated and the protein will be retained. As a consequence the protein concentration at the membrane surface will rise to much higher values than in the bulk flow. See Figure 5.21.

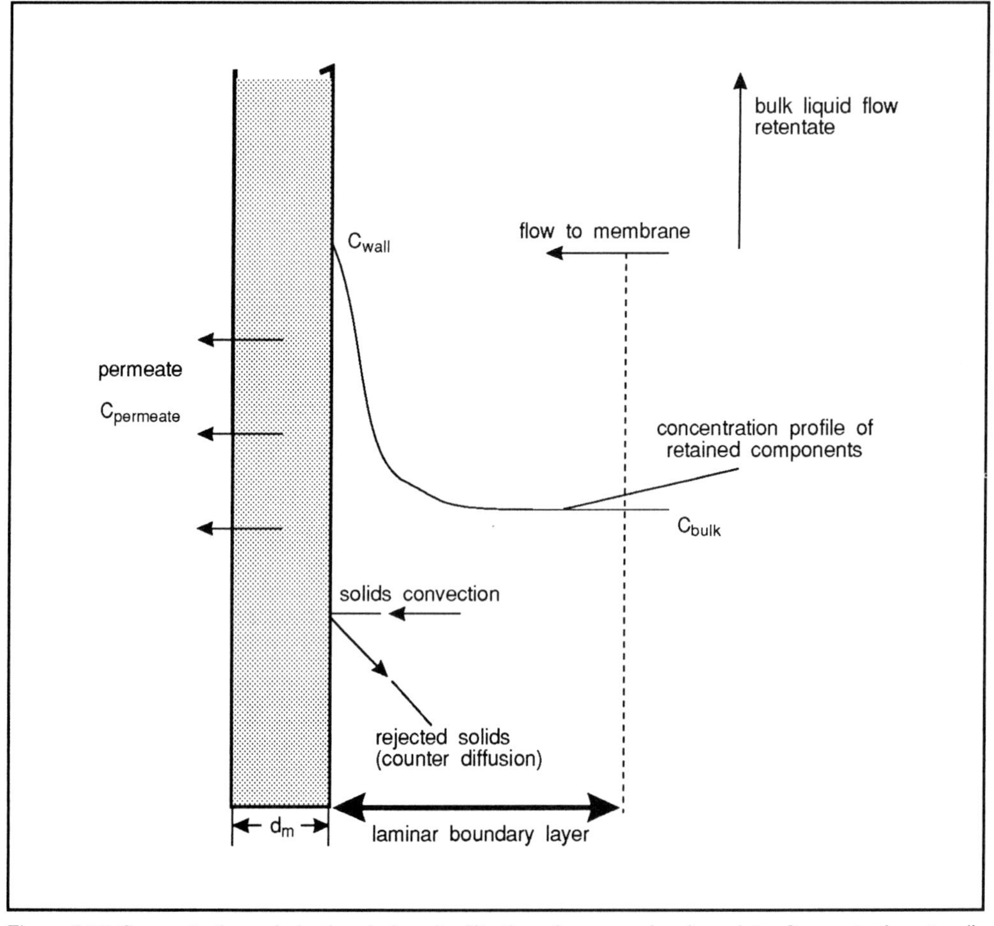

Figure 5.21 Concentration polarisation during ultrafiltration of macromelecular solute. Concentration at wall (C_{wall}), concentration in bulk (C_{bulk}), concentration in permeate ($C_{permeate}$).

concentration polarisation reduces flux

This phenomenon is known, as concentration polarisation. Here the flux is limited by the balance of convective flow to the membrane surface of the protein and its counter diffusion to the bulk as a result of a concentration difference.

Applying a differential mass balance the limited membrane flux can be derived as:

$$j = K' \ln \frac{C_{wall} - C_{permeate}}{C_{bulk} - C_{permeate}} \quad \text{(E - 5.47a)}$$

where:

K' = mass transfer coefficient (m s^{-1});

C_{wall} = concentration at wall (kg m^{-3});

C_{bulk} = concentration in bulk (kg m^{-3});

$C_{permeate}$ = concentration in permeate (kg m^{-3}).

Since $C_{permeate} << C_{bulk}$ we obtain:

$$j = K' \ln \frac{C_{wall}}{C_{bulk}} \qquad \text{(E - 5.47b)}$$

So, if concentration polarisation occurs the flux will be independent of the transmembrane pressure but dependent on the ratio C_{wall} / C_{bulk}.

Figure 5.22 shows this phenomenon for haemoglobin solutions.

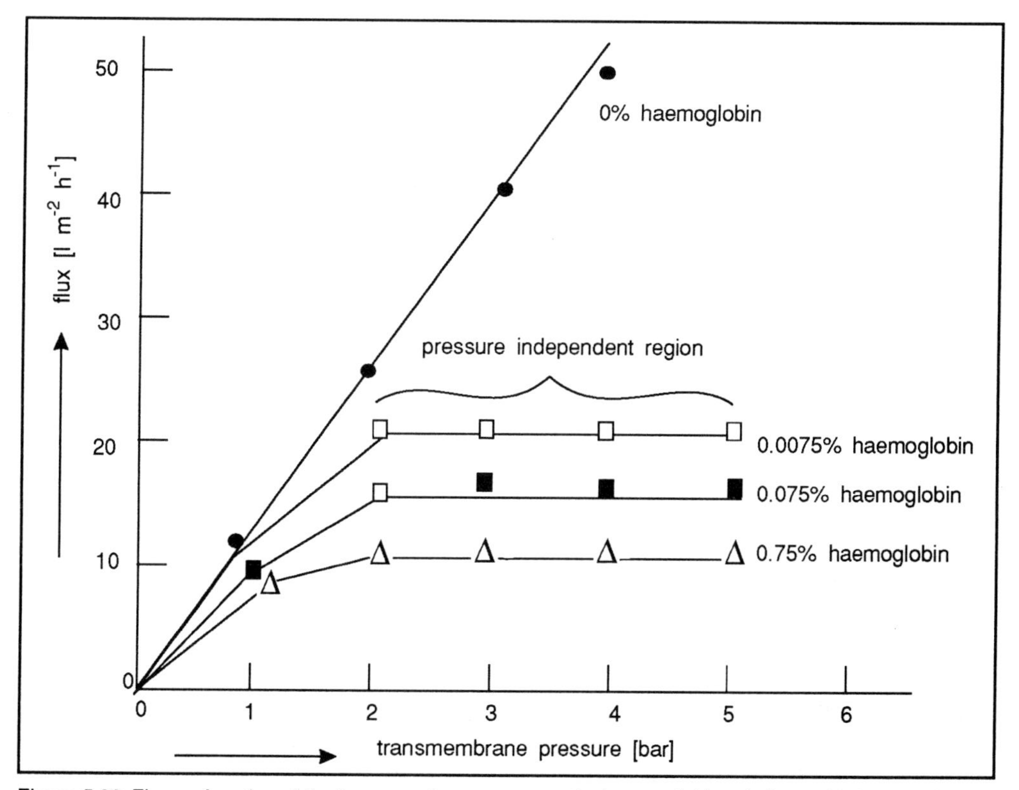

Figure 5.22 Flux as function of the transmembrane pressure for haemoglobin solutions with the protein concentration as parameter.

In Equations 5.45 and 5.45b K' is known as the 'mass' transfer coefficient. K' is defined here as the volume of material (usually water m^3) transferred across an interface (membrane) per surface (interface) area (m^2) per unit of time (s or h). Thus K' is a velocity. This coefficient can be found in the literature and depends on the flow conditions and the fluid properties.

Π Write, as fully as possible, expressions that define the following parameters: 1) Permeability coefficient; 2) Membrane resistance; 3) Membrane flux, with concentration as the driving force.

We hope that you would have recalled the following:

1) Permeability coefficient (K) $= \dfrac{\varepsilon R^2}{8\tau d_m}$ (see Equation 5.43)

2) Membrane resistance $\dfrac{1}{K} = \dfrac{1}{\left(\dfrac{\varepsilon R^2}{8\tau d_m}\right)}$ (see Equations 5.43 and 5.45)

3) Membrane flux (j) $= K' \ln \dfrac{C_{wall}}{C_{bulk}}$ (Equation 5.47b)

many factors influencing flux

We have already seen that many factors affect the flux and the separation characteristics of a membrane. The theory presented might have suggested that all of these influences can be predicted, but in general that is not the case. Experimental work is of crucial importance in the design of membrane processes. However the theory can be used to explain certain influences and, moreover, it can be used to set up experiments in a systematical way. As we have seen the driving force is the main influence on the flux but the fluid velocity along the membrane, temperature, pH of the protein solution and the concentration can also determine the flux to a great extent.

A well known flux model for the concentration of protein solutions using a UF membrane can be derived from the concentration polarisation model.

For concentration in batch mode the following equation has been derived.

$$j = K'' - K' \ln \alpha = K'' - K' \ln \frac{1}{(1-\Delta)} \qquad \text{(E - 5.48)}$$

membrane flux for concentration in batch mode

K'' is the initial flux in the absence of concentration polarisation. As soon as the concentration increases, the flux declines according to the concentration polarisation theory. K' is the 'mass' transfer coefficient.

Figure 5.23 shows the flux decline as function of the concentration factor.

increase in solute concentration can decrease flux

The decline in flux will continue until the maximum concentration ratio has been reached. At that point the flux becomes zero and the protein will be partially deposited at the membrane surface as a gel.

It is very easy to obtain the curve shown in Figure 5.23 experimentally. While the batch process is running, the actual flux and the volume will be recorded. A plot of j as a function of Ln α yields a straight line. By curve fitting K' and K'' can be obtained.

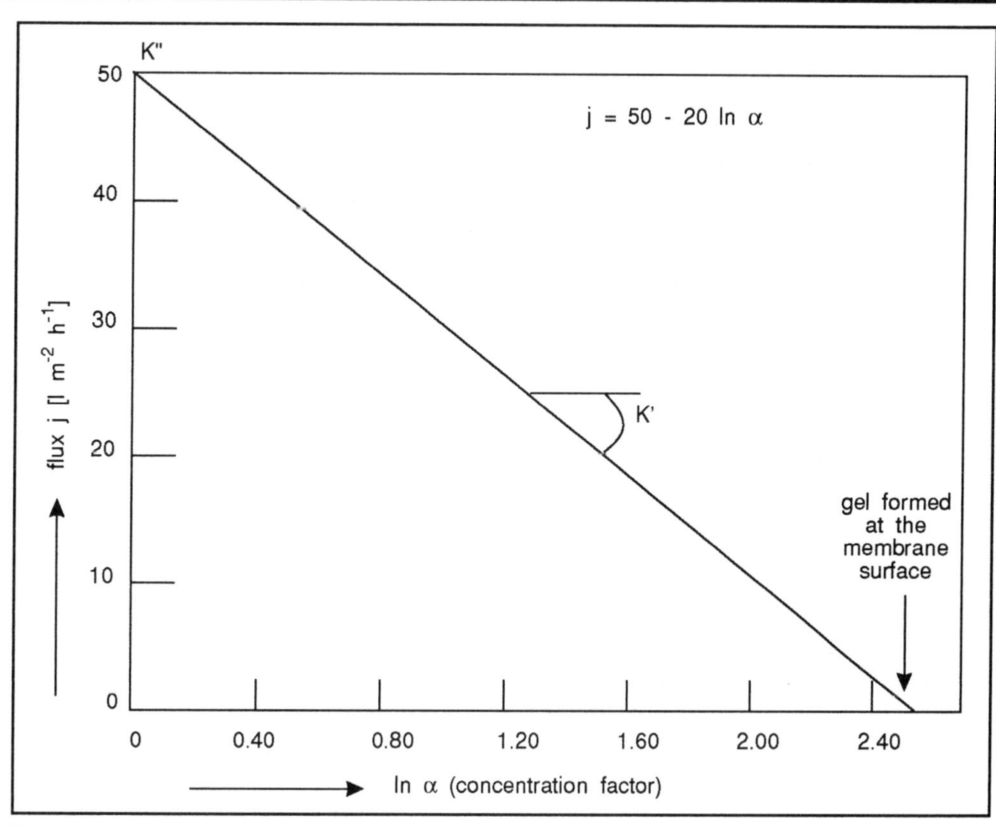

Figure 5.23 Effect of solute concentration on flux for macromolecules. Note K'' = intercept on j axis, K' = slope.

SAQ 5.11

1) Explain how a very small pore radius would influence membrane flux.

2) What limits flux in concentration polarization?

3) Explain how an increase in solute concentration at the wall, relative to that in the bulk liquid, would influence membrane concentration.

4) Define each of the components in the following equation:

   $j = K'' - K' \ln\alpha$

5) Explain how you would obtain values for the initial flux (in the absence of concentration polarization) and the mass transfer coefficient for batch concentration of protein solution using a UF membrane.

5.4.6 Applications

Concentration in batch mode operation

Earlier, we described several definitions for concentration factor (α), permeate yield (Δ) and rejection coefficient (R) - see Equations 5.26, 5.27 and 5.29. We can use these definitions to describe a relationship between the feed concentration (C_f) and the final concentration (C_r) of a filtration carried out in batch mode. The system used is illustrated in Figure 5.24.

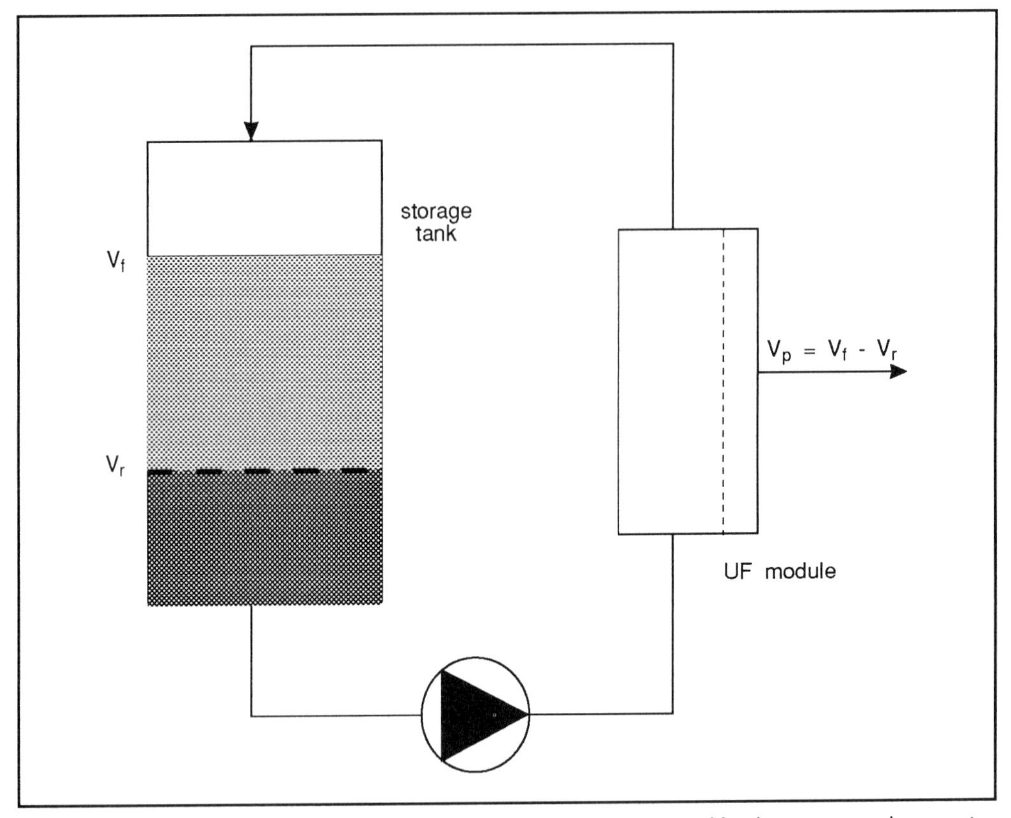

Figure 5.24 Concentration in batch mode. Vf, Vr and Vp represent volume of feed, retentate and permeate respectively.

From Figure 5.24 note that $V_p = V_f - V_r$ (the volume of the permeate = volume of the feed - volume of the retentate).

We remind you that:

rejection coefficient $R = 1 - \frac{C_p}{C_r}$

concentration factor $\alpha = \frac{V_f}{V_r}$

permeate yield $\Delta = \frac{V_p}{V_f}$

and the non-permeate fraction $\psi = \frac{V_r C_r}{V_f C_f}$

Since $\psi = \frac{V_r}{V_f} \frac{C_r}{C_f}$ and $\alpha = \frac{V_f}{V_r}$ it follows that:

$$\psi = \frac{C_r}{\alpha C_f}$$

or $C_r = \psi \alpha C_f$ (E - 5.49)

But in a batch system:

$\psi = (1 - \Delta)^{1-R}$ (see Equation 5.36)

Thus $C_r = (1 - \Delta)^{1-R} \alpha C_f$ (E - 5.50)

We can also write α in terms of Δ.

Since $\Delta = \frac{V_p}{V_f}$ and $V_p = V_f - V_r$

Then $\Delta = \frac{V_f - V_r}{V_f} = 1 - \frac{1}{\alpha}$

Therefore $\frac{1}{\alpha} = 1 - \Delta$

or $\alpha = (1 - \Delta)^{-1}$

Thus substituting this into Equation 5.50 gives:

$$C_r = \left(\frac{1}{1-\Delta}\right)^R C_f \quad \text{(E - 5.51a)}$$

or $C_r = C_f\ \alpha^R$ (E - 5.51b)

R = rejection coefficient

α = concentration factor

Π Calculate C_r in terms of C_f when R = 1 and R = 0.

If R = 1 then $C_r = C_f/1\text{-}\Delta$

If R = 0 then $C_r = C_f$. This answer makes sense since if there is no retention (ie R = 0) then the concentration on both sides of the membrane will be equal.

Equation 5.51b is graphically presented for $\alpha > 1$ in Figure 5.25 as the solid lines (we will discuss the dotted lines a little later).

C_r increases with increasing R

As can be seen from Figure 5.25 the higher the rejection coefficient the more C_r increases. At low R-values the product will be partly permeated and will be considered as lost. The instantaneous permeate flux can be calculated from Equation 5.48. Because α varies with time, the average flux should be taken into account.

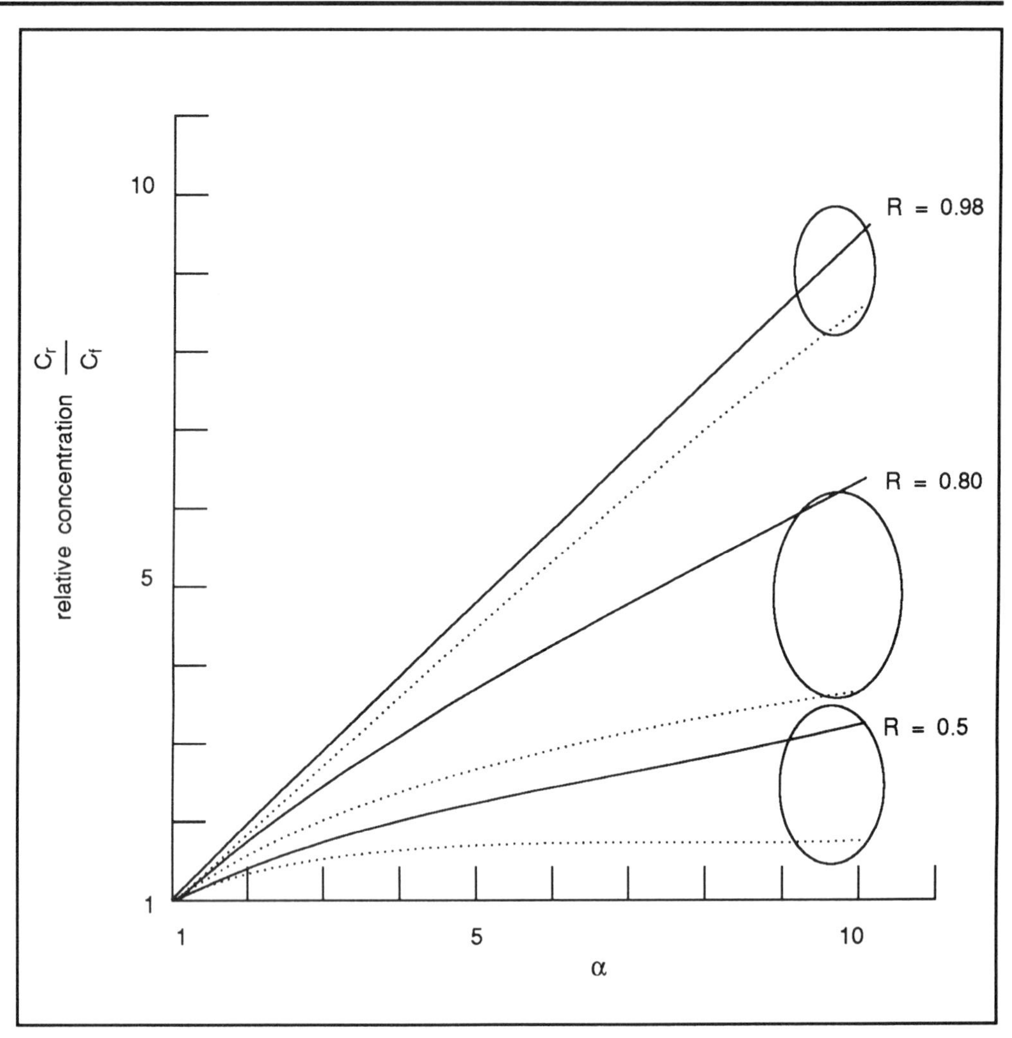

Figure 5.25 Relative concentration as a function of the concentration factor for different values of R (see text for details). Solid line = batch operation, dotted line = continuous operation.

Concentration in single-stage continuous operation

Substitution of the appropriate ψ-value (see Equation 5.35) into Equation 5.49 gives the retentate concentration in a single stage.

Since $\psi = \frac{1-\Delta}{1-R\Delta}$ (Equation 5.35) and $\alpha = \frac{1}{1-\Delta}$ and $C_r = \psi\,\alpha\,Cf$ (Equation 5.49)

$$\text{then } C_r = C_f \cdot \frac{1}{1-R\Delta} \text{ or } = \frac{\alpha}{\alpha - R\,(\alpha - 1)} \qquad \text{(E - 5.52)}$$

single-stage continuous operations have high permeate losses

Equation 5.52 is plotted in Figure 5.25 as the dotted lines. The figure shows that batch operation is more effective because the permeate losses are lower at all rejection and concentration values. The permeate flux for single-stage continuous operation can be calculated by substitution of the desired α-value into Equation 5.46.

Concentration in multi-stage continuous operation

If for some reason a continuous process is required a multi-stage operation can be a solution to prevent high permeate losses. The mathematics of this is a little more complex.

In each stage (i) of the cascade a permeate yield Δ_i will be attained, see Figure 5.26.

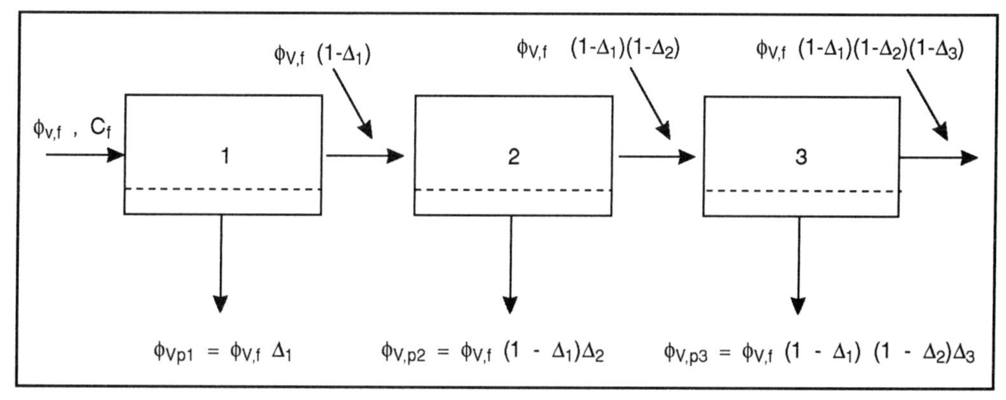

Figure 5.26 Definition of a multi-stage membrane operation.

From this figure, it can be shown that the concentration factor can be obtained as:

$$\alpha_N = \frac{1}{\prod_{i=1}^{N} (1 - \Delta_i)} \qquad \text{(E - 5.53)}$$

where:

π = continued multiplication;

i = multiplication variable;

N = number of cascades.

The retentate concentration C_r in stage N is given by Equation 5.49 ($C_r = \psi\ \alpha\ C_f$). However, the non-permeated fraction for a single-stage membrane operation with ideal mixing (Equation 5.35):

$\psi = \frac{1 - \Delta}{1 - R\Delta}$ should now be substituted by:

$$\psi_N = \prod_{i=1}^{N} \frac{(1 - \Delta_i)}{(1 - R\Delta_i)} \qquad \text{(E - 5.54)}$$

Substitution of Equation 5.53 and 5.54 into Equation 5.49 and rearranging gives:

$$C_r = \frac{C_f}{\prod_{i=1}^{N}} \qquad \text{(E - 5.55a)}$$

If $\Delta_1 = \Delta_2 = \Delta_3 = \Delta_{N'}$ then:

$$C_r = C_f \frac{1}{(1-R\Delta_1)(1-R\Delta_2)\,..........\,(1-R\Delta_N)} = \frac{1}{(1-R\Delta)^N} \qquad \text{(E - 5.55b)}$$

The flux in each stage can be determined using Equations 5.48 with the appropriate permeate yield factor Δ.

∏ For a multi-stage continuous membrane concentration, write an expression for:

1) the flow rate of feed from the 2nd to 3rd stage.

2) the flow rate of permeated fraction for the 3rd stage.

From Figure 5.26: 1) $\phi_{V,f}(1-\Delta_1)(1-\Delta_2)$, 2) $\phi_{V,f}(1-\Delta_1)(1-\Delta_2)\Delta_3$

Process times and required membrane area

For a continuous single-stage process the required membrane area can be calculated from the process time t, the actual membrane flux j and the volume to be processed V_p.

$$A = \frac{V_p}{j\ t_p} = \frac{V_f}{j\ t_p}\Delta \qquad \text{(E - 5.56)}$$

A is calculated from , j and V_p

In case of a multi-stage process we obtain:

$$A_1 = \frac{V_{p1}}{j_1\ t_p}\ ,\quad A_2 = \frac{V_{p2}}{j_2\ t_p}\ \ldots\ldots\ A_N = \frac{V_{pN}}{j_N\ t_p}$$

The total area to be installed is then:

$$A_{tot} = \sum_{i=1}^{N} A_i \qquad \text{(E - 5.57)}$$

for batch processes j is used

Prediction of the membrane area and process time for a batch process is more complicated because the flux varies in time.

Therefore Equation 5.56 should be slightly modified:

$$A = \frac{V_p}{\bar{J}\ t_p} = \frac{V_f}{\bar{J}\ t_p}\Delta \qquad \text{(E - 5.58)}$$

$\bar{J}$ is the average flux and may be calculated in a number of ways.

We will demonstrate one way here. It is rather demanding mathematically. Even if you cannot fully grasp the mathematics, try to understand what the stages are and how they can lead to determining either process time or the required membrane area.

For convenience we rewrite the flux equation for batch mode operation (Equation 5.48). Rewriting gives:

$$j = A \ln B\ (1 - \Delta) \qquad \text{(E - 5.59)}$$

where:

$A = K'$ and $B = e^{K''/K'}$

It can be shown mathematically that for the average flux the following equation holds:

$$\bar{J} = \frac{A\,B\,(\Delta_f - \Delta_i)}{E_i\,[\ln B\,(1 - \Delta_i)] - E_i\,[\ln B\,(1 - \Delta_f)]} \qquad \text{(E - 5.60)}$$

where Δ_i = initial permeate yield (usually 0), Δ_f = final permeate yield.

E_i (x) is the exponential integral, defined as follows:

$$E_i(x) = -\int_{-x}^{\infty} \frac{e^{-v}}{v}\,dv = \int_{-\infty}^{x} \frac{e^{-v}}{v}\,dv \qquad \text{(E - 5.61)}$$

Tables exist which test the exponential integral for different values of x (see Table 5.4).

Although this equation looks difficult, the contrary is the case as shown in Table 5.4.

Example:

We wish to calculate the membrane area required to concentrate a protein solution (volume = 150 m^3) by 11 times in 10 hours. We would like to do this for a batch and for continuous, single stage and three stage processes. For the three stage process, the following holds: $\Delta_1 = 0.5$, $\Delta_2 = 0.5$ and $\Delta_3 = 0.64$

Batch

The types of processes we are visualising using are like those illustrated in Figures 5.12 - 5.14. We will use the data presented in Figure 5.23 for values of K′ and K″.

membrane area for a batch system

We start by calculating the average flux using Equation 5.60, therefore Δ_i (initial Δ) and Δ_f (final Δ) is needed. We have already shown:

$$\Delta = 1 - \frac{1}{\alpha}$$

$\Delta_i = 0$ because there is no inital concentration ie $\alpha_i = 1$.

$$\Delta_f = 1 - \frac{1}{\alpha_f} = 1 - \frac{1}{11} = 0.909$$

x	$E_i(x)$	x	$E_i(x)$	x	$E_i(x)$	x	$E_i(x)$
.00	-	.50	.45 422	1.0	1.89 512	6.0	85.9 898
.01	-4.01 790	.51	.48 703	1.1	2.16 738	6.1	93.0 020
.02	-3.31 476	.52	.51 953	1.2	2.44 209	6.2	100.626
.03	-2.89 912	.53	.55 173	1.3	2.22 140	6.3	108.916
.04	-2.60 126	.54	.58 365	1.4	3.00 721	6.4	117.935
.05	-2.36 788	.55	.61 529	1.5	3.30 129	6.5	127.747
.06	-2.17 528	.56	.64 668	1.6	3.60 532	6.6	138.426
.07	-2.01 080	.57	.67 782	1.7	3.92 096	6.7	150.050
.08	-1.86 688	.58	.70 873	1.8	4.24 987	6.8	162.707
.09	-1.73 866	.59	.73 941	1.9	4.59 371	6.9	176.491
.10	-1.62 281	.60	.76 988	2.0	4.95 423	7.0	191.505
.11	-1.51 696	.61	.80 015	2.1	5.33 324	7.1	207.863
.12	-1.41 935	.62	.83 023	2.2	5.73 261	7.2	225.688
.13	-1.32 866	.63	.86 012	2.3	6.15 438	7.3	245.116
.14	-1.24 384	.64	.88 984	2.4	6.60 067	7.4	266.296
.15	-1.16409	.65	.91 939	2.5	7.07 377	7.5	289.388
.16	-1.08 873	.66	.94 878	2.6	7.57 611	7.6	314.572
.17	-1.01 723	.67	.97 802	2.7	8.11 035	7.7	342.040
.18	- .94 915	.68	1.00 712	2.8	8.67 930	7.8	372.006
.19	- .88 410	.69	1.03 608	2.9	9.28 602	7.9	404.701
.20	-.82 176	.70	1.06 491	3.0	9.93 383	8.0	440.380
.21	-.76 187	.71	1.09 361	3.1	10.6 263	8.1	479.322
.22	-.70 420	.72	1.12 220	3.2	11.3 673	8.2	521.831
.23	-.64 853	.73	1.15 068	3.3	12.1 610	8.3	568.242
.24	-.59 470	.74	1.17 906	3.4	13.0 121	8.4	618.919
.25	-.54 254	.75	1.20 733	3.5	13.9 254	8.5	674.264
.26	-.49 193	.76	1.23 551	3.6	14.9 063	8.6	734.714
.27	-.44 274	.77	1.26 360	3.7	15.9 606	8.7	800.749
.28	-.39 486	.78	1.29 161	3.8	17.0 948	8.8	872.895
.29	-.34 820	.79	1.31 954	3.9	18.3 157	8.9	951.728
.30	-.30 267	.80	1.34 740	4.0	19.6 309	9.0	103 7.88
.31	-.25 819	.81	1.37 518	4.1	21.0 485	9.1	113 2.04
.32	-.21 468	.82	1.40 290	4.2	22.5 774	9.2	123 4.96
.33	-.17 210	.83	1.43 056	4.3	24.2 274	9.3	1347.48
.34	-.13 036	.84	1.45 816	4.4	26.0 090	9.4	147 0.51
.35	-.08943	.85	1.48 571	4.5	27.9 337	9.5	160 5.03
.36	-.04926	.86	1.51 322	4.6	30.0 141	9.6	175 2.14
.37	-.00979	.87	1.54 067	4.7	32.2 639	9.7	191 3.05
.38	.02901	.88	1.56 809	4.8	34.6 979	9.8	208 9.05
.39	.06718	.89	1.59 547	4.9	37.3 325	9.9	228 1.58
.40	.10 477	.90	1.62 281	5.0	40.1 853	10.0	249 2.23
.41	.14 179	.91	1.65 013	5.1	43.2 757		
.42	.17 828	.92	1.67 741	5.2	46.6 249		
.43	.21 427	.93	1.70 468	5.3	50.2 557		
.44	.24 979	.94	1.73 192	5.4	54.1 935		
.45	.28 486	.95	1.75 915	5.5	58.4 655		
.46	.31 950	.96	1.78 636	5.6	63.1 018		
.47	.35 374	.97	1.81 356	5.7	68.1 350		
.48	.38 759	.98	1.84 075	5.8	73.6 008		
.49	.42 108	.99	1.86 793	5.9	79.5 382		
.50	.45 422	1.00	1.89 512				

Table 5.4 Exponential integral $E_i(x)$ for values of x - 0 $\rightarrow$ 10.

Now we need to know A and B. The flux equation is given in Figure 5.23, so K′ = 20 l m^{-2} h^{-1} and K″ = 50 l m^{-2} h^{-1}, so A = K′ = 20 and B = $e^{K''/K'}$ = $e^{2.5}$ = 12.18.

Now we can rewrite Equation 5.59 as follows:

$$j = 20 \ln 12.18 (1 - \Delta)$$

So for each value of Δ, the actual flux can be calculated. For a batch system however Δ (or α) varies with time so an average flux $\bar{J}$ is needed for calculations. We may now substitute the values just found into Equation 5.60.

$$\bar{J} = \frac{20 * 12.18 * (0.909 - 0.0)}{E_i [\ln 12.18 (1 - 0)] - E_i [\ln 12.18 (1 - 0.909)]}$$

$$\bar{J} = \frac{221.4}{E_i [2.5] - E_i [0.103]}$$

In Table 5.4 we may find the next values for the exponential integral:

$$E_i [2.5] = 7.07$$

$$E_i [0.103] = -1.6$$

Substitution gives:

$$\bar{J} = \frac{221.4}{7.07 - (-1.6)} = \frac{221.4}{8.67} = 25.5 \text{ l} m^{-2} h^{-1}$$

The membrane area can now be calculated using Equation 5.58:

$$A = \frac{150 . 10^3 . 0.909}{25.5 . 10} = 535 \text{ m}^2$$

Single stage process

membrane area for a single stage system

In a single stage continuous system the actual flux is very low because of the high value for Δ = 0.909. Applying Equation 5.59 we find:

j = 20 ln 12.18 (1 - 0.909) = 2.06 l m^{-2} h^{-1} - which is a very low value!

Substitution in Equation 5.56:

$$A = \frac{150 . 10^3 . 0.909}{2.06 . 10} = 6625 \text{ m}^2$$

Three stage process:

membrane area for a three stage system

For a three stage process the total membrane area is the sum of the areas required in each stage. We calculate the fluxes in the three stages first using Equation 5.59. The permeate yield in the first stage is 0.5 (given) so the flux will be here:

$$j_1 = 20 \ln 12.18 (1 - 0.5) = 36.2 \text{ l } m^{-2} h^{-1}$$

In the second stage the permeate yield factor is:

$$(1 - \Delta_1)(1 - \Delta_2) = 1 - \Delta_{overall}$$

$$\Delta_{overall} = 0.75$$

So: $j_2 = 20 \ln 12.18 (1 - 0.75) = 22.2 \text{ l m}^{-2} \text{h}^{-1}$

The third stage:

$$(1 - \Delta_1)(1 - \Delta_2)(1 - \Delta_3) = (1 - \Delta_{overall})$$

$$\Delta_{overall} = 0.909$$

So: $j_3 = 20 \ln 12.18 (1 - 0.909) = 2.06 \text{ lm}^{-2} \text{h}^{-1}$

The membrane areas can be calculated now using Equation 5.56.

First stage:

$$A_1 = \frac{V_f \Delta_1}{j_1 t} = \frac{150 . 10^3 . 0.5}{36.2 . 10} = 207 \text{ m}^2$$

Second stage:

$$A_2 = \frac{V_f (1 - \Delta_1) \Delta_2}{j_2 t} = \frac{150 . 10^3 . 0.25}{22.2 . 10} = 169 \text{ m}^2$$

Third stage:

$$A_3 = \frac{V_f (1 - \Delta_1)(1 - \Delta_2) \Delta_3}{j_3 t} = \frac{150 . 10^3 . 0.16}{2.06 . 10} = 1165 \text{ m}^2$$

The total area to be installed is now according to Equation 5.57:

$$A = 207 + 169 + 1165 = 1541 \text{ m}^2$$

summary of systems

We have summarised the results in Table 5.5.

	Batch	Single stage	Three stage
concentration factor	11 overall 3.4 average	11	2 4 11
flux $l\ m^{-2}\ h^{-1}$	25.5 average	2.06	36.2 22.2 2.06
area (m^2)	535	6625	207 169 1165

Table 5.5 Summary of the data derived from the calculation.

As can be seen from the table, the batch system is favourable with respect to membrane area to be installed. The single stage system is most unfavourable because it operates at the highest concentration and so the lowest permeate flux.

SAQ 5.12

Calculate the average concentration factor ($\overline{\alpha}$) for the batch operation described in the previous example (you will need to use the data on Figure 5.23 and Table 5.5).

Crossflow microfiltration

In contrast with ultrafiltration, microfiltration deals with suspended solids. Crossflow microfiltration is frequently used as an alternative to dead-end filtration.

Figure 5.27 shows the main differences in the filtration performance.

in crossflow microfiltration cake thickness is very thin

With dead end filtration the cake is deposited on the membrane. The thickness of the filter cake grows with time. In crossflow microfiltration the cake thickness is limited to a very thin layer. As a consequence the flux should remain constant in time.

In practice, however, the membranes in crossflow microfiltration are subject to heavy fouling since the feed contains cells, cell constituents and, in most cases, anti-foam, which deposit on the membrane surface and block the pores.

Frequent backflushing (reversing the feed stream) is necessary to keep the flux sufficiently high.

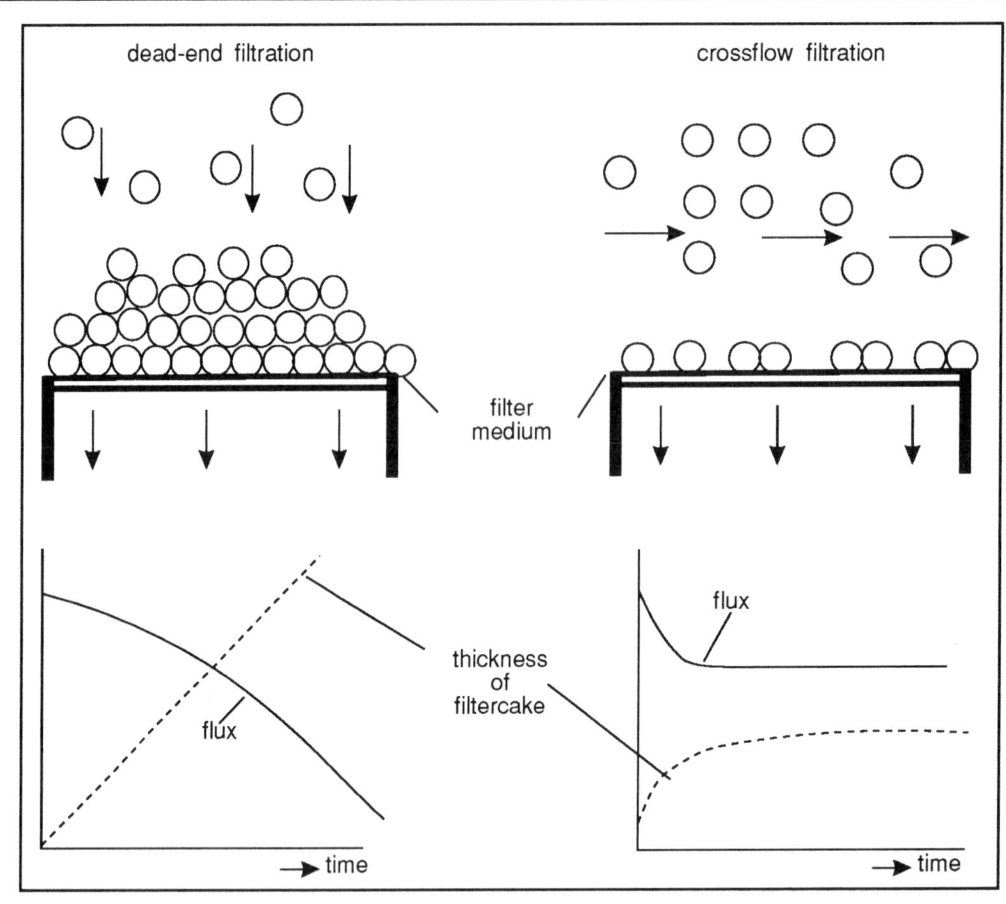

Figure 5.27 Performance of dead-end and crossflow microfiltration.

Summary and objectives

Evaporation is a very simple and versatile concentration technique which can be applied in both small and large scale processes. To avoid thermal degradation, evaporators are capable of operating at relatively low temperatures (30-40°C). The energy consumption for single effect evaporators is high. Applying more evaporation effects will reduce the energy consumption but will substantially increase the residence time and consequently the thermal damage to the product.

Because of the complexity of the fluid, the heat transfer in evaporators is based on experimental results rather than on the standard calculations that can be found in many textbooks and handbooks.

In industry, membrane systems are currently being applied in large scale processes (up to 500 m^2). Because of their modular set-up, membrane systems are very flexible and easy to expand. Many membrane types are available for a wide range of applications. One of the major drawbacks of membranes is their sensitivity to fouling which may considerably reduce the flux. The design of a membrane system will always be based on experimental results, preferably on pilot plant scale.

Now that you have completed the chapter you should be able to:

- describe the principle of operation of different types of evaporators;
- calculate specific steam consumption;
- use mass and enthalpy balances to evaluate energy consumption in evaporators;
- list advantages/disadvantages for different types of evaporators;
- distinguish micro-, ultra- and hyperfiltration according to particle size and pressure difference;
- compare and contrast different modes of operation of filtration processes;
- define rejection, concentration factor, permeate yield and non-permeate yield for batch and continuous modes of operation;
- write equations that describe how various factors influence water flux through an ideal membrane;
- derive expressions for final concentration and required membrane area for different modes of operation of membrane processes.

Concentration of products - II

Concentration of products - II

6.1 Introduction

concentrating procedures

In this chapter we continue the study of the concentration of products by examining liquid-liquid extraction and precipitation.

Liquid-liquid extraction is applied on a large scale in biotechnology. It is not only used as a concentration step but to a small extent as a purification step. The same also holds for precipitation. A possible disadvantage of both liquid-liquid extraction and precipitation, when compared with evaporators and membrane processes, is that they require an auxiliary phase (solvent, salt) which has to be recovered.

6.2 Extraction processes

Solvent extraction is one of the classical and most versatile unit operations for the recovery of bioproducts. It can be applied for products with a broad range of physicochemical properties.

recovery of lipophilic secondary metabolites

In biotechnology however it is predominantly applied in the recovery of lipophilic (fat loving) secondary metabolites.

Here we consider solvent extraction as an enrichment technique rather than a purification technique, although in practice a substantial purification can be achieved.

Solvent extraction of charged products is most efficient if they can be re-extracted into an aqueous phase (buffer). In this way extractable neutral impurities are easily separated. In a single stage extraction the degree of extraction is limited and for economic reasons multistage extraction is most frequently employed.

6.2.1 Extraction fundamentals

The basic principle of solvent extraction is the treatment of the (aqueous) feed with a non miscible organic solvent. The solute originally present in the feed will distribute itself over both phases. The more solute present in the organic phase, the better the extraction efficiency. In many cases the desired efficiency cannot be reached in a single stage and the aqueous phase will have to be treated with solvent again.

The repetitious treatment of the aqueous phase can be done in different modes as shown in Figure 6.1.

Four different modes of extraction are common as shown in Figure 6.1.

batch:

- single stage;
- multi stage.

continuous:

- cocurrent;
- countercurrent.

It depends on the nature of the process which of these modes are used. Batch and countercurrent modes are most commonly used.

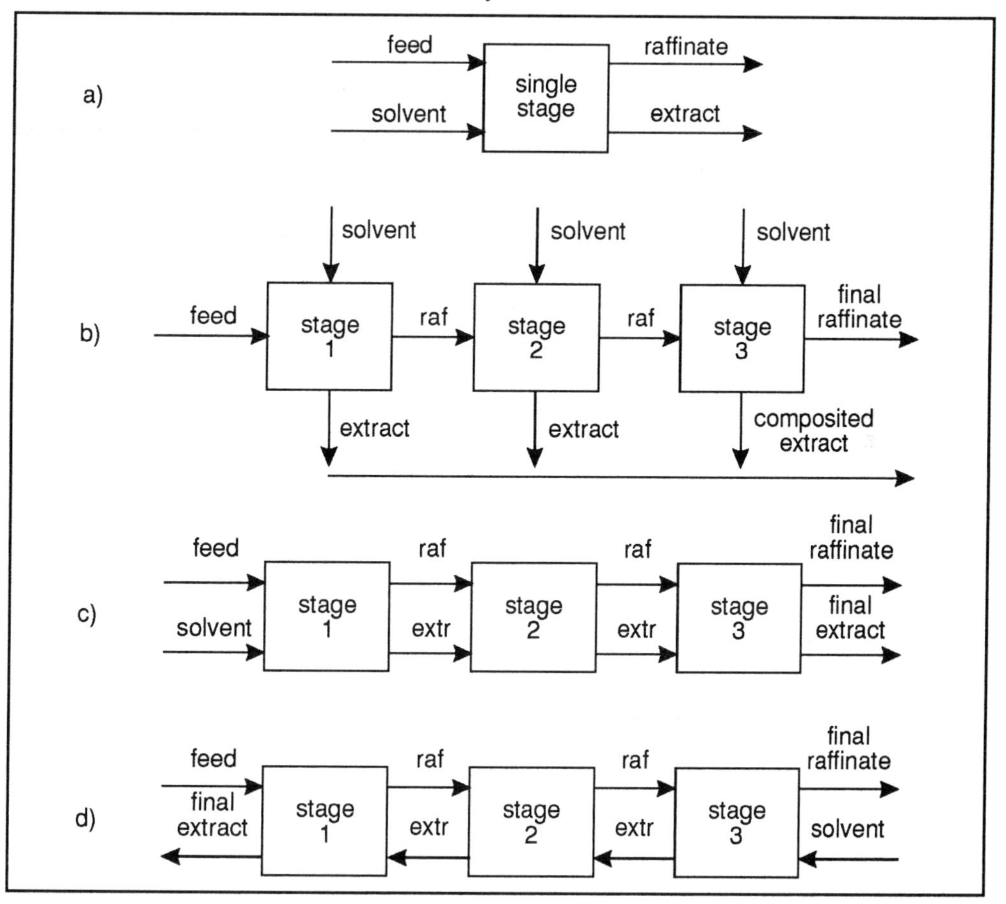

Figure 6.1 Modes of extraction. a) batch: single stage, b) batch: multi stage (also called cross-current), c) continuous: cocurrent, d) continuous: countercurrent. Note, raffinate is the remaining material left after solvent extraction (raf = raffinate, extr = extract).

modelling of extraction processes

For modelling extraction processes two subjects are of great importance. These are:

- equilibrium (thermodynamics);
- kinetics (mass transfer).

Equilibrium is relevant to the consideration of the number of subsequent extraction stages required to achieve a certain degree of separation (efficiency).

Kinetics is relevant to the consideration of the size of each extraction stage.

The dimensions of the extraction unit will be determined by the number of stages required, the size of each stage and the volume flow rates of both phases.

Phase equilibrium and distribution

determination of distribution coefficient

Extraction involves the use of systems composed of at least three components and in most cases all three components appear in both mutually soluble phases. In some cases this ternair equilibrium can be determined thermodynamically. The phase equilibrium thermodynamics is, however, outside of the scope of this book. Here we restrict ourselves to two immiscible liquids (feed, which in biotechnology is usually aqueous and a solvent). The product to be extracted is soluble in both phases but is preferentially soluble in the organic phase. The ratio of solubilities (or concentration) in the aqueous phase and organic phase is called the distribution or partition coefficient. Let us now see how extraction works.

Consider a small vessel which contains water and a low fraction of phenol, (see Figure 6.2).

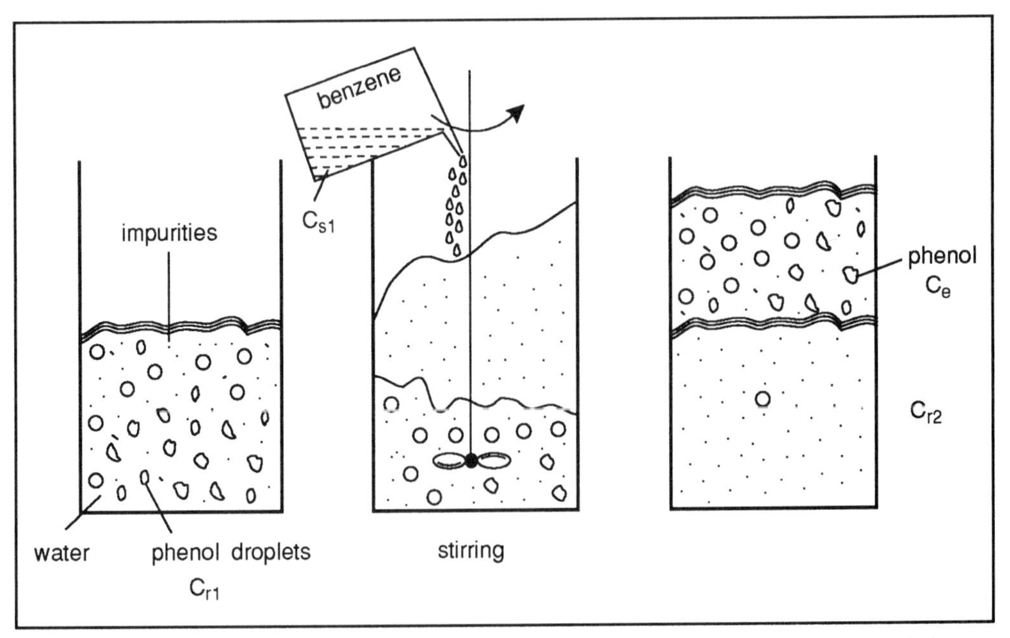

Figure 6.2 Single-stage batch solvent extraction. Concentration in aqueous phase before (C_{r1}) and after (C_{r2}) extraction; concentration in solvent phase before (C_{s1}) and after ($C_{s2} = C_e$) extraction.

determination of the distribution coefficient

Under thorough stirring we add benzene as an extractant. Because benzene does not dissolve in water and has a lower density it will form a top layer after the stirring has been stopped.

The distribution coefficient (α) is defined as:

$$\alpha = \frac{C_e}{C_r} \qquad \text{(E - 6.1)}$$

provided that the volumes of both phases are equal and equilibrium has been reached.

Note C_e is the concentration of the solute in the extract phase and C_r is the concentration in the raffinate phase at equilibrium. In general, the distribution coefficient depends on the temperature and the concentration.

Now lets consider a theoretical ideal single-stage extraction in continuous mode in which after a certain contact time of both phases the effluent phases are in thermodynamical equilibrium (see Figure 6.3).

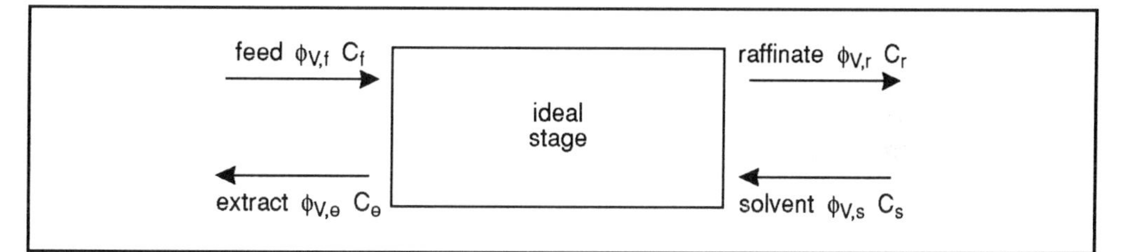

Figure 6.3 Theoretical equilibrium stage for a single stage continuous countercurrent extraction. Flow rates for feed etc ($\phi_{V,f}$; $\phi_{V,r}$; $\phi_{V,s}$ and $\phi_{V,e}$). Concentration of solute in the aqueous feed (C_f), aqueous raffinate (effluent stream; C_r), fresh solvent (C_s) and extract (C_e).

A solute balance gives:

$$\phi_{V,f}\, C_f + \phi_{V,s}\, C_s = \phi_{V,r}\, C_r + \phi_{V,e}\, C_e \qquad \text{(E - 6.2)}$$

Because the fresh solvent is free of solute ($C_s = 0$) the feed is divided into two streams:

$\phi_{V,r}\, C_r$ which flows forward, and

$\phi_{V,e}\, C_e$ which flows backward

(ie countercurrent extraction).

The ratio:

$$\frac{\text{backflow}}{\text{forward flow}} = \frac{\phi_{V,e}\, C_e}{\phi_{V,r}\, C_r} = \alpha\, \frac{\phi_{V,e}}{\phi_{V,r}} = S \qquad \text{(E - 6.3)}$$

Note for a single stage batch process $S = \alpha \dfrac{V_s}{V_r}$ where V_s = volume of solvent, V_r = volume of raffinate.

separation factor related to distribution coefficient

S is called the separation factor. Once S is known the non-extracted fraction (C_r/C_f) can be calculated applying the following assumptions, $C_s = 0$; $C_e = \alpha C_r$; $\phi_{V,e} = \phi_{V,s}$. and finally $\phi_{V,r} = \phi_{V,f}$

So, a material balance becomes:

$$1 - f = \frac{C_r}{C_f} = \frac{1}{1+S} \qquad \text{(E - 6.4)}$$

where: f = extracted fraction

Π It would be useful for you to derive Equation 6.4 for yourself. Begin with Equation 6.2. Remember that $C_e = \alpha C_r$, $S = \alpha \frac{\phi_{V,e}}{\phi_{V,r}}$, $C_s = 0$ and $\phi_{V,r} = \phi_{V,f}$

Here is our solution:

Since:

$\phi_{V,f} C_f + \phi_{V,s} C_s = \phi_{V,r} C_r + \phi_{V,e} C_e$ (Equation 6.2) and $C_s = 0$, $\phi_{V,r} = \phi_{V,f}$ and $C_e = \alpha C_r$

then:

$$\phi_{V,r} C_f + 0 = \phi_{V,r} C_r + \phi_{V,e} \alpha C_r$$

Divide through by $\phi_{V,r} C_f$

$$1 = \frac{C_r}{C_f} + \frac{\phi_{V,e}}{\phi_{V,r}} \alpha \frac{C_r}{C_f}$$

We defined the non-extracted fraction as C_r/C_f.

But we can define the non-extracted fraction (C_r/C_f) as 1 - f where f is the extracted fraction. We also defined $S = \alpha \frac{\phi_{V,e}}{\phi_{V,r}}$.

Thus $1 = (1 - f) + S(1 - f)$

Dividing through by 1 - f gives:

$$\frac{1}{1-f} = 1 + S$$

Thus $(1 - f) = \frac{1}{1+S}$

Thus the non-extracted fraction $1 - f = \frac{C_r}{C_f} = \frac{1}{1+S}$

In Figure 6.4 the non-extracted fraction (1 - f) is plotted as function of the separation factor (dotted line).

This line shows the relative concentration of the effluent stream (raffinate) after a single stage countercurrent extraction.

The extracted fraction or the degree of extraction is defined as follows:

$$f = \frac{\text{amount extracted}}{\text{amount in feed}} = \frac{\phi_{V,e}\, C_e}{\phi_{V,f}\, C_f} \qquad \text{(E - 6.5)}$$

Using the same assumptions as before we will find:

$$f = \frac{S}{S+1}$$

(this is simply a rearrangement of Equation 6.4). (E - 6.6)

In Figure 6.4 the solid line represents the fraction of extracted solute.

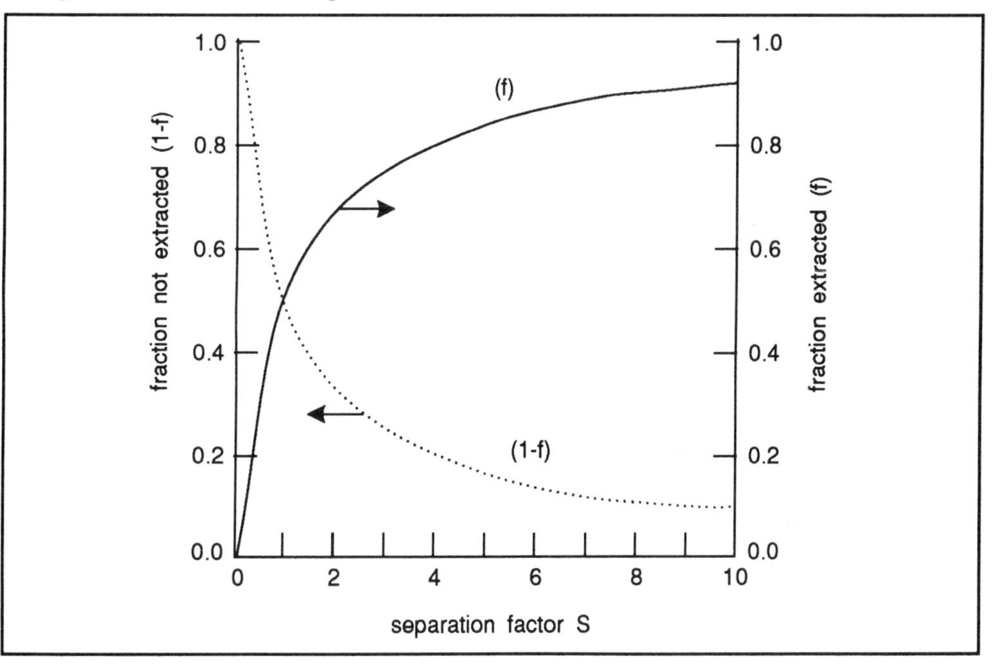

Figure 6.4 Non-extracted fraction (- - - -) and extracted fraction (——) as functions of the separation factor.

high S values essential for single stage extractions

As can be seen, a high degree of extraction can only be achieved in a single stage at high values for S. A high distribution coefficient is preferable because a high solvent/raffinate ratio is expensive as all the solvent has to be recovered for the next cycle.

If a high distribution coefficient cannot be attained further treatment with solvent in a second or third stage may have to be the solution.

Π **Write an expression for 1) the distribution coefficient, 2) the separation factors and 3) the degree of extraction, using Figure 6.2 as a model. Try to do this without using the text.**

You should have come to the following conclusions:

1) $$\alpha = \frac{C_e}{C_r}$$

2) $$S = \alpha \frac{\phi_{V,e}}{\phi_{V,r}}$$

3) $$f = \frac{\phi_{V,e} C_e}{\phi_{V,f} C_f}$$

Π If the distribution coefficient α is 10, and the concentration of the solute in the raffinate is 1 mol m^{-3}, what will be the concentration in the extraction solvent?

The answer is 10 mol m^{-3} as $\alpha = \frac{C_e}{C_r}$

Π If the flow rate of the extraction solvent in this system is $1 m^3 h^{-1}$, and the flow rate of the feed is $10 m^3 h^{-1}$, what is the extracted fraction (f).

f is 1 because $f = \frac{\phi_{V,e} C_e}{\phi_{V,f} C_f} = \frac{1}{10} \times \frac{10}{1} = 1$. That is, all of the solute is extracted, a result which is virtually impossible to achieve in practice.

Let us now consider a two stage extraction unit with a feed flow and a solvent feed flow of 1 $m^3 s^{-1}$. The solute mass flow of the raffinate is r (see Figure 6.5).

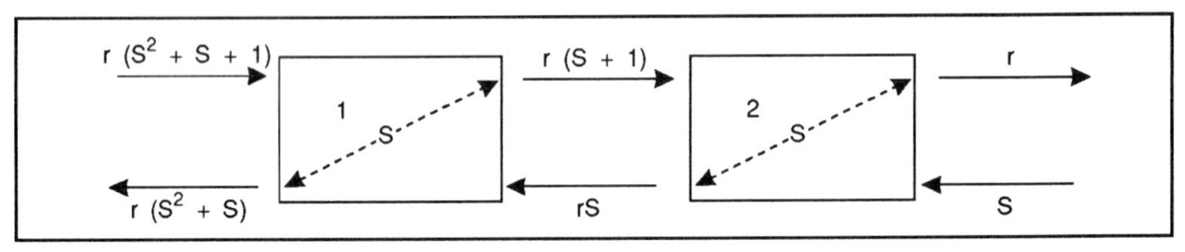

Figure 6.5 Multi-stage extraction (see text for details).

Let us explain the terms displayed on Figure 6.4

We will assume that the process is in a steady state and the amount of solute in the raffinate from stage 2 is r and the separation factor (S) is the same at each stage. The amount of solute in the solvent effluent from stage 2 is therefore Sr.

We can apply a material balance for stage 2:

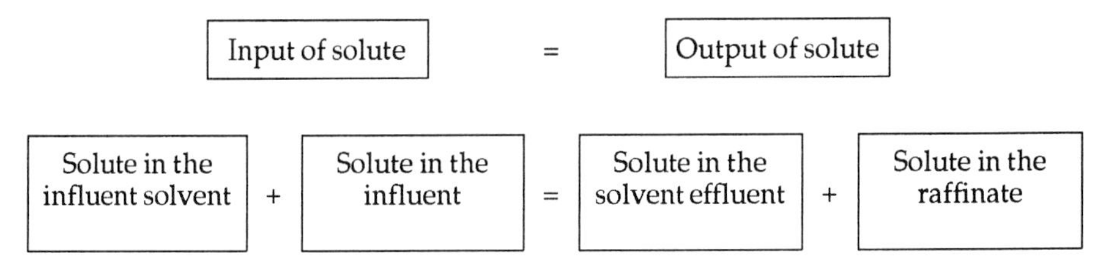

If the influent solvent contain no solute, then:

Solute in the influent	$= rS + r$
	$= r\ (S + 1)$

For stage 1, if the effluent solute contains r (S + 1), then at equilibrium, the solute in the solvent effluent will be r(S + 1) multiplied by the separation factor S and therefore $= r\ (S^2 + S)$.

Using a similar material balance to that described for stage 2:

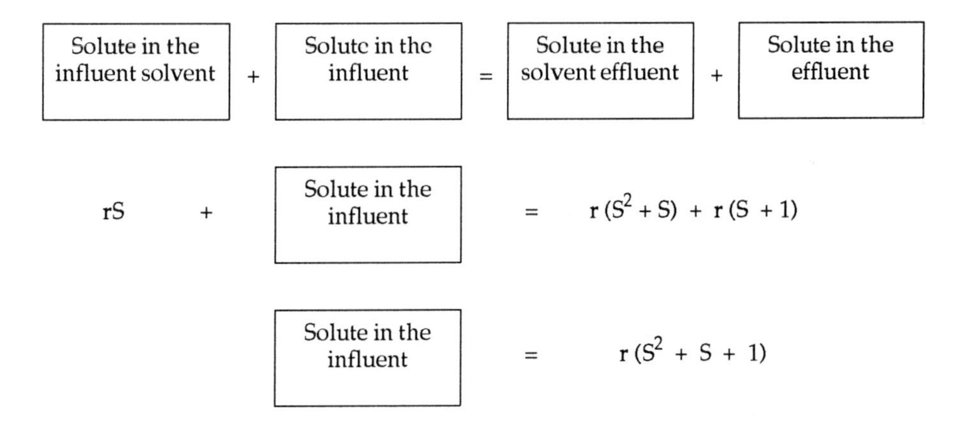

Π What is the non-extracted fraction for this two stage process?

r is the non-extracted material and the input (feed) is $r\ (S^2 + S + 1)$. Thus the non-extracted fraction is:

$$\frac{r}{r\ (S^2 + S + 1)}$$

We can extend this argument for a multistage process.

For N stages, the non-extracted fraction will be defined as in Equation 6.4 as:

$$\frac{C_N}{C_f} = \frac{r}{r\ (S^N + S^{N-1}_{+} \ldots\ldots + S + 1)} \qquad \text{(E - 6.6a)}$$

where C_N is the concentration of the solute in the raffinate after N stages and $\frac{C_N}{C_f}$ is the non-extracted fraction.

We can write this in another way:

$$\frac{C_N}{C_f} = \frac{1}{\sum_{i=1}^{N} S^i + 1} \qquad \text{(E - 6.6b)}$$

This is a so called geometric series.

Kremser equation

Equation 6.6b can be simplified further to:

$$1 - f = \frac{C_N}{C_f} = \frac{S - 1}{S^{N+1} - 1} \qquad \text{(E - 6.7a)}$$

This equation is known as the Kremser equation. Mathematically, we can re-write Equation 6.7a as:

$$N = \frac{\ln\left[\frac{(S-1)}{(1-f)} + 1\right]}{\ln S} - 1 \qquad \text{(E - 6.7b)}$$

Based on this equation we are able to predict the number of stages in series given the non-extracted fraction (1 - f) and the separation factor S.

Alternatively, if we know the separation factor we can predict the number of stages we will need to achieve a particular value for the non-extracted fraction.

Figure 6.6 shows the effect of the number of stages at different S-values on the non-extracted fraction. Compared with Figure 6.4 the efficiency can be enhanced significantly by applying more than one stage. Examine Figure 6.6 before attempting SAQ 6.1.

SAQ 6.1

Predict the number of stages in series required in an extraction process given the following data.

Distribution coefficient = α = 5.

Flow rate of extract = $2m^3\ h^{-1}$.

Flow rate of raffinate = $5m^3\ h^{-1}$.

Extracted fraction = 95%.

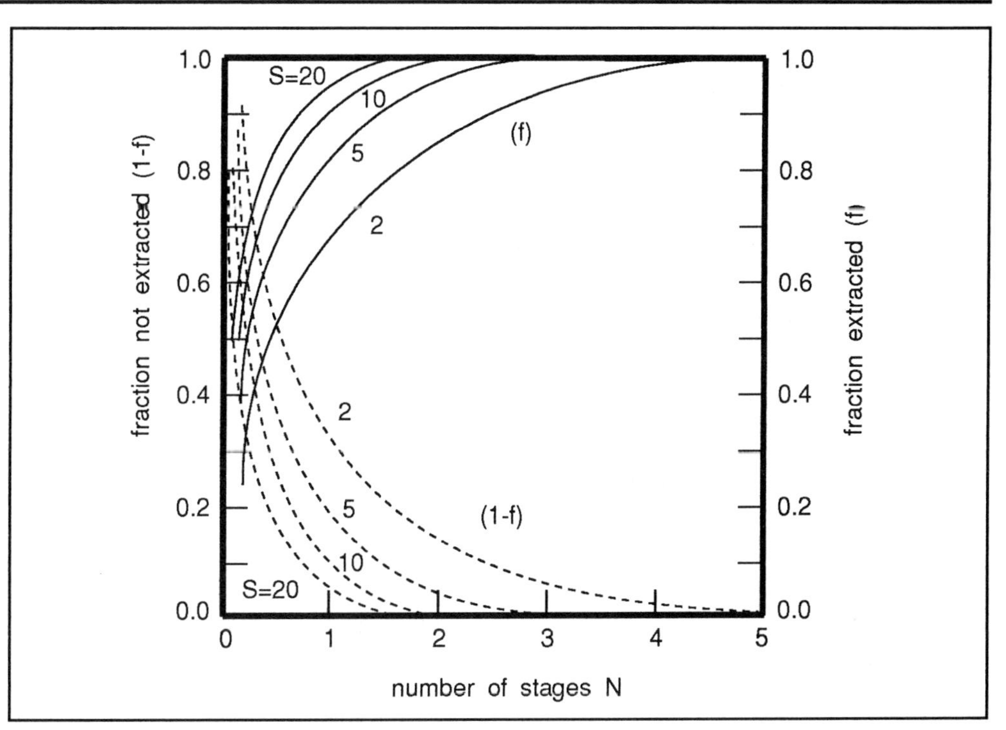

Figure 6.6 Non-extracted fraction (.......) and extracted fraction (——) as a function of the number of stages with various separation factors (S) as a parameter.

SAQ 6.2

Identify each of the following statements as True or False. If False give a reason for your response.

1) The higher the separation factor the lower the number of stages required to obtain a given fraction extracted.

2) In a two-stage extraction process the raffinate flow between the stages is given by rS.

3) The non-extracted fraction is given by $\frac{S}{S+1}$.

4) Fraction extracted increases as the separation factor increases.

5) It is reasonable to assume that $C_r = \frac{C_e}{\alpha}$.

Having considered equilibrium and the number of stages required for a particular degree of extraction, we will now turn our attention to the second of the two subjects important for modelling extraction processes: kinetics of mass transfer.

Kinetics

The kinetics of the mass transfer is determined to a great extent by the contact between the two immiscible phases. Figure 6.7 shows a single stage extraction unit used in batch operation. Note the symbols shown on this figure, they are important in the subsequent discussion. Note also that filling/mixing and phase separation takes place in the same vessel.

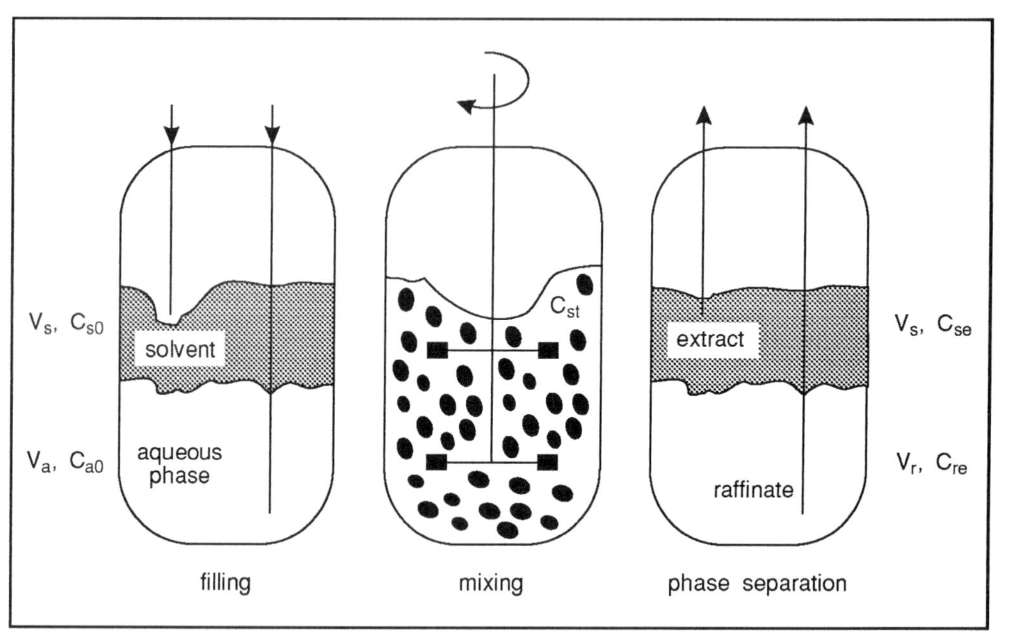

Figure 6.7 A single stage extraction unit. V_s = volume of solvent, V_a = volume of aqueous phase, V_r = volume of raffinate, C_{s0} = initial concentration of solute in the solvent at time t = 0, C_{a0} = initial concentration of solute in the aqueous phase, C_{se} = final concentration in the extract, C_{re} = final concentration in the raffinate, C_{st} = concentration of solute in the solvent at time t.

separation of phases relies on density difference

Figure 6.8 shows a mixer-settler, a single stage continuous extraction unit which is probably the most widely used extraction device. The mixer is filled with the continuous phase, while the dispersed phase is added near the impeller. Both phases are intensively mixed, until equilibrium has been reached. Separation of the two phases occurs in the settler provided there is a significant density difference.

The mass transfer for batch contacting of two immiscible phases, may be discribed by the following rate equation in terms of solvent concentration (we will not derive this equation here, it is derived in the BIOTOL text 'Bioprocess Technology: Modelling and Transport Phenomena').

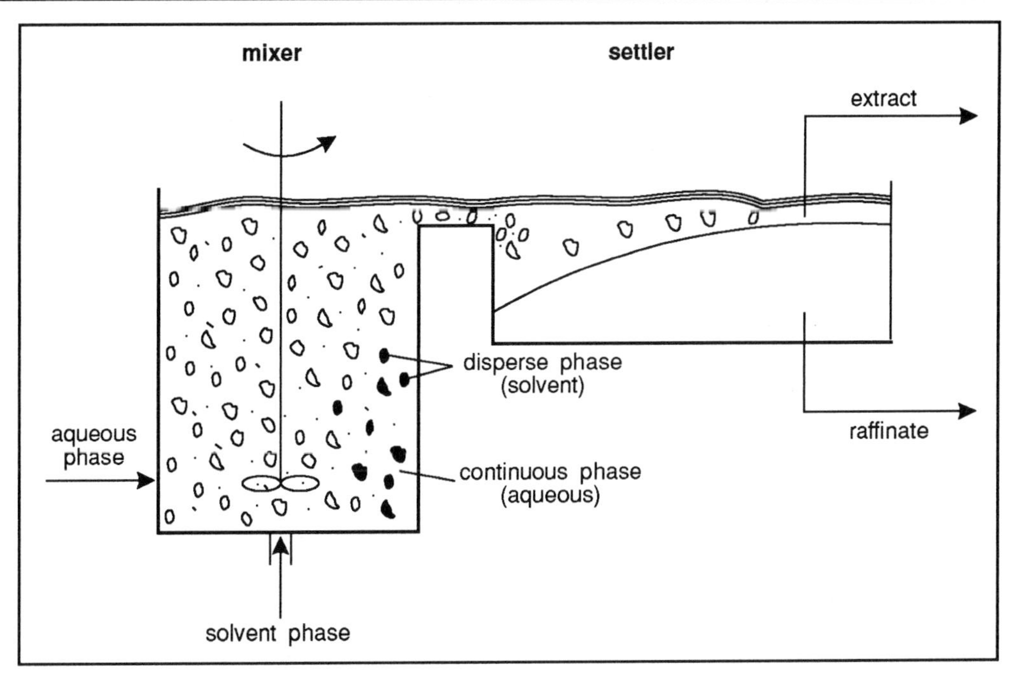

Figure 6.8 Mixer-settler.

$$V_s \frac{dC_s}{dt} = K_s\, a\, V_m (C_s^* - C_s) \qquad \text{(E - 6.8)}$$

where:

V_s = solvent volume (m^3)

V_m = mixer volume (m^3)

K_s = average mass transfer coefficient based on the solvent phase (s^{-1})

C_s = concentration of solute in solvent phase (kg m^{-3})

C_s^* = concentration of solute in solvent phase in equilibrium with concentration in (aqueous) raffinate phase (kg m^{-3})

a = specific area (m^2 m^{-3})

What Equation 6.8 is telling us is that the rate of change in the amount of solute in the solvent $\left(V_s \frac{dC_s}{dt}\right)$ is related to the mass transfer coefficient (K_s), the specific surface area (a) between the solvent and aqueous phase, the total volume of the mixture (V_m) and the difference between the concentration of the solute (C_s) and its equilibrium concentration in the solvent (C_s^*).

The specific area a is defined as follows:

$$a = \frac{\text{surface}}{\text{volume}}$$

For spherical droplets the specific area can be estimated as follows:

$$a = \frac{\pi d^2}{\frac{\pi}{6} d^3} = \frac{6}{d} \qquad \text{(E - 6.9a)}$$

where d = droplet diameter (m)

We can calculate the specific area of a dispersion as follows

$$a = \frac{6\varepsilon}{d} \qquad \text{(E - 6.9b)}$$

in which ε is the volume fraction droplets of the dispersion.

It is common in biochemical engineering to express processes, which tend to equilibrium, in a fractional approach to that equilibrium.

$$\text{So: } E_b = \frac{C_{St} - C_{S0}}{C_{Se} - C_{S0}} \qquad \text{(E - 6.10)}$$

where:

E_b = fractional approach to equilibrium for a batch system, the subscript b refers to batch.

C_{St} = concentration of solute in solvent phase at time t (kg m^{-3})

C_{S0} = concentration at t = 0 (initial concentration) (kg m^{-3})

C_{Se} = concentration at the end of the process (final concentration) (kg m^{-3})

Thus if C_{St} = 50% of C_{Se} and $C_{S0} = 0$, then $E_b = 0.5$

Using the geometrically observed relation:

$$\alpha \frac{V_s}{V_r} = \frac{C_S^* - C_{Se}}{C_{Se} - C_S} \qquad \text{(E - 6.11)}$$

where $C_S^* = \alpha C_r$

Integration of Equation 6.8 and substitution of Equation 6.10 and 6.11 into Equation 6.8 yields:

$$E_b = 1 - \exp\left(\underbrace{\frac{-K_s\, a\, V_m\left(1 + \alpha\dfrac{V_s}{V_r}\right)}{V_s}}_{-\,b}\, t\right) \qquad \text{(E - 6.12a)}$$

So: $E_b = 1 - \exp(-bt)$ (E - 6.12b)

Equation 6.12 can be used in experimental design. A plot with experimental data of $\ln(1 - E_b)$ versus t should give a straight line, indicating that b = constant. Because of the limited amount of published data, it will be necessary to conduct small scale tests and pilot plant work in equipment which is similar in design to the intended full scale plant.

Π Let us try a numerical example.

Given $K_s = 10^{-6}\ ms^{-1}$, $V_s = 3\ m^3$, $\alpha = 25$, $V_r = 10\ m^3$

Calculate S, 1 - f, b and the diameter of the droplets if 99% equilibrium has been reached in one minute (try this before reading on). You can first find S which should enable you to calculate 1 - f. Then you can determine b in two ways one from values of K_s, V_s, V_r and a in terms of a and also from $E_b = 1 - \exp(-bt)$ using the equilibrium data. This should enable you to calculate a and then the diameter of the droplets.

Here is our solution:

We can use Equation 6.3 to determine the separation factor (S).

$$S = \alpha\frac{V_s}{V_r} = 25 \cdot \frac{3}{10} = 7.5$$

The non-extracted fraction (1 - f) can be determined using Equation 6.4:

$$(1 - f) = \frac{1}{1 + S} = \frac{1}{1 + 7.5} = 0.118$$

So 11.8% of the component to be extracted remains in the raffinate. A value as high as this is generally unacceptable and additional stages are required.

Now we wish to calculate b (using Equation 6.12a):

$$b = \frac{K_s \cdot a\, V_m\left(1 + \alpha\dfrac{V_s}{V_r}\right)}{V_s}$$

$$= \frac{10^{-6} \cdot a \cdot 13\left(1 + 25\dfrac{3}{10}\right)}{3}$$

Note $V_m = V_s + V_r$

$= 3.68 \cdot 10^{-5}\ a$

But b can be calculated from Equation 6.12b

Since $E_b = 1 - \exp(-bt)$ and $E_b = 0.99$ (ie 99%) at t = 60s

Thus $0.99 = 1 - \exp(-b.60)$

$b = 7.67 \cdot 10^{-2}$

Equating both results gives:

$$a = \frac{7.67 \cdot 10^{-2}}{3.68 \cdot 10^{-5}} = 2.1 \cdot 10^{3}\ m^2\ m^{-3}$$

But the droplet diameter d and the specific surface area a are related by Equation 6.9b.

$a = \frac{6\varepsilon}{d}$ where ε is the volume fraction of droplets of the dispersion.

In this case $\varepsilon = \frac{V_s}{V_r + V_s} = \frac{3}{13}$

Thus $d = \frac{6}{a}\ \frac{3}{13} = \frac{18}{2.1 \cdot 10^{3} \cdot 13} = 6.6 \cdot 10^{-4}\ m$

$= 0.6$ mm

Note that the power input necessary to obtain droplets of this size can also be calculated. We will not examine this in detail here since this topic is covered in the BIOTOL text 'Operational Modes of Bioreactors'. Generally the power input per unit mass is related to density (ρ), viscosity (η) and the surface tension (σ).

We may extend our calculation. Using the calculated b value and Equation 6.12b, we may produce a plot of E_b as a function of time.

This is done in Figure 6.9 for different values of b. You will notice that the higher the value of b, the faster the extraction process proceeds.

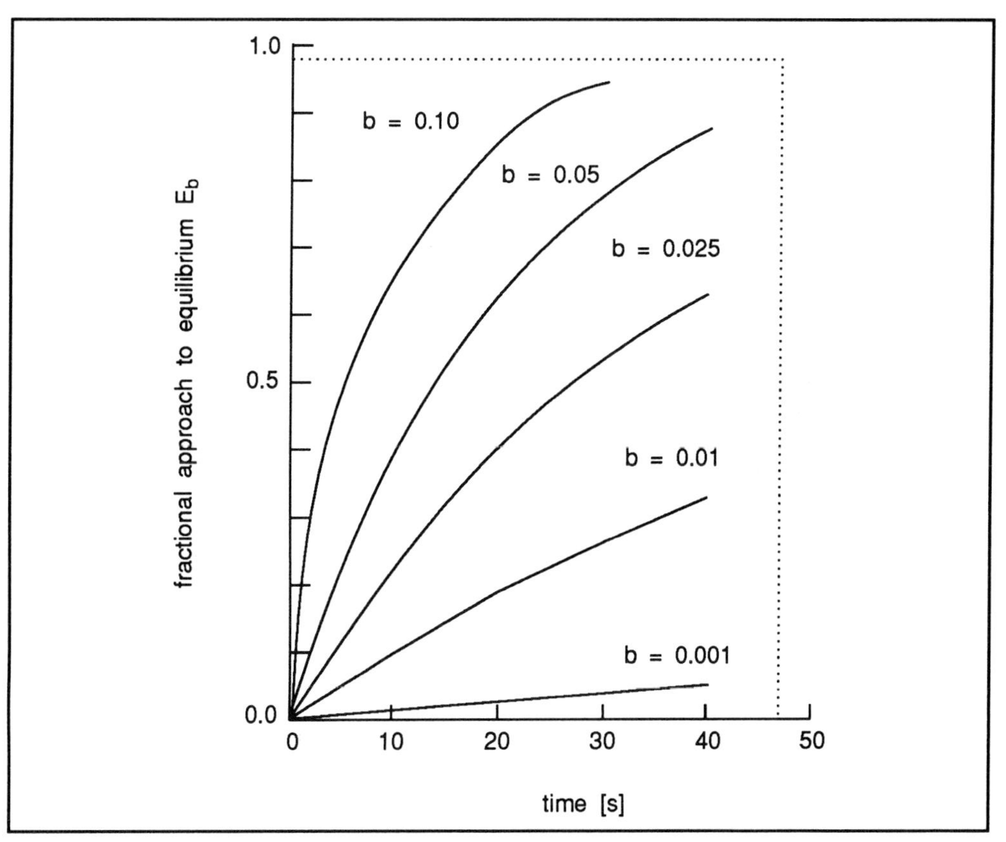

Figure 6.9 Plot of functional approach to equilibrium against times for different values of b.

We have reported the time taken to reach equilibrium in batch and continuous flow processes for different values of E_b and b in Table 6.1.

	E_b (E_f)	b 0.1	0.05	0.025	0.01	0.001
	0.99	46	92	184	460	4600
A	0.95	25	60	120	300	2495
	0.90	23	46	92	230	2300
	0.99	990	1980	3960	9890	98900
B	0.95	190	380	760	1900	19000
	0.90	90	180	360	900	9000

Table 6.1 Equilibrium time in s for different b values and efficiencies. A) for a batch process, B) for a continuous flow process (see text for discussion).

Π Which type of system (batch or continuous) reach equilibrium the quickest?

The data in Table 6.1 clearly indicates that equilibrium is reached quicker in batch, rather than continuous flow systems. In a flow system such as that illustrated in Figure 6.8, we may assume that the flow ratio $\phi_{v,s} / \phi_{v,r}$ equals the volume ratio (V_s / V_r). In this case, the fractional approach to equilibrium is given by:

$$E_f = \frac{bt}{1 + bt} \qquad \text{(E - 6.13)}$$

where E_f is the fractional approach to equilibrium for a flow system.

Π In an earlier example, we calculated that the value of b = 7.67 . 10^{-2} for a batch system which reached 99% equilibrium in 1 minute. How long would it take a flow system with a similar b value to reach 99% equilibrium.

$$\text{Since } E_f = \frac{bt}{1 + bt} \text{ then } E_f = \frac{7.67 \,.\, 10^{-2}\, t}{1 + 7.67 \,.\, 10^{-2}\, t} = 0.99$$

t = 1238 s (approximately 20 min)

We can conclude that batch systems reach equilibrium faster than flow systems with the same b value.

E_f can however be reached very rapidly providing the value of b is high enough.

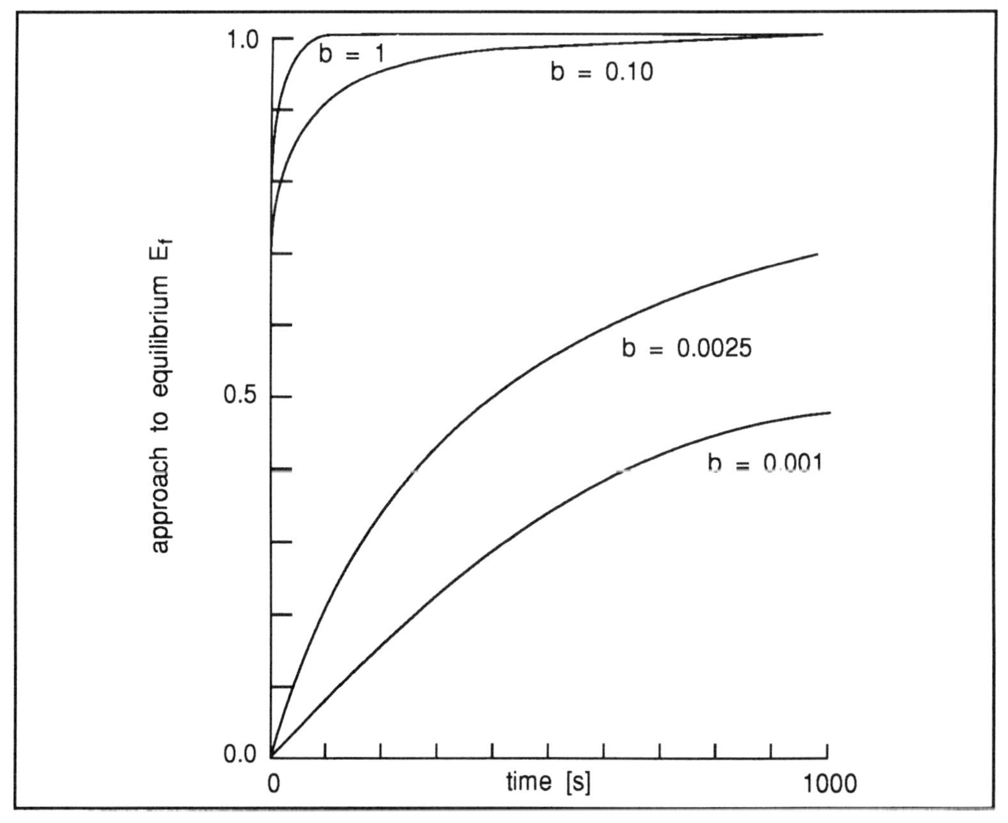

Figure 6.10 Relationship between the fractional approach to equilibrium (E_f) and time for a flow system at diffrent values of b. Note E_f is given by $\frac{bt}{1+ bt}$.

We have illustrated the relationship between E_f and time for a flow system in Figure 6.10. Some values of E_f at different b values are reported in Table 6.1.

The important conclusion is that E_b is reached more rapidly than E_f. The explanation for this is that the driving force for mass transfer (the difference between C_s^* and C_{St}) is small for a continuous flow system. In a batch system the concentration difference is much larger. The following illustrates this diagrammatically.

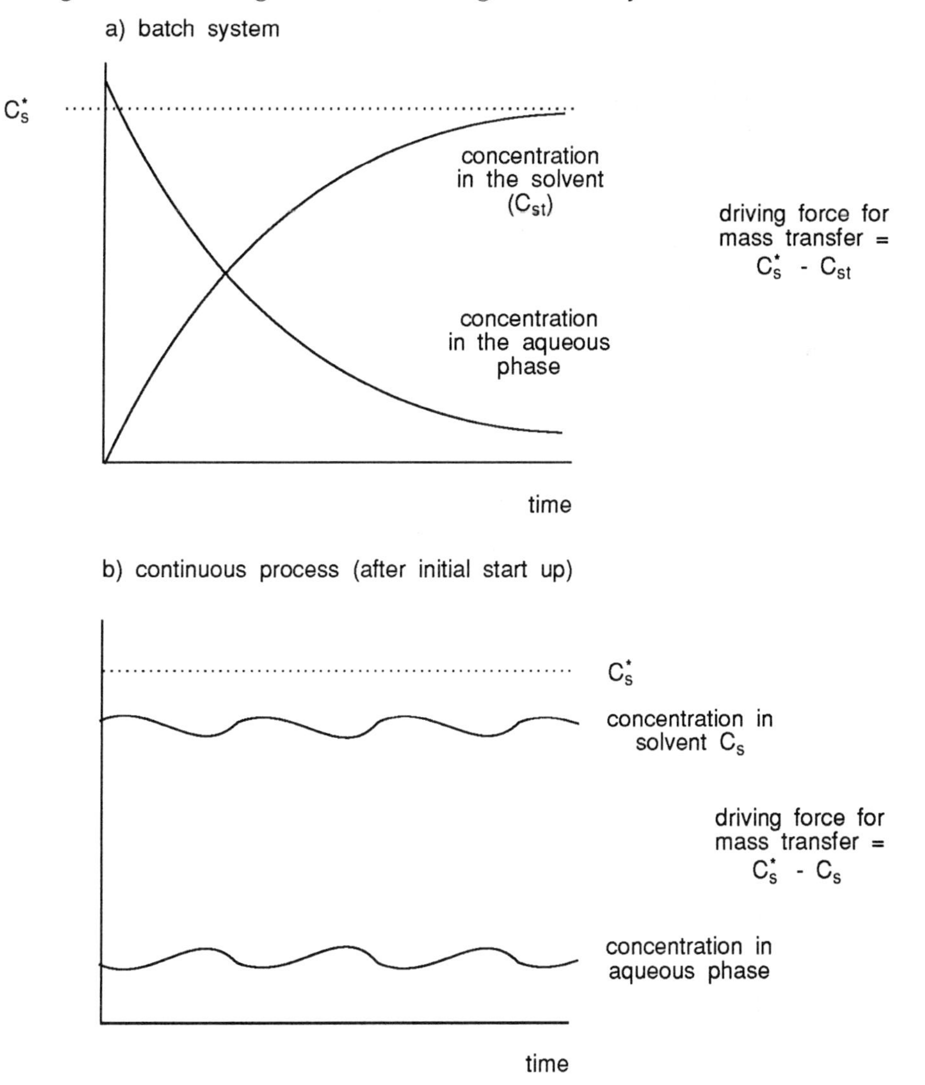

Π Make a list of the factors which will influence b and E_f (Equation 6.12a may help you).

We can see from Equation 6.12a that:

b is influenced by the $\frac{V_s}{V_r}$ ratio and α.

constant s and α

But , much more important to the engineer is the value of a (specific surface). This can be changed by changing the power input (eg by using a larger stirrer or high stirrer speed). This also alters the value of K_s.

In a solvent extraction both the equilibrium and the distribution coefficients can be determined thermodynamically. In many cases the raffinate and solvent phases are completely immiscible and the solute concentrations are low (1-10% w/w). Here, we may consider the distribution as a constant which can simply be determined (measured). However, in dissociation extraction (see next section) and aqueous two phase extraction the distribution coefficient can be a function of pH, pK_a, etc. These relations will be explained in the next sections.

6.2.2 Extraction applications

In the application of liquid-liquid extraction two distinctions can be made as to type of product, depending on molecular weight. Products such as carboxylic acids (citric acid, butyric acid, lactic acid, etc.) are considered as low molecular products. These products are stable and will not deteriorate when organic solvents are used, which makes them suitable for solvent extraction.

Products such as proteins (pharmaceuticals, plasma proteins, enzymes) are considered to be high molecular weight products (5,000 - 500, 000 Daltons). They are sensitive to many solvents and relatively unstable. These products are suitable for aqueous two phase extraction.

Extraction of low molecular components

Low molecular components can be extracted in one of three ways:

- physical extraction;
- dissociative extraction;
- reactive extraction.

physical, dissociative and reactive extraction

In physical extraction the component involved, distribute themselves over the two phases according to their physical preference (hexane/water/phenol).

In dissociation extraction, differences in the dissociation constant of the components are exploited to achieve separation.

Reactive extraction is characterised by the addition of a carrier, for example an aliphatic amine, to the organic solvent which carries the component from the aqueous phase to the organic phase.

We will now consider dissociation extraction in more detail, using extraction of simple organic bases and acids as an illustrative example.

To predict the number of stages necessary to achieve a certain degree of extraction, we need to know the distribution coefficient of the solute between the aqueous and the organic phase.

Here we will derive a general equation for the distribution coefficient for simple organic bases and acids, and apply it to penicillin as an example.

The distribution coefficient can be derived based on the following assumptions:

- the product is treated as a monobasic acid;
- no association occurs;
- the product does not dissociate in the organic phase;
- undissociated product dissolves to a large extent in the organic phase.

The dissociation of a weak acid can be written as follows:

$$HP \rightleftarrows H^+ + P^-$$

At very low pH (high H^+) the equilibrium shifts to the left hand side. In this case we can define the intrinsic distribution coefficient (α^o):

$$\alpha^o = \frac{[HP]_o}{[HP]_w} \qquad \text{(E - 6.14)}$$

at low pH there is no dissociation

in which o and w refer to the organic and aqueous (water) phases respectively. Here, all P molecules are undissociated.

At higher pH values the product is partly dissociated and therefore not available for extraction.

The effective distribution coefficient (α) is then defined as:

$$\alpha = \frac{[HP]_o}{[HP]_w + [P^-]_w} \qquad \text{(E - 6.15)}$$

The dissociation constant (K_a) is given by:

$$K_a = \frac{[H^+]\ [P^-]}{[HP]_w} \qquad \text{(E - 6.16)}$$

Substitution of Equation 6.16 in Equation 6.15 gives:

$$\alpha = \frac{[HP]_o}{[HP]_w}\left(1 + \frac{K_a}{[H^+]}\right)^{-1} \qquad \text{(E - 6.17)}$$

Substitution of Equation 6.11 in Equation 6.14 gives:

$$\alpha = \frac{\alpha^o}{1 + \dfrac{K_a}{[H^+]}} \qquad \text{(E - 6.18)}$$

Applying the definitions of:

$pH = -\log_{10} [H^+]$ and

$pK_a = -\log_{10}K_a$

and substitution in Equation 6.18 gives:

$$pH - pK_a = \log_{10}\left(\frac{\alpha^o}{\alpha} - 1\right) \text{ for acids} \qquad \text{(E - 6.19a)}$$

and

$$pK_b - pH = \log_{10}\left(\frac{\alpha^o}{\alpha} - 1\right) \text{ for bases} \qquad \text{(E - 6.19b)}$$

Equation 6.19a is graphically presented in Figure 6.10.

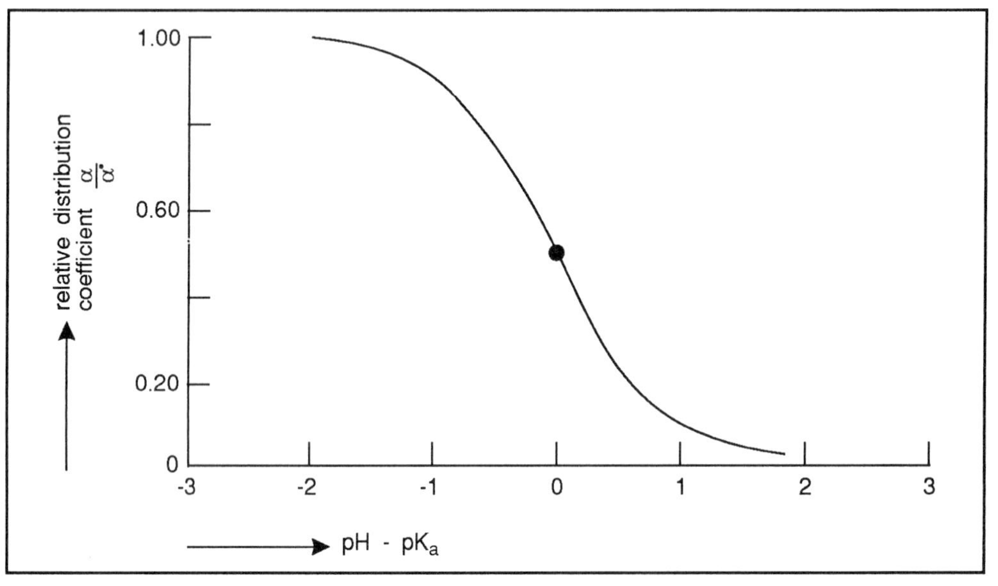

Figure 6.11 Relative distribution coefficient as function of pH - pK_a for an acid.

α^o depends on product and solvent used

These equations hold for many acids and bases. The value of α^o depends on the product and the solvent used.

Π For the reaction $OHP \rightleftarrows OH^- + P^+$ write expressions for: 1) The intrinsic distribution coefficient. 2) The effective distribution coefficient. 3) The dissociation coefficient.

The relationships we anticipate you would write are:

1) $$\alpha^o = \frac{[OHP]_o}{[OHP]_w}$$

2) $$\alpha = \frac{[OHP]_o}{[OHP]_w + [P^+]_w}$$

3) $$K_b = \frac{[OH^-]\ [P^+]_w}{[OHP]_w}$$

As an example, some of the intrinsic distribution coefficients for penicillin (pK = 2.75) are given in Table 6.2.

Solvent	α^o
Butyl acetate	48
i-Amyl acetate	22
Chloroform	12.3
Di-isopropylether	2.4
n-Hexane	0.2

Table 6.2 Intrinsic distribution coefficients for different solvents in penicillin extraction.

Figure 6.9 shows the distribution coefficient for penicillin as function of the pH using the α^o-values from Table 6.1.

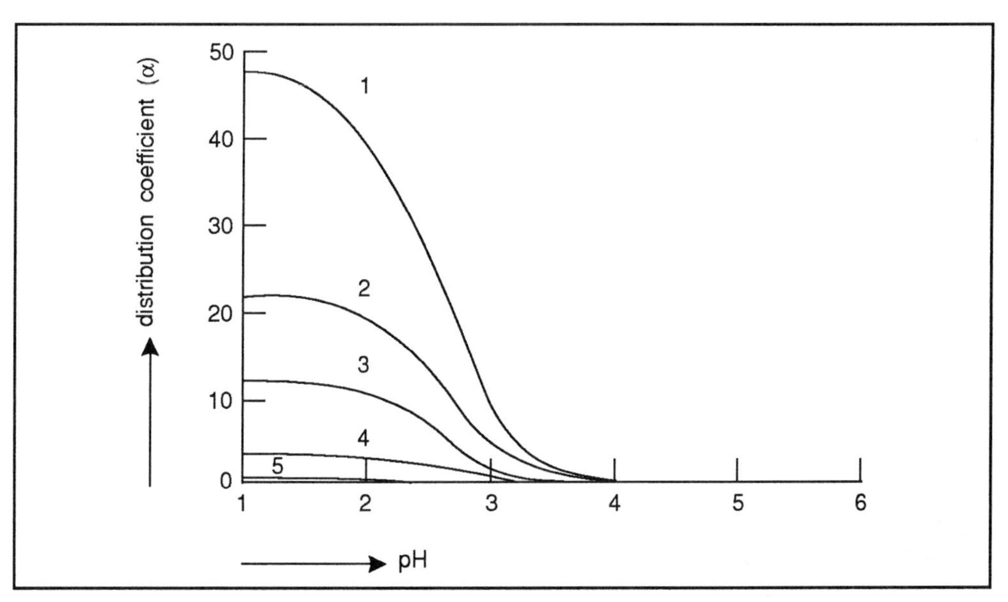

Figure 6.12 Distribution coefficient as function of the pH for different solvents (pK_a = 2.75 penicillin). 1-5 refer to the solvents described in Table 6.2 (see text for details).

high α at low pH for acids

From this figure it can be seen that high distribution coefficients will be obtained at low pH values (for bases it is the other way around, of course).

Π Name the solvents for plots 1 and 3 in Figure 6.12, using the information in Table 6.2.

Since plot 1 has the highest distribution coefficient at low pH, the solvent must have the highest intrinsic distribution coefficient; from Table 6.2 we can see that this is butyl acetate. The same approach indicates that plot 3 must be chloroform.

A severe problem associated with the extraction of antibiotics is their stability at low and/or high pH values. Clearly it is pointless to use a low or high pH to improve extraction if these bring about destruction of the material we are intending to extract!

The residence time in an extractor should therefore be very short (minutes!). This is the main reason why centrifugal extractors are exclusively applied in the extraction of antibiotics. Thus the compromise is to use an extreme of pH which improves extraction, and to reduce the time of exposure of the antibiotics to pH stress to a minimum.

Figure 6.13 shows a typical unit for the extraction of antibiotics from filtered broths. Two centrifugal extractors are operating in a countercurrent mode (see also Figure 6.1). Another solution to this problem is the application of carriers (reaction extraction) which allows extraction at higher pH values. In this case, centrifugal extractors can be avoided. Essentially how these systems work is to add a carrier which interacts with the product we wish to extract. Providing the carrier : product complex has a favourable distribution coefficient, this will lead to improved extraction. We will not however, discuss the systems further at this stage.

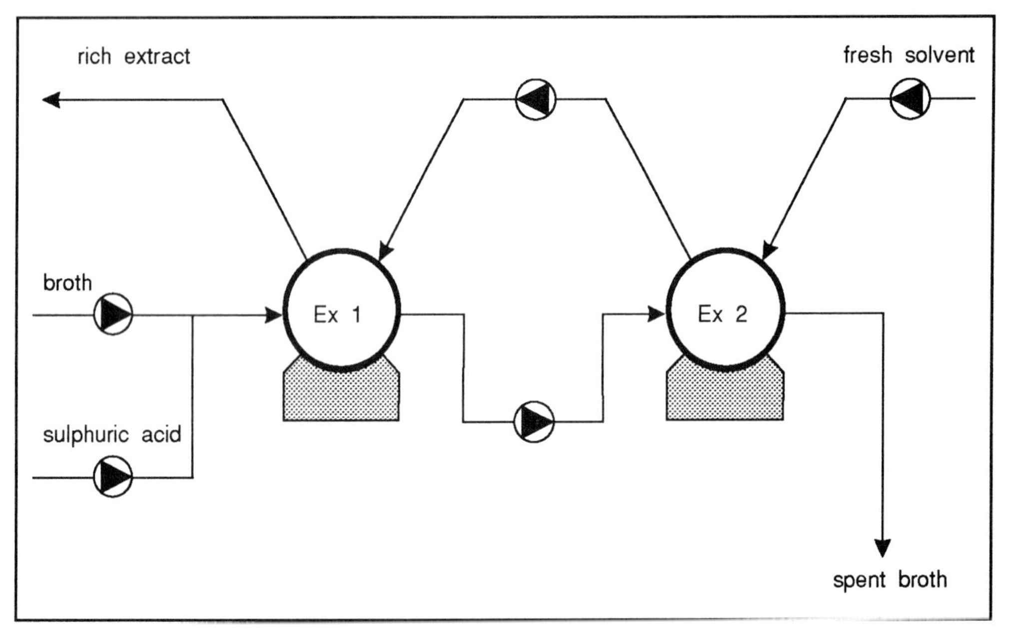

Figure 6.13 Extraction of filtered antibiotic broth. Ex 1 and Ex 2 are centrifugal extracts.

Knowing the distribution coefficient α and by choosing a proper pH value and aqueous/solvent ratio the extraction factor can be calculated. Together with an extraction efficiency (fraction not extracted) we may calculate the number of stages required. Let us work through an example:

Penicillin extraction is carried out a pH = 2 with butyl acetate in an aqueous/solvent ratio of 5. We wish to calculate the number of stages required to achieve 98% extraction. Remember pKa for penicillin = 2.75. The data in Table 6.2 will be used.

For convenience we rewrite Equation 6.19a as follows:

$$\alpha = \frac{\alpha^{\circ}}{1 + 10^{(pH - pK_a)}}$$

Substitution of values given yields.

$$\alpha = \frac{48}{1 + 10^{(2 - 2.75)}} = 40$$

The separation factor is now according to Equation 6.3.

$$S = \alpha \frac{\phi_{ve}}{\phi_{vr}} = 40 * \frac{1}{5} = 8$$

The number of stages required for 98% efficiency can be calculated using Equation 6.7b.

$$N = \frac{\ln\left[\frac{(S-1)}{(1-f)} + 1\right]}{\ln s} - 1$$

$$= \frac{\ln\left[\frac{7}{0.02} + 1\right]}{\ln 8} - 1 = 1.8 \text{ stages}$$

In practice 2 stages will be chosen.

SAQ 6.3

1) Explain why you would expect the distribution coefficient for a base to increase with increasing pH.

2) An acid has a pK_a of 4. If the pH is 5, calculate the intrinsic distribution coefficient α° in terms of α.

3) Repeat the calculation done in 2) but this time use pH 6.

Extraction of high molecular weight components

aqueous two phase systems have several advantages

A technique recently developed for extracting labile high molecular weight components such as enzymes is to use an aqueous phase extraction. For example, proteins can be extracted into water containing a polymer, such as dextran. The major advantages of aqueous two phase systems include high protein capacity, non denaturing solvent environment and high selectivity.

many factors influence α of a protein

Many factors contribute to the distribution of a protein between two phases. It is a complex function involving hydrophobic, electrical, conformational and biospecific features of the proteins.

To a first approximation, these factors can be treated as independent, so the distribution coefficient can be written as:

$$\ln\alpha = \ln\alpha_o + \ln\alpha_{el} + \ln\alpha_{biosp} + \ln\alpha_{hydro} + \ln\alpha_{conf} \quad \text{(E - 6.20)}$$

α_o contains factors not specifically accounted for in the other coefficients including polymer properties, molecular weight, concentration and protein structural aspects such as size, charge and charge distribution.

In practice it is far too difficult to distinguish the various contributions to the overall distribution coefficient since these parameters cannot always be determined experimentally. Therefore the Brønsted equation can serve as a first approximation:

$$\alpha = \exp(M\lambda/KT) \qquad \text{(E - 6.21)}$$

where:

M = relative molecular weight (mass);

K = Boltzman constant ($1.38 * 10^{-23}$ JK^{-1});

T = absolute temperature (K);

λ = characteristic of the system dependent on all factors other than those represented in the Brønsted equation (J m^{-2}). Since molecular distribution is determined by the molecular surface energy and the surface area, these should appear in the expression and not in the molecular weight (mass). (Note: surface area is proportional to $M^{2/3}$).

λ may be negative or positive

Figure 6.14 shows the dependence of the distribution coefficient on the molecular weight. The larger the molecular weight, the smaller α. This is due to the negative value of λ. Larger entities prefer to stay in the aqueous phase while small entities prefer to move to the polymer phase.

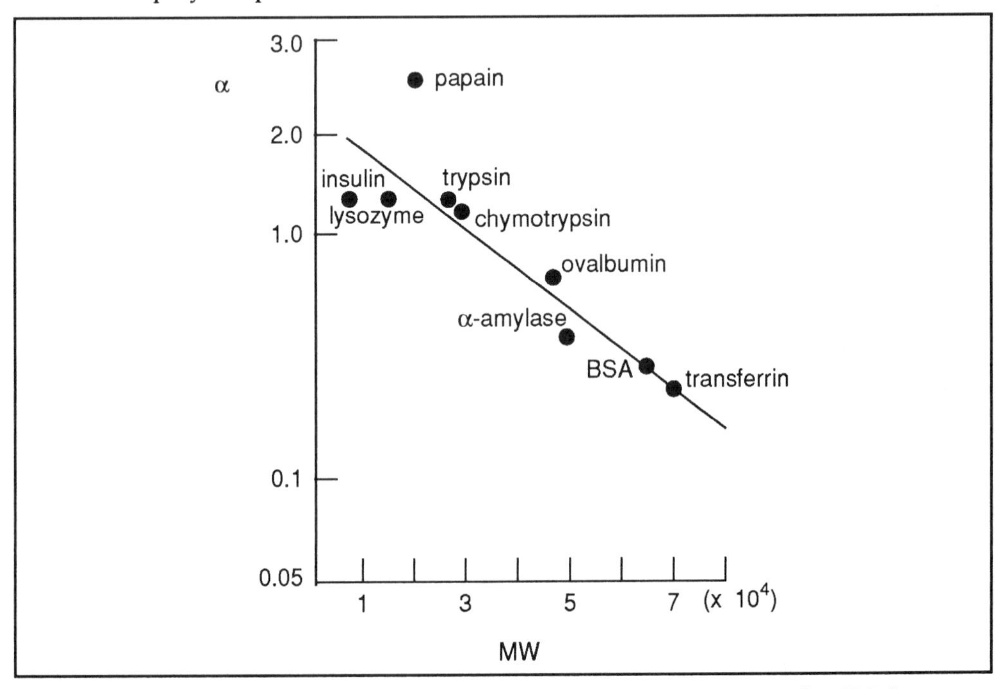

Figure 6.14 Distribution coefficient as function of the molecular weight for various proteins. (7% dextran 500, 10 mmol^{-1} phosphate 20°C). Solid line represents the Brønsted equation. BSA = bovine serum albumin.

You must note however that λ may vary strongly depending on the types of polymers and salts used and can even be a positive number.

According to Equation 6.20 the distribution coefficient may also be affected by an electrochemical potential difference. This can be the result of either unequal phase distribution of cations and anions of the (buffer) salts used or charge on the protein which depends on its iso-electric point and on the pH value of the solution.

Furthermore the distribution coefficient may be affected by:

- the molecular weight of the phase forming polymers;
- the phase composition;
- the type of buffer salts (ions).

For the moment there is no general theory available to predict the distribution coefficient in a polymer/water or polymer/polymer system.

Figure 6.15 shows a possible scheme for the large scale recovery of an enzyme. This is not meant to represent all the various schemes that have been developed. It does however provide you with a model. It would be worthwhile reading through the scheme represented in Figure 6.15 and to remember the sequence.

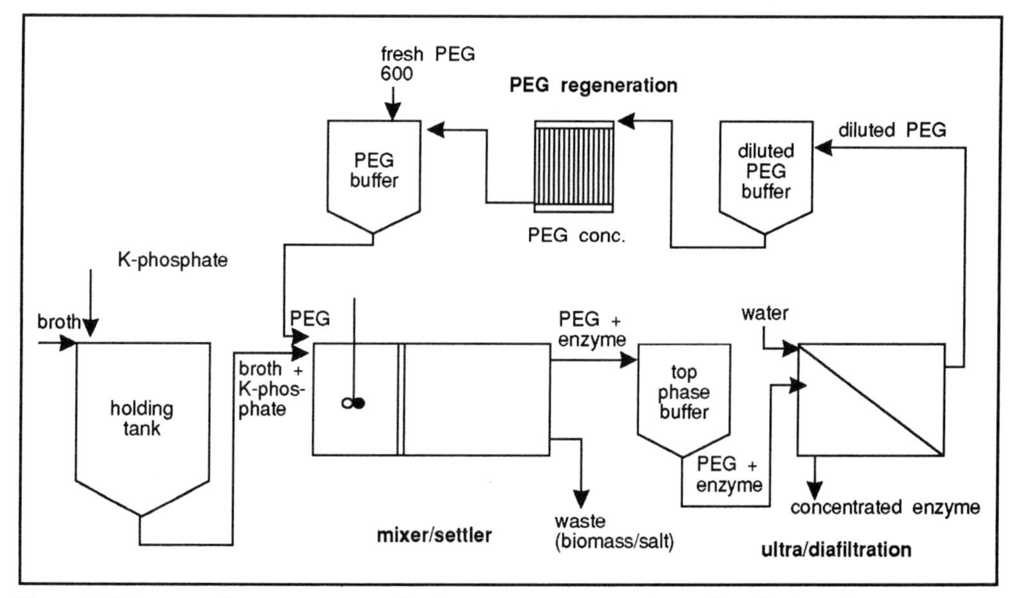

Figure 6.15 Large scale recovery system for the recovery (concentration and purification) of an enzyme using polyethylene glycol (PEG).

Π List factors that influence the value of the distribution coefficient of a protein in an aqueous two phase system. You should be able to list at least eight factors. If you are unable to do this refer back to the text. As you do so, write down each factor mentioned on a sheet of paper so that you have your own comprehensive list.

SAQ 6.4

Complete the following statements.

1) According to the Brønsted equation, the distribution coefficient is given by [ ].

2) If λ is positive, an increase in molecular weight will [ ] the distribution coefficient.

3) Aqueous two phase systems offer three major advantages, these are:

   a)

   b)

   c)

6.3 Precipitation

Precipitation is a well established method for the recovery of proteins. It is widely used for the recovery of bulk proteins.

In precipitation, the solubility is reduced by addition of salts (salting out) or organic solvents.

It can be applied to fractionate proteins (separate different types), as done in blood plasma processing, or as a volume reduction method.

Precipitation can also be used as a purification step prior to high resolution methods, for example to achieve the removal of undesired by-products such as nucleic acids, pigments and other residual components from the broth. These methods, include protamine treatment, salt treatment, iso-electric precipitation and precipitation using non-ionic polymers. We will not discuss precipitation as a purification step at this stage but focus on precipitation as a volume reduction method.

6.3.1 Colloidal stability of protein solutions

A protein can be characterised as a globular polymer with a large number of non-uniformly distributed titratable surface charges. Figure 6.16 shows a stylised version of the distribution of charges and hydrophobic patches over the surface of a protein.

DLVO theory

The stability of protein solutions can be explained by the Deryagin-London-Verweij-Overbeek (DLVO) theory. The attractive and repulsive forces between colloidal particles of uniform charge can be added. The height of the energy barrier compared with the thermal energy of collisions determines whether aggregation will occur or not and at what rate.

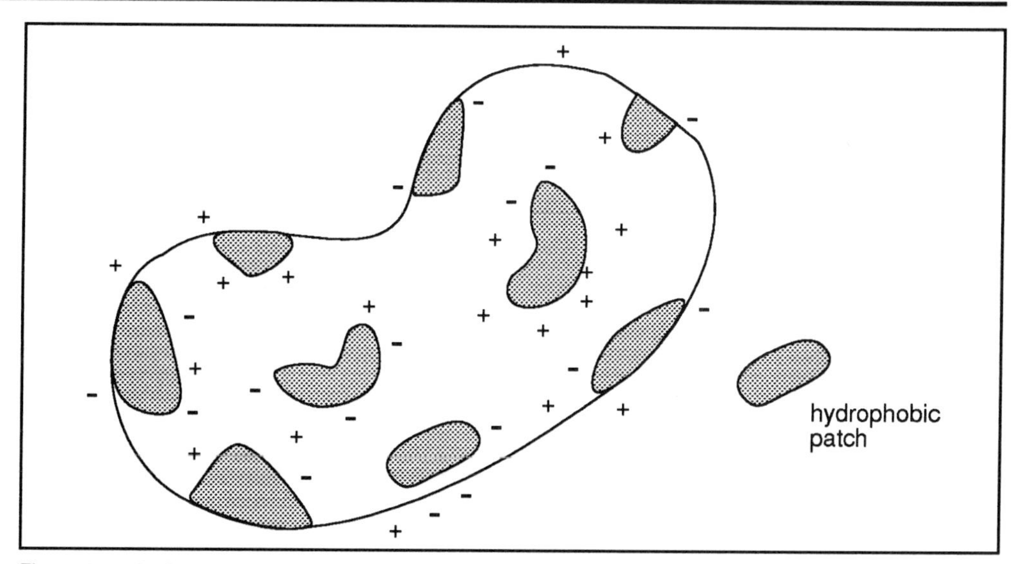

Figure 6.16 Stylised distribution of charges and hydrophobic patches over the surface of a typical protein.

proteins attract positive ions

Stern layer

sum of forces determines stability

In their 'natural' environment, proteins usually exhibit a net negatively charged surface. When they are suspended in an electrolyte solution, they will attract positive ions from the solution to form a so called 'Stern layer' of counter ions, close to the protein surface. Next to it, there is a more diffuse (Gouy-Chapman) layer of mobile counter ions with a gradually decreasing concentration (see Figure 6.17). It has been proven that the Stern layer indirectly controls the stability of the colloid by its control of the thickness of the diffuse layer. When two proteins are brought together, repulsion is induced by the double layer (Stern and Gouy-Chapman). At the same time colloidal particles attract one another because of the ionic interactions. Whether colloidal particles will precipitate or not, depends on the sum of the repulsive and attractive forces.

It has theoretically and experimentally been proven that the double layer can be 'compressed' by electrolytes and organic solvents. The extent of destabilisation of colloids (proteins) depends on the ionic strength and the dielectric constant of the solution.

Industrial precipitation techniques are mainly based on the compression of this double layer by the addition of salts and organic solvents.

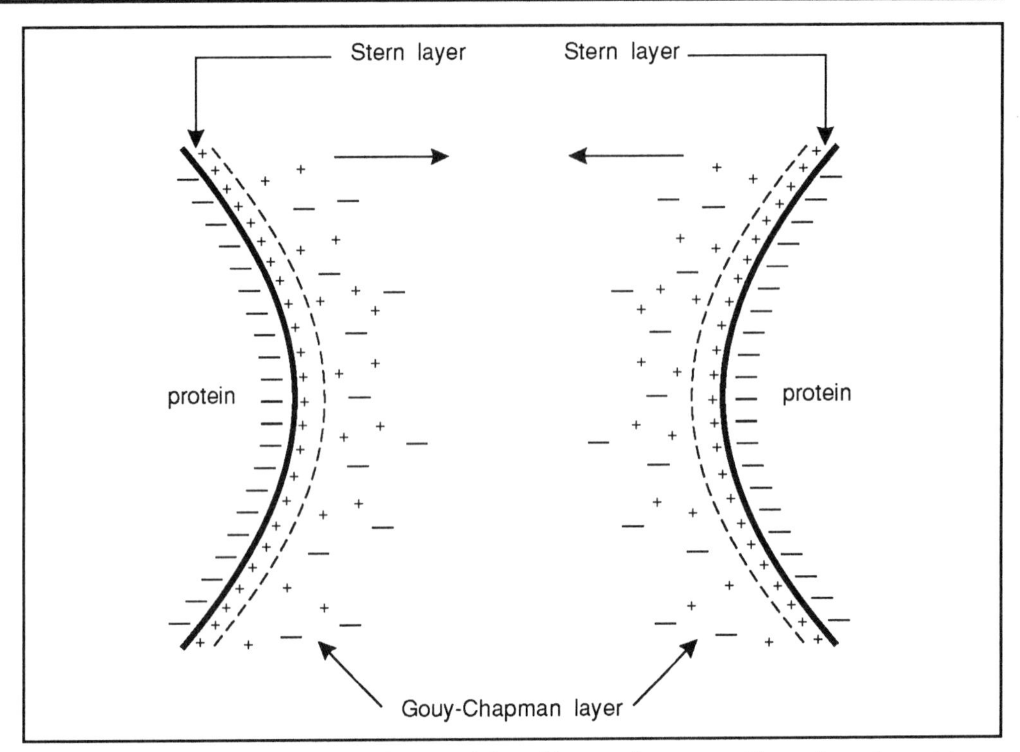

Figure 6.17 Electric double layer (Stern layer) and Gouy-Chapman layer around the surface of a protein. Compressing the Gouy-Chapman layer allows collision and subsequent precipitation of proteins.

6.3.2 Precipitation techniques

Salting out

Salting out is one of the oldest and most important techniques for the recovery of proteins (enzymes) from fermentation broths. However, little has been published on this subject.

Precipitation is achieved here by adding high concentrations of inorganic salts, for example ammonium sulphate, sodium sulphate or magnesium sulphate.

The solubility (S) as a function of the ionic strength (Γ) can be described in a semi-empirical way:

$$\log_{10} S = \beta - K_1\Gamma \qquad \text{(E - 6.22)}$$

β and K_1 are constants. β is strongly dependent on the pH and temperature, while K_1 depends on the salt employed, and they are characteristic of each protein. An example is given in Figure 6.18, which shows the salting out of fumarase by ammonium sulphate.

polyvalent anions are most effective

Not every salt is equally effective in its salting out capacity. Salts with polyvalent anions, such as sulphates (SO_4^{2-}) and phosphates result in higher K_1-values than univalent salts.

Polyvalent cations such as calcium (Ca^{2+}) or magnesium (Mg^{2+} reduce the K_1 -value. For that reason, ammonium sulphate and sodium sulphate have been widely used.

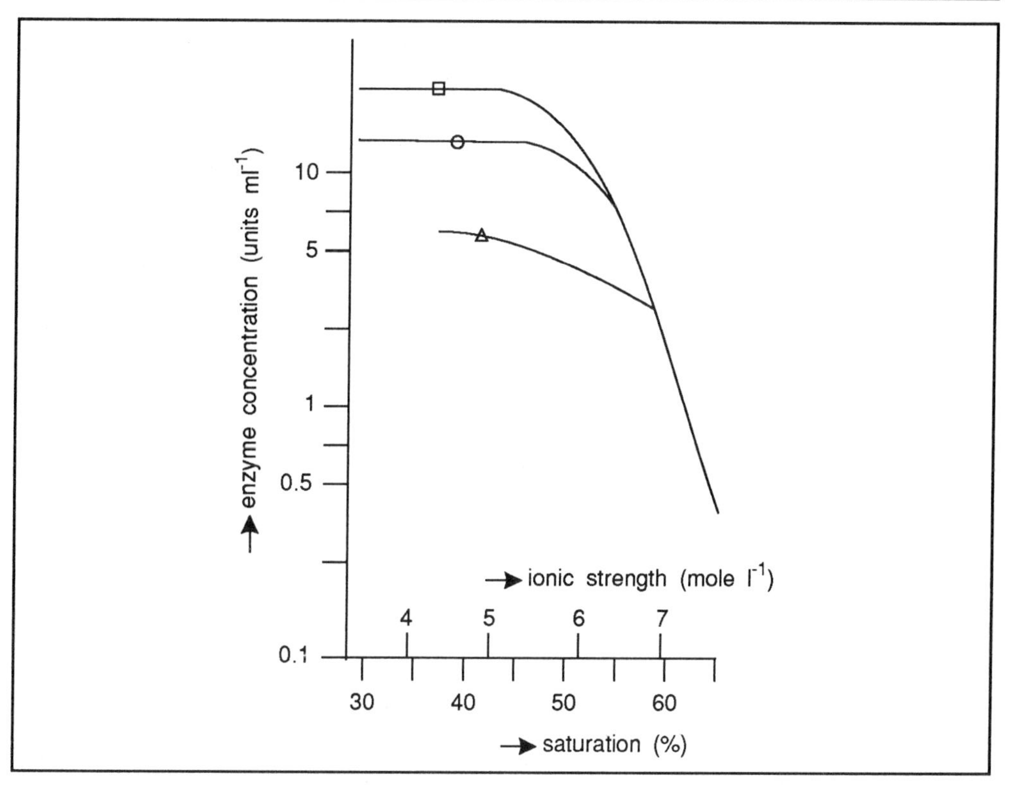

Figure 6.18 The soluble enzyme concentration during the salting out of fumarase by ammonium sulphate at 6°C. Initial enzyme concentrations (units/ml^{-1}) were (Δ) 5.4; (o) 15.2; (□) 25.5.

Hofmeister series of neutral salts

The relative effectiveness of neutral salts in salting out, especially of anions, gives rise to a series in which citrate > phosphate >sulphate > acetate > chloride > nitrate > thiocyanate. This series is known in the literature as the Hofmeister series.

The tendency of a salt to cause structural damage to a protein appears to be inversely related to its position in the series. Sulphate ions are associated with structural stabilisation and thiocyanate with destabilisation.

Precipitation using organic solvents

Water soluble, neutral, weakly polar organic solvents can be used for the precipitation of proteins. Reduction of the dielectric constant increases the electrostatic interaction between opposite charges on the protein surface. Solvents such as ethanol, methanol, isopropyl alcohol or acetone can be used.

Since organic solvents tend to inactivate biologically active proteins, operations will generally have to be performed at low temperatures, typically at 0-5°C (for human plasma even lower: -10°C).

SAQ 6.5

1) Complete the following statements:

   a) Proteins usually exhibit a net [   ] charge on their surface.

   b) The Stern-layer comprises [   ] ions.

   c) The [   ] layer is a diffuse layer of mobile counter ions.

2) The Hofmeister series is given: citrate > phosphate > sulphate > acetate > chloride > nitrate > thiocyanate. Now select the correct response to the following statements.

   a) Nitrate salts are more effective than phosphate salts for salting out - True/False.

   b) Nitrate salts tend to cause more structural damage to proteins than phosphate salts - True/False.

The change in solubility of a protein at its iso-electric point with a change in effective dielectric constant of the solvent can be described by the following empirical equation:

$$\log_{10} S = \frac{K_2}{D_\varepsilon^{\ 2}} + \log_{10} S_o \qquad \text{(E - 6.23)}$$

where:

S = solubility (arbitrary units);

K_2 = constant (-);

D_ε = dielectric constant (-)

S_o = original solubility

K_2 depends on the temperature and the protein employed.

The dielectric constant depends on the type of solvent used.

Table 6.3 shows D_ε values for various solvents.

The attraction between polar groups (eg Na^+ and Cl^-) is weakest in solvents of high dielectric constant, and highest in solvents of low dielectric constant. For hydrophobic (non-polar) interactions, the reverse is true; that is attraction is greatest in solvents of high dielectric constant. Let us focus on ionic attraction.

Table 6.3 shows that the attractive forces between for example Na^+ and Cl^- ions at a given distance is forty times higher in hexane than in water. Hence NaCl will precipitate (crystallise) more quickly in hexane than in water.

Solvent	D_ε	Formula
Water	80	H_2O
Methanol	33	CH_4O
Ethanol	24	C_2H_6O
Acetone	22	C_3H_6O
Benzene	2.3	C_6H_6
Hexane	1.9	C_6H_{14}

Table 6.3 Dielectric constant of various solvents.

Π A protein has a isoelectric point (pI) at pH 7. Does this mean that in a salt solution, solubility of the protein will be highest or lowest at pH 7? Give reasons for your choice.

At the isoelectric point, a protein carries equal amounts of positive and negative charges (ie there is no net charge). Thus, at pH = pI, repulsion will be minimised. Any hydrophobic areas of the protein molecules would tend to be strongly attracted to each other in the high dielectric environment. The protein molecules would therefore tend to be attracted (ie aggregate) to each other and thus be relatively insoluble.

If we lowered the pH, the protein molecules would then carry a net positive charge and the individual molecules would repell each other. Likewise, at higher pHs, the molecules would carry a net negative charge and thus repell each other. Under these two conditions, the net repulsion of the molecules would reduce their chances of aggregating and they would not therefore precipitate.

6.3.3 Kinetics of protein aggregation

In the previous sections the equilibrium between solute (protein) and solvent was described. However, for sizing a production plant we need to know how fast the equilibrium will be reached, and whether aggregates will be formed which can be handled further, for example by filtration or centrifugation.

Unfortunately very little has been published on the kinetics of precipitation, and what has been published is not directly applicable to industrial processes. Therefore we will restrict ourselves to a qualitative description of the kinetics of protein aggregation.

diffusion controls perikinetic growth

If a precipitant is added to a protein solution the double layer (Stern and Gouy-Chapman) will be compressed. As soon as the attractive forces have become stronger than the repulsive forces, the proteins will start to aggregate. This aggregation takes place on the smallest possible scale. Two protein molecules will collide with each other resulting in aggregation. This collision called perikinetic growth is determined by the number of molecules or small aggregates, and the diffusion coefficient, which in its turn is determined by the temperature. Diffusion is the controlling mechanism in perikinetic growth. Intensive stirring has no effect on aggregate formation.

stirring controls orthokinetic growth

Once an aggregate with a particle size of approx 1 μm has been formed, fluid motion will cause particles to collide and hence aggregate further, a process called orthokinetic growth. Here the effectiveness of collision is dependent on the power dissipation per unit of volume, ie effect of stirring.

6.3.4 Equipment used in precipitation

Precipitation is carried out to a great extent in batch mode in stirred tanks. Depending on the precipitation characteristics the precipitant is added to the protein solution or vice versa.

After the aggregates have been formed the stirrer will be stopped to allow settling of the aggregates onto the bottom of the tank. The mother liquor is then removed and the aggregate slurry is fed to a centrifuge or a filter.

Batch mode precipitation does not provide optimal conditions for precipitation since the two types of aggregate growth (peri- and orthokinetic) are affected by different mechanisms.

In continuous precipitation, perikinetic growth under optimal conditions (no stirring) and high precipitant concentration (mixed, for example, in a static mixer) occurs first. Then the orthokinetic growth can be performed in a tank reactor under continuous stirring.

SAQ 6.6

Identify each of the following statements as True or False.

1) In general, an increase in the dielectric constant increases the solubility of protein.

2) Perikinetic growth is greatly influenced by temperature.

3) Perikinetic growth is influenced by diffusion.

4) Addition of non-polar solvent to a protein solution compresses the Stern and Gouy-Chapman layers.

5) Attractive forces for protein molecules are inversely proportional to the dielectric constant.

Summary and objectives

Solvent extraction is one of the classical and most versatile unit operations used in the recovery of bioproducts. It is still by far the most important operation in the pharmaceutical industry, and in particular in the manufacturing of antibiotics. Its use in biotechnology is however limited, mainly because of the poor stability of the bioproducts under extraction conditions. For large molecules such as enzymes and pharmaceutical proteins, aqueous two phase extraction could be a solution to the stability problem. It is applied on a relatively small scale, because of the high costs associated with the recovery of the polymers and salts used. For solvent extraction the theory fits well with the experimental work. However, aqueous two phase extraction is relatively new and a comprehensive theory for solvent extraction has not yet been developed.

Precipitation is one of the oldest recovery methods for proteins. Several precipitation methods can be used, but salt and solvent precipitation occupy a prominent position. To some extent it is possible to describe the precipitation process by semi empirical correlations. The available theories give only an indication of the parameters important to the aggregation process.

Now that you have completed this chapter you should be able to:

- define separation factor and explain how extraction equilibrium influences this factor;
- describe the kinetics of extraction processes;
- predict the number of stages required for an extraction process;
- describe how pH and pK influences extraction of a low molecular weight compound;
- interpret the Brønsted equation and list factors that influence the distribution coefficient of a high molecular weight compound in solvent extraction;
- describe how the electric double layer surrounding a protein molecule influences precipitation;
- interpret the Hofmeister series;
- explain how the dielectric constant influences precipitation;
- broadly describe the kinetics of protein aggregation.

Purification

Purification

Purification is the final and in many cases essential step in the recovery chain.

Depending on the degree of purification this type of process is generally costly and operates with relatively low yields.

Here discussions of purification is restricted to sorptive processes. In sorption the interaction between a solid phase and a liquid phase is exploited to separate components. We can identify two types of sorptive processes; adsorption and chromatography.

In adsorption, the preferential concentration of a component on the solid surface is the separating phenomenon, while in chromatography separation is caused by the difference in affinity or interaction of components with the solid phase.

adsorption and chromatography differ in mode of operation

Both processes are based on the same principles but differ in mode of operation. In adsorption a step change in the feed and its response at the end of the operation are considered, while in chromatography the response to a pulse in the feed is considered.

Adsorption has been applied as a concentration and purification technique on a large scale for commodity chemicals such as antibiotics and other small molecules for many years. Adsorption has been mathematically described in detail. Equilibria are known and mass transfer rates can be calculated or can be found in literature.

Chromatography, however, is a relatively new technique and is applied for the fractionation of bio-macromolecules on a relatively small scale. Far less is known about chemical equilibria and kinetics of this technique. We will discuss both techniques in this chapter. We begin by describing sorption and sorption mechanisms.

7.1 Sorption

7.1.1 Sorption mechanisms

The mechanisms of sorption processes can be divided into five categories as shown in Figure 7.1

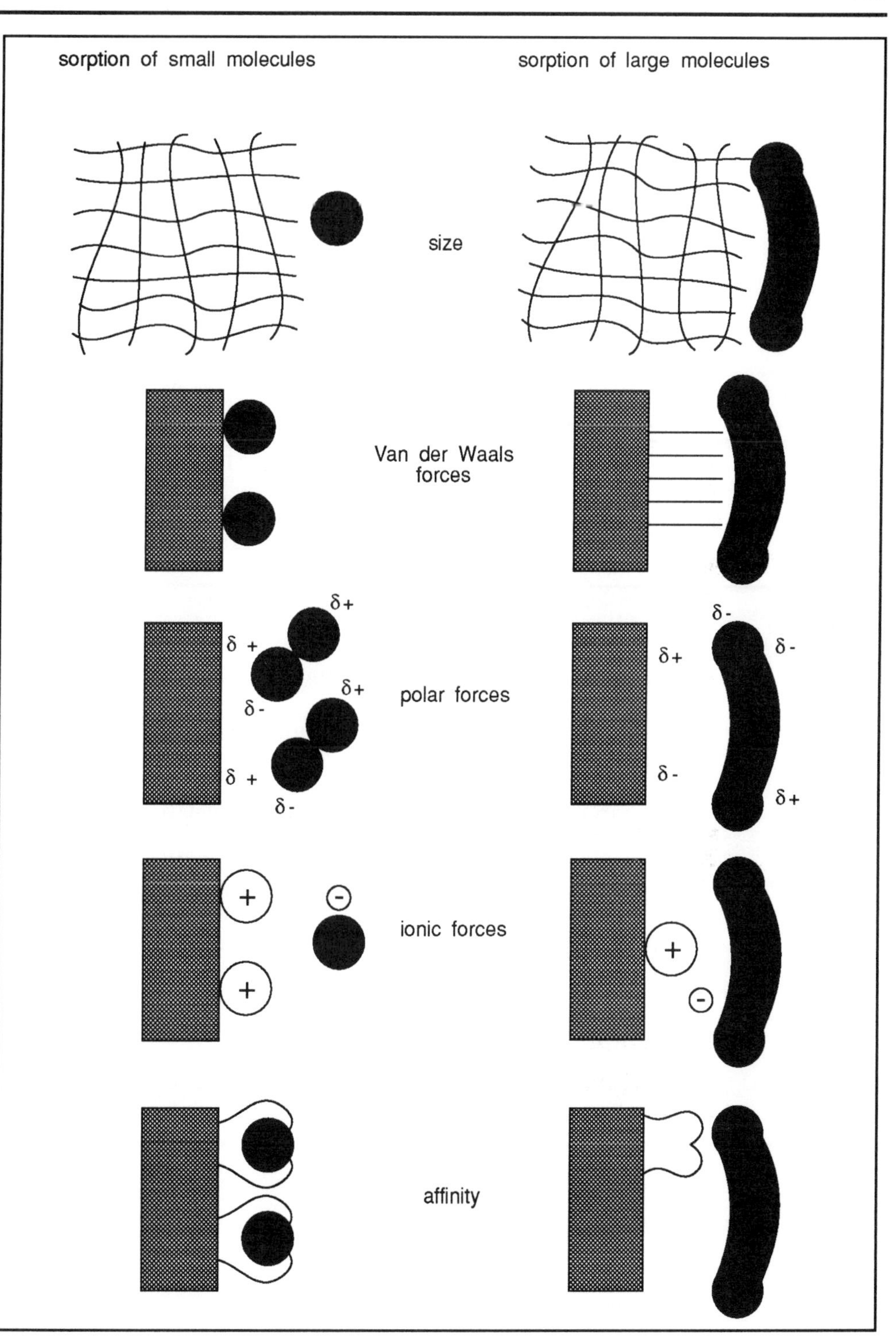

Figure 7.1 Sorption mechanisms.

The first one is more or less a form of size exclusion. Gel permeation is based on this phenomenon. Large molecules do not enter the pores while small molecules do so. The result is a velocity difference between large and small molecules. In fact separation is achieved on the basis of kinetics. We can represent this by a simple diagram.

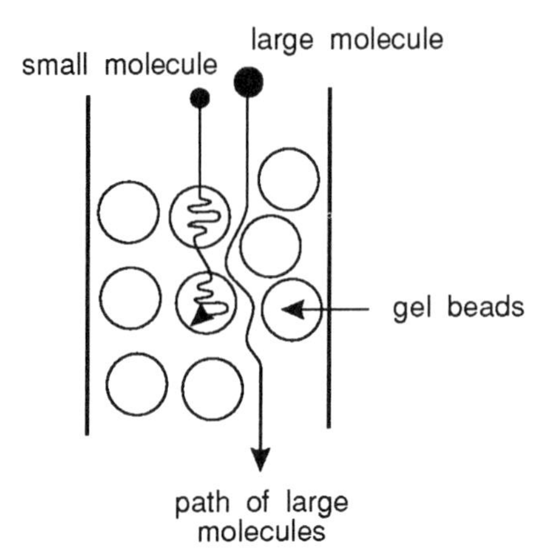

Large molecules cannot enter the beads and pass quickly through the bed. Small molecules penetrate the beads and their progress through the layer takes much longer.

In all other cases of sorption surface phenomena plays a dominant role.

The Van der Waals attractive force causes the weakest bounding to a surface of a (macro) molecule.

Differences in polarity of the surface and the (macro) molecule causes polar binding forces.

Charged surfaces are able to bind molecules with a net overall opposite charge.

Finally (bio) specific binding can be accomplished by shape of molecules that attaches on the surface of a carrier; in the same way as a key fits into a lock.

7.1.2 Sorbtion materials

Different sorbtion materials (also called medium, immobile phase, matrix, sorbent etc) utilise different binding mechanisms. All sorbtion materials can be classified in three main groups.

They may be:

- inorganic (silica, carbon);
- organic (dextran, polystyrene);
- composite (silica, dextran).

The inorganic sorbents, such as activated carbon, silica and alumina, are generally rigid, stable and available in many shapes and sizes. The selectivity, however, is rather low. Also, regeneration is difficult and their lifetime is limited. They are applied in decolourisation and in the removal of low molecular weight impurities.

composition determines properties

The polymeric sorbents, such as dextran and polystyrene, are available in a gel type and a macroreticular, sponge-like structure. They are applied in a wide rage of application such as in the recovery of antibiotics. Their mechanical strength, especially that of the gel type, limits their applicability in large scale processes, and they are mainly used in chromatography.

The composite sorbents combine the advantages of the inorganic rigid carrier with the favourable adsorption behaviour of the polymeric adsorbents. In fact, the inner surface of an inorganic carrier is covered with a very thin layer of organic material. Composite sorbents can be applied in the recovery or proteins.

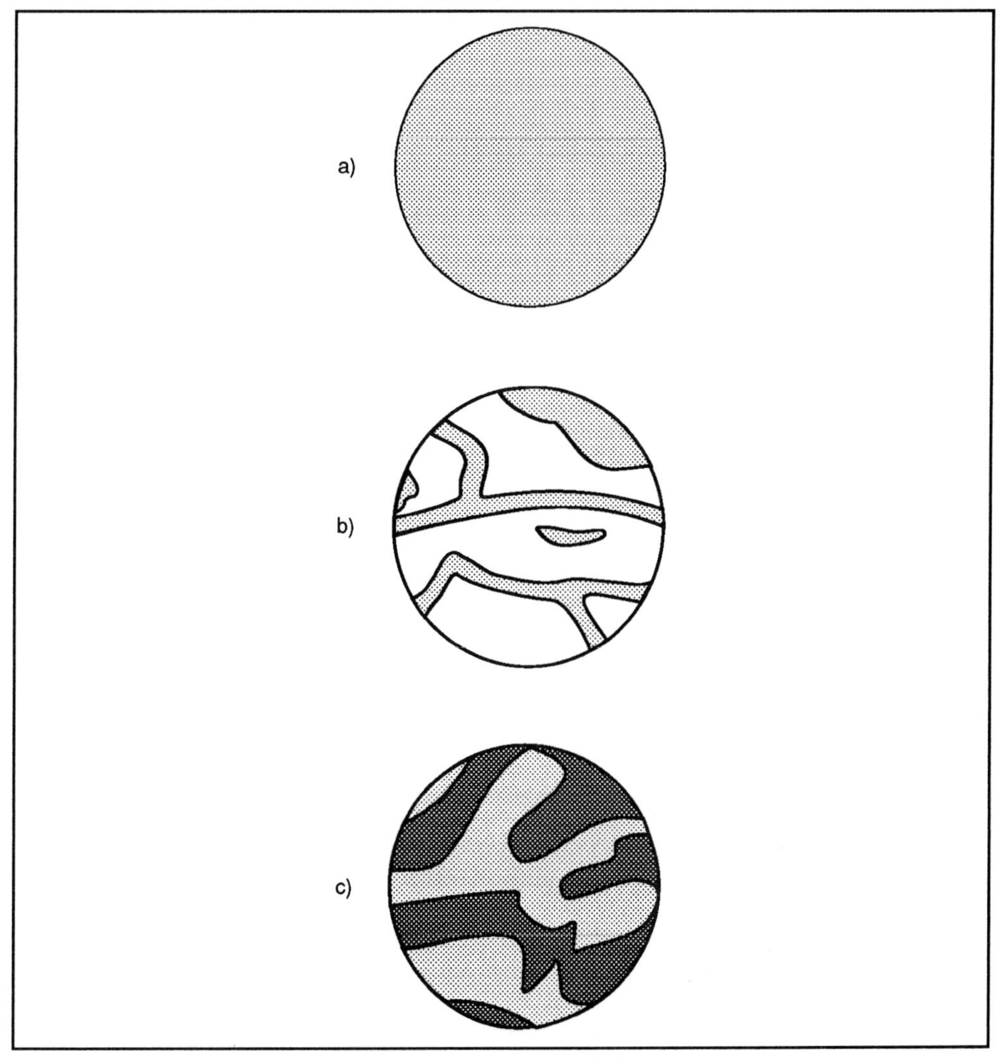

Figure 7.2 Sorbtion materials a: gel, b: macroreticular (sponge-like), c: composite.

SAQ 7.1

Complete the following statements on sorption

1) Affinity adsorption is [ ] specific than physical sorption and ion exchange.

2) The main binding forces in physical sorption are [ ] and [ ].

3) An adsorbent that is an inorganic material covered with a thin layer of organic material is termed a [ ].

The usefulness of an absorbent in a purification process is a function of:

- composition;
- porosity and surface area;
- hydrophicility/hydrophobicity;
- ligand or other functional groups;
- mechanical and chemical stability.

In addition, sorbent should also be:

- biocompatible;
- chemically stable;
- resistant to micro-organisms;
- sterilisable;
- of constant quality.

and have:

- high selectivity.

They should preferably be cheap and available in large quantities from different suppliers.

Π Think of reasons why they should be available from different suppliers.

Probably uppermost in your mind is that if only one supplier was available it would be difficult to get a competitive price. Also if the single supplier failed to deliver, then your process could become inoperative. Multiple suppliers usually means both competitive process and better service.

Table 7.1 show some sorbent materials used in protein purification.

medium	application
cellulose	gel permeation
agarose	gel permeation/ion exchange
poly acrylamide	gel permeation
dextran	gel permeation/ion exchange
porous glass	HPLC
silica	adsorption of proteins
polystyrene	ion exchange

Table 7.1 Media used in protein purification (HPLC = high performance liquid chromatography).

7.2 Sorption fundamentals

The basic concept for modelling and design of sorption processes is the same as that for extraction. Sorption can even be considered as an extraction process using a solid as an auxiliary phase instead of a liquid (solvent)!

Modelling of sorption processes requires a detailed description of equilibrium and kinetics.

7.2.1 Equilibrium

Consider three beaker glasses filled with water and an organic acid, say acetic acid, varying in concentration from 0.5-3 mmol l^{-1}. Then active carbon is added, say 50 gl^{-1} (5% w/v). During gentle mixing, the acetic acid concentration in all three beaker glasses is monitored and recorded.

When the acetic acid concentration stops changing, equilibrium has been reached. To avoid artifacts or systematic errors, three more beaker glasses are filled with a water acetic acid mixture of constant composition to which different amounts of active carbon (2-4% w/v) are added.

The amount of adsorbed acetic acid per gram of active carbon in equilibrium can be calculated from the decrease in acid concentration in the water phase.

equilibrium curve or isotherm

Plotting the amount of adsorbed acetic acid as a function of its equilibrium concentration in the liquid phase yields an equilibrium curve, usually designated as an isotherm because the experiment has to be carried out a certain fixed temperature. Figure 7.3 shows the result of this experiment.

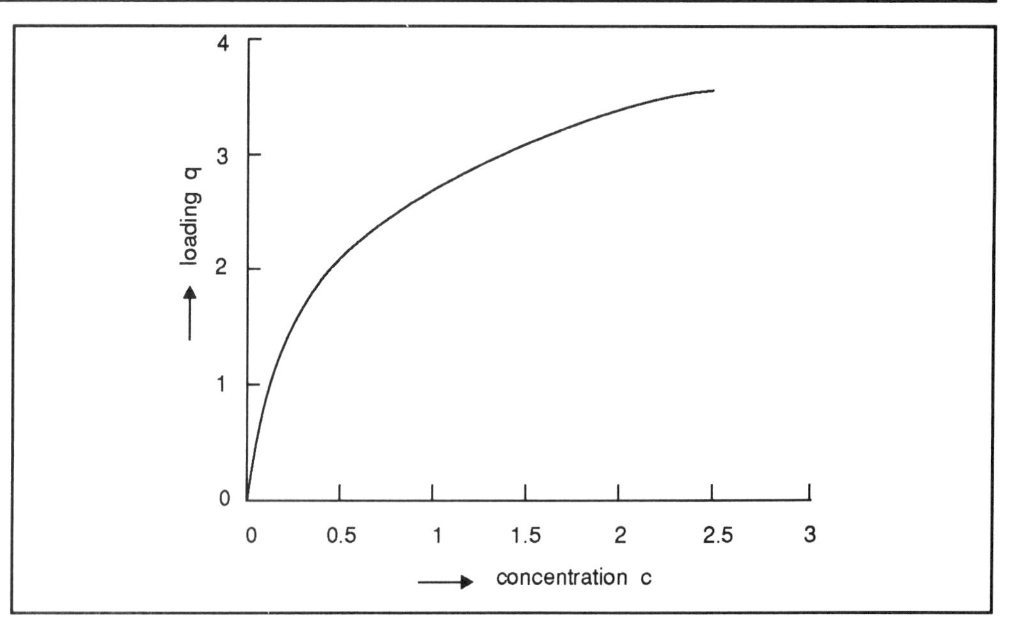

Figure 7.3 Sorption of acetic acid on activated carbon (K_F = 2.65, n = 0.35). See text for details.

Freundlich equation

The curve can be described mathematically by the Freundlich equation:

$$q = K_F C^n \qquad \text{(E - 7.1)}$$

where:

q = solid loading (kg m^{-3}; mol m^{-3});

K_F = Freundlich equilibrium constant;

C = concentration (kg m^{-3}; mol l^{-1});

n = exponent.

K_F and n are empirical parameters and can be used as fit parameters as well.

Freundlich equation limited to dilute solutions and simple molecules

The Freundlich equation can be derived from the Gibbs sorption equation combined with the description of the free energy of the surface. The Freundlich sorption isotherm, however, is limited to highly diluted solutions and simple molecules.

Based on molecular sorption as a monolayer, Langmuir derived an equation giving a good description for many cases of protein sorption:

$$q = q_{max} \frac{C}{C + K_L} \qquad \text{(E - 7.2)}$$

where:

K_L = Langmuir equilibrium constant

Since q_{max} is the maximum loading of the sorbent and is the number of moles or grams of solute adsorbed per unit of weight of sorbent, it depends on the sorbent (specific area) and solute (size of molecule) types.

Langmuir equation more suited to complex molecules

K_L depends on the strength of binding forces on the surface and its value is always greater than zero. In the case of proteins, K_L depends on the isoelectric point of the protein, the pH and the ionic strength of the solution.

In practice, K_L and q_{max} are often used as fit parameters in non-linear computer programs.

According to the Freundlich equation, the nature of an isotherm can be linear, convex or concave, as illustrated in Figure 7.4.

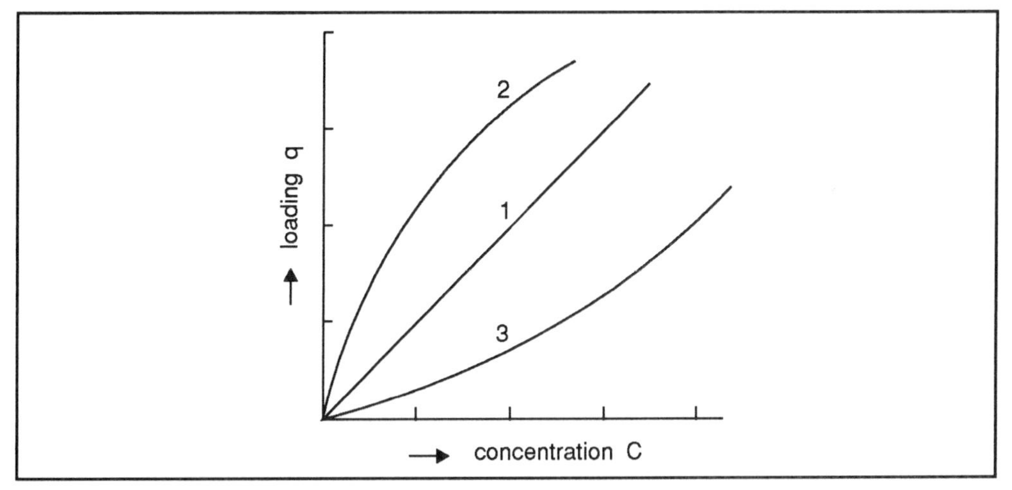

Figure 7.4 Types of sorption isotherm: 1) linear; 2) concave; 3) convex.

If n = 1 we obtain a linear relationship, and if n > 1 or n < 1 we obtain a convex or concave curve respectively.

The significance of the different types of curves in relation to the column performance will be outlined in the next sections.

Π Why do you think that q_{max} is sometimes referred to as a theoretical maximum? Examine Equation 7.2 before you answer this.

We can see from Equation 7.2 that q will approach q_{max} at high concentrations. However, since K_L has some positive value, q will never actually equal q_{max}. Because q_{max} is theoretically unattainable, it is referred to as a theoretical maximum.

Π What shape of isotherm do you think the Langmuir equation will show? Give a reason for your answer.

The Langmuir equation shows a concave isotherm. As the concentration increases K_L becomes less significant in the $(C/C + K_L)$ ratio.

7.3 Modes of operation in sorption processes

There are basically two ways to contact the fluid with the sorbent: batchwise and continuous. In biotechnology, however, only the batch modes are employed. There are two types:

- stirred tank;
- packed bed.

7.3.1 Stirred tank sorption

Stirred tank sorption, is by far the simplest configuration. In this mode of operation, measured quantities of sorbent are added to the fermentation broth or filtrate. Under gentle stirring, the sorbent is left in contact with the fluid until equilibrium has been reached.

The sorbent is then removed by screening, washed to remove undesired impurities and after that repeated with an eluting solvent. After elution the sorbent can be reused for the next batch.

stirred tank adsorption is simplest configuration

The antibiotic bacitracin, for example, can be recovered in this way using a porous polystyrene resin. After six hours the resin is then screened for separation. The resin is placed in a column and eluted with methanol. Water is added to the elute, after which it is distilled under vacuum. Bacitracin is then crystallised and dried with a 86% yield.

The design and operation of a stirred tank process are simple. As long as only one stage is needed on a small scale, the manual handling is limited. If a larger number of stages is required on a large scale, packed bed adsorption is preferred.

7.3.2 Packed bed sorption

The most common way to carry out sorption processes, is in a packed bed. A packed bed is a column with a diameter/height ratio of 1/5 to 1/20 filled with rigid sorbent material. Usually the feed enters at the top of the column.

The upper part of the packed bed is contacted with fresh feed and the sorbent in this part is saturated first. A sorption front will slowly move downwards through the column.

When the front has reached the exit of the column the sorbate 'breaks through'.

Once a column has been loaded it needs to be eluted. This is generally done in the same mode as in sorption, ie downwards. For details see Equilibrium break through (Section 7.4.3).

In most cases the elution is the critical step, and is particularly sensitive to the behaviour of the sorption isotherm and the dispersion behaviour of the elution fluid in the column (we will discuss this later).

7.4 Performance of sorption columns

The performance of a sorption column depends on many factors such as medium used, superficial velocity of eluent, column porosity, permeability of the medium, distribution coefficient between stationary and mobile phase, capacity factor, etc.

Two phenomena play an important role during the sorption process: zone separation and peak broadening.

These two phenomena will be illustrated on a chromatographic process. Before we start to do this we will first remind you of the main differences in operation mode of sorption processes: adsorption and chromatography.

7.4.1 Adsorption

adsorbate, adsorbent

Adsorption is the phenomenon in which a component (the adsorbate) concentrates on a solid surface (the adsorbent) without chemical change.

The adsorbent usually is a porous material with a high internal surface area. The main binding forces in physical adsorption are the Van der Waals forces (short range electrostatic forces) and polarity. In ion exchange the electrostatic interaction between a charged molecule and an opposite charge on the surface of the adsorbent is the governing binding force. Ion exchangers usually are crosslinked polymers (resins) to which charged groups are attached.

The performance of an adsorptive process depends heavily on the capacity of the adsorbent. Consequently, adsorption is in fact a concentration process. The purification level, however, is much higher than achieved by using membranes or by precipitation. Adsorption, therefore, can be considered as a purification technique.

description involved a change in conditions

After adsorption the component needs to be eluted from the solid surface sorbent. This operation is called desorption or elution.

7.4.2 Chromatography

In chromatography, differences in migration velocity for the different components present in the feed are exploited to cause separation.

In nearly every application separation is carried out in a packed column filled with spherical particles or a gel as a stationary phase. The sample is introduced at the top of the column and allowed to migrate through the column as illustrated in Figure 7.5.

Depending on the interaction of one or more components with the solid phase they will migrate slower or faster through the column, thus affecting resolution.

Π Which of the components labelled a), b) and c) in Figure 7.5 migrates through the column the fastest and which interacts with the stationary phase the most?

Components a) migrates through the column the fastest and component c) interacts with the stationary phase the most.

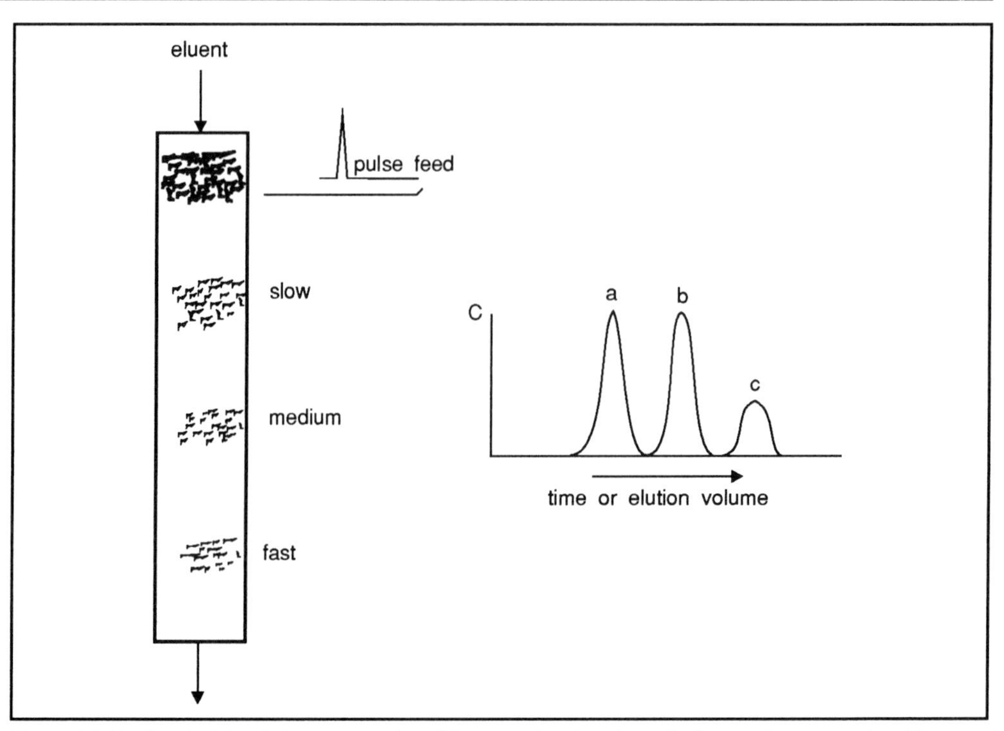

Figure 7.5 Basic principle of chromatography: differences in migration velocity causing separation (C = concentration).

As the components migrate through the packed bed, they tend to spread our, a process known as peak or zone broadening.

If the migration process proceeds sufficiently fast the zone separation is faster than the zone broadening. Separation can be achieved but dilution of the effluent is the result. This effect is depicted in Figure 7.6.

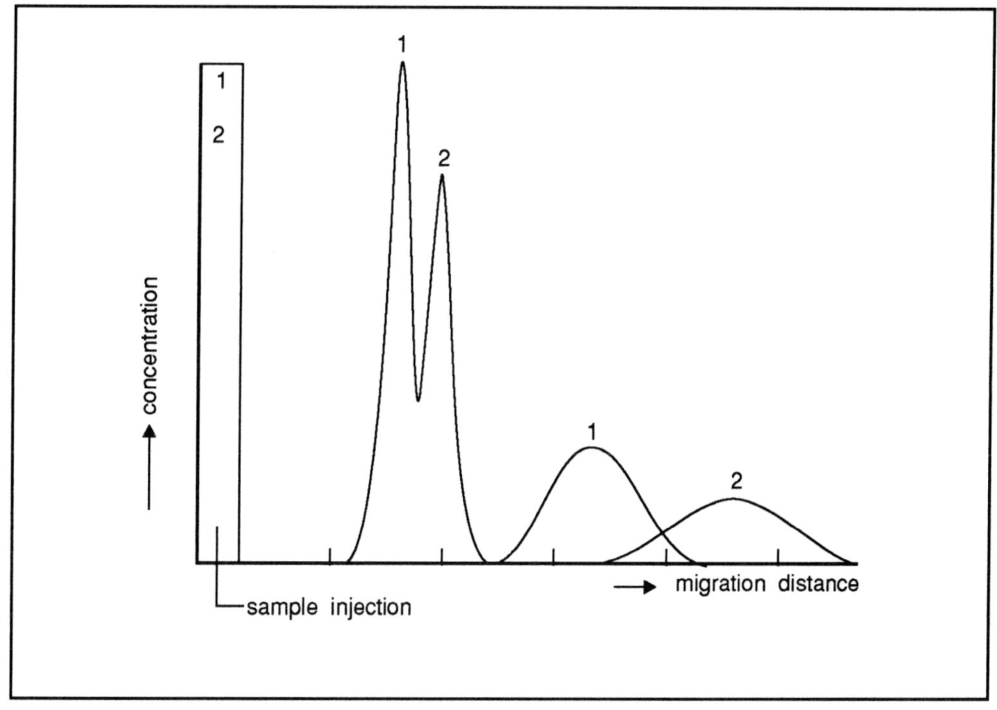

Figure 7.6 Peak broadening in sorption columns.

We may conclude that peak or front broadening is an important phenomenon that contributes negatively to the separation. Peak broadening can be suppressed by a proper choice of the sorption isotherm. However, there are also non-equilibrium effects on front or peak broadening. Both effects will be explained in more details in the next sections.

7.4.3 Equilibrium breakthrough and peak broadening

As already mentioned in the previous sections a relation exist between the shape (type) of the isotherm (see Figure 7.7) and the mode of operation with respect to the performance. This is especially true in the case of column operation.

Let us consider the behaviour of breakthrough curves for the various types of adsorption isotherms.

Based on a differential mass balance we are able to calculate the 'concentration velocity' through the column (see Figure 7.7).

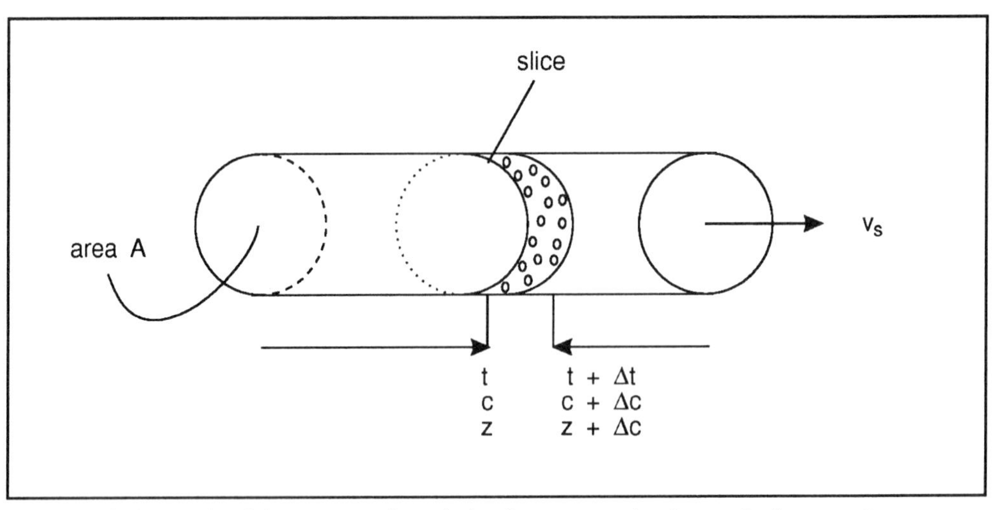

Figure 7.7 Definition sketch for concentration velocity. A = cross-sectional area of column, t = time, v_s = superficial velocity , C = concentration, ε = porosity (liquid between particles). See text for further discussion.

From a mass balance over a thin slice Δ it follows:

Input:	$v_s A \varepsilon C \Delta t$
Output:	$v_s A \varepsilon (C + \Delta C) \Delta t$
Accumulation:	$(1 - \varepsilon) A \Delta L \Delta \bar{q} + \varepsilon A \Delta L \Delta C$
	bound on medium / present in the pores

Algebraic manipulation yields:

$$\frac{\Delta L}{\Delta t} = \frac{v_s}{1 + \frac{1-\varepsilon}{\varepsilon} \frac{\Delta \bar{q}}{\Delta C}} \qquad \text{(E - 7.3)}$$

where

L = column length (m)

v_s = superficial liquid velocity (based on empty column) m s^{-1}

ε = porosity (liquid between particles)

$\frac{\Delta \bar{q}}{\Delta C}$ = slope of the adsorption isotherm

$\frac{\Delta L}{\Delta t}$ = propagation velocity m s^{-1}

Based on Equation 7.3 the local propagation velocity of a concentration through the column depends on the slope of the adsorption isotherm ie $\Delta \bar{q} / \Delta c$. It can be concluded that a sharp breakthrough at the discharge outlet of the column can be achieved if the isotherm is concave. This type of isotherm is, therefore, called 'favourable'. Convex isotherms on the other hand causes 'front spreading' and is called unfavourable. The result is an efficient used of the sorbent in case of concave isotherms and inefficient use in the case of convex isotherms. This is illustrated in Figure 7.8.

Linear isotherms causes constant concentration velocities and no change in front shape in time will result.

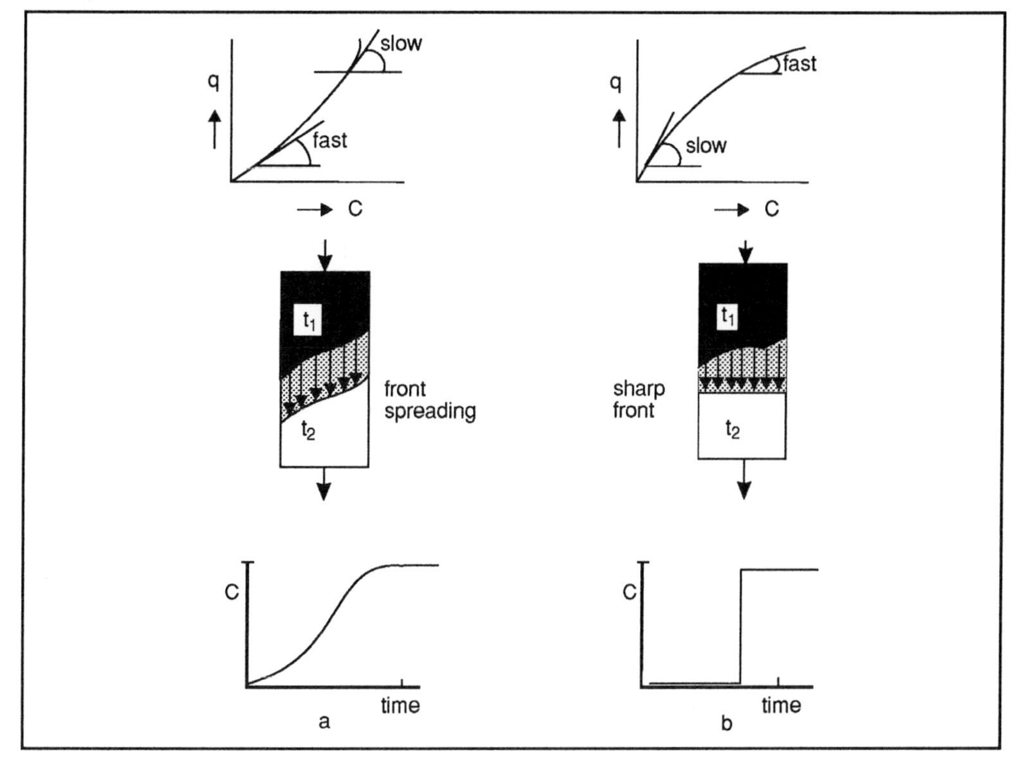

Figure 7.8 Relation between type of sorption isotherm and breakthrough curve, a: unfavourable isotherm; b: favourable isotherm (t = time), during column loading.

So far we dealt with (ad)sorption or loading the column. The same holds for unloading (called elution) of the column.

Convex isotherms causes front spreading with a (strong) dilution of the effluent as a result, while concave isotherms causes a sharp elution front (see Figure 7.9).

7.4.4 Non-equilibrium breakthrough

The non-equilibrium effects in sorption processes causes additional effects on peak or front broadening.

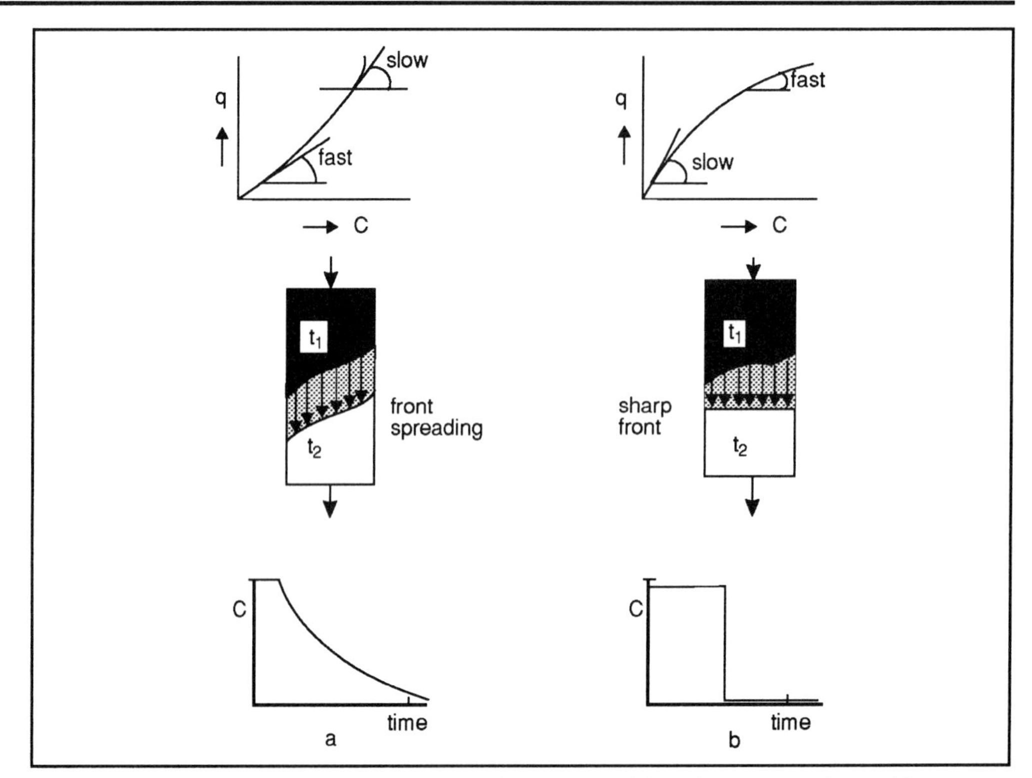

Figure 7.9 Relation between type of adsorption isotherm and breakthrough curve. a: unfavourable isotherm; b: favourable isotherm (t = time) for elution.

Three main effects can be distinguished here:

- axial dispersion;
- longitudinal diffusion;
- mass transfer limitation.

Before we start to explain these effects in more detail, according to the Van Deemter theory, we need to introduce the theoretical plate concept.

According to this concept the column is considered to be subdivided into a number of theoretical plates. The mobile phase is transported from one plate to another. For each plate, solid and liquid are assumed to be in equilibrium. In fact the number of plates for a certain column is a measure of efficiency.

The length of the column is determined by the number of plates multiplied by their height H or HETP (Height Equivalent to a Theoretical Plate).

$$H = \frac{L}{N} \qquad \text{(E - 7.4)}$$

where:

H = plate height (m);

L = column length (m);

N = number of plates.

The more plates over a certain length, the lower the HETP, the higher the efficiency. In other words efficiency is inversely proportional to plate height.

efficiency increases with increase in plate number

If the isotherm is linear, and there are no mass transfer limitations, it can mathematically be proven that symmetrical peaks occur in which about 68% of the component is within a maximum of one standard deviation (a Gaussian shaped curve is assumed).

The number of plates (N) is then defined as follows:

$$N = \left(\frac{t_r}{\sigma}\right)^2 \qquad \text{(E - 7.5)}$$

where:

t_r = residence time (s);

σ = standard deviation (s).

Figure 7.10 shows the distribution and peak parameters of a symmetrical peak.

It can also be proven mathematically that w_b (the width at the base of the peak) = 4σ. Substitution of this value into Equation 7.5 gives:

$$N = \left(\frac{t_r}{w_b}\right)^2 . \ 16 = \left(\frac{t_r}{w_{\frac{1}{2}}}\right)^2 . \ 5.55 \qquad \text{(E - 7.6)}$$

Π Determine the number of plates from the peak shown in Figure 7.10 (dimension of peak should be measured on meters).

$$\text{Number of plates (N)} = \left(\frac{t_r}{w_b}\right)^2 . \ 16$$

From Figure 7.10 $w_b = 34.10^{-3}$ m, $t_r = 500$ s

$$N = \left(\frac{500}{34.10^{-3}}\right)^2 . \ 16 = 3.46 . 10^9$$

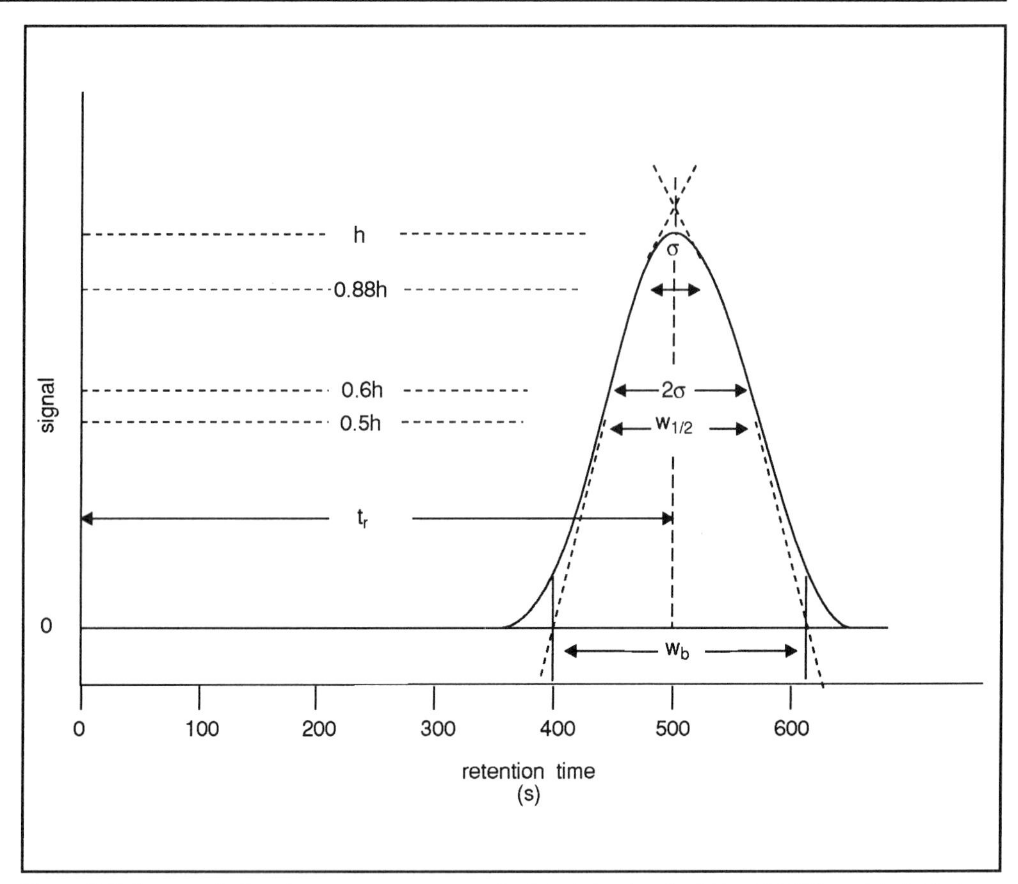

Figure 7.10 Symmetrical peak parameters: σ = standard deviation (peak width at 0.8882 peak height); t_r = residence time; w = width of peak (m); w_b = width of peak at base. $w_{\frac{1}{2}}$ width at half peak height.

Figure 7.11 shows the relation of the width of the peak and the number of plates.

Peak broadening reflects a lower number of theoretical plates in a given column and this makes the column less efficient.

Now we will continue the explanation of the non-equilibrium effects in relation to operational variables such as flow rate and medium properties.

As a first approximation to the relation of plate height and operational variables such as flow rate and medium properties, the Van Deemter equation can be used:

$$H = A + \frac{B}{v_s} + Cv_s \qquad \text{(E - 7.7)}$$

where:

H = plate height (m);

v_s = superficial velocity (flow rate) (ms^{-1});

A, B and C are constant and will be explained further below.

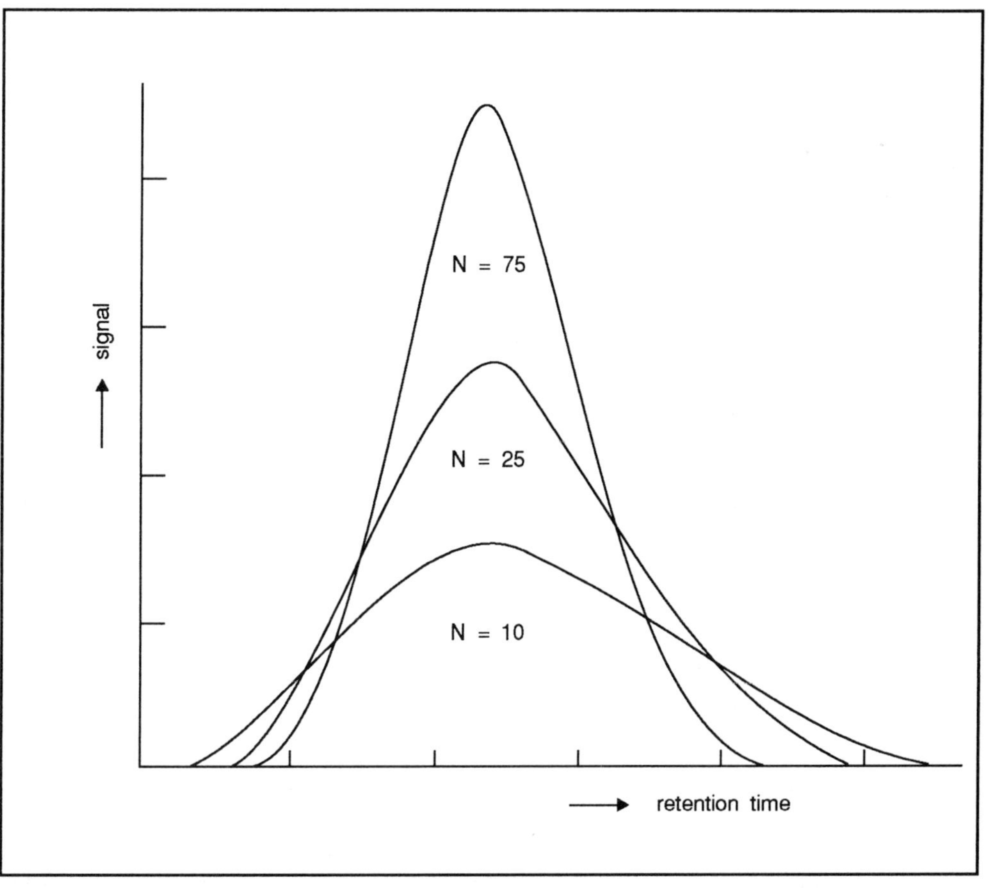

Figure 7.11 Peak broadening as function of the number of plates (N).

We will learn that A relates to axial dispersion, B relates to longitudinal dispersion and C relates to resistance to mass transfer.

Axial dispersion or Eddy diffusion

Van Deemter equation related plate height to operational variables

Factor A in Equation 7.7 represents the axial dispersion, also called Eddy diffusion. Axial dispersion is phenomenon caused by differences in the distance covered for different molecules as shown in Figure 7.12. Some molecules have a more or less linear path others take a more random used route. Dispersion causes peak broadening.

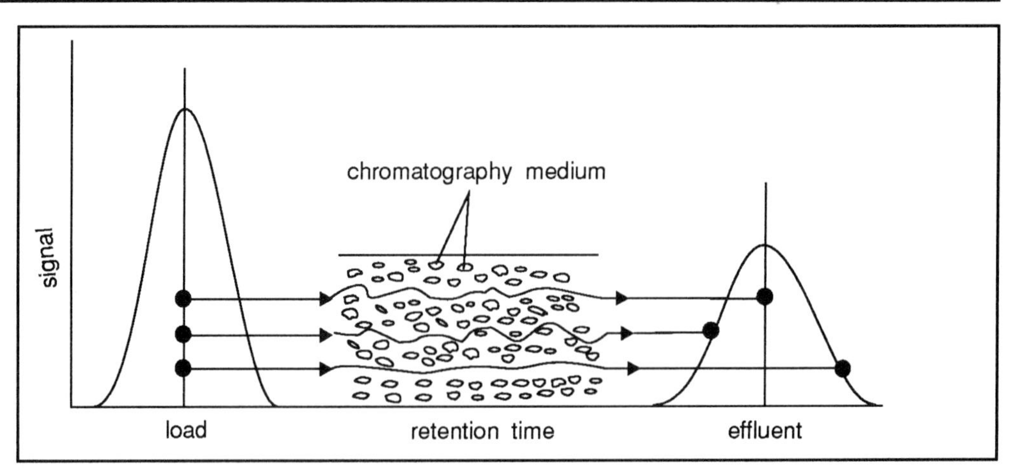

Figure 7.12 Peak broadening caused by axial dispersion (see text).

It can be mathematically expressed as follows:

$$A = 2\lambda d_p \qquad \text{(E - 7.8)}$$

where:

λ = particle ordering factor;

d_p = particle diameter (m).

λ is a factor which takes the particle ordering (ie the way in which the immobile particles are arranged) into account. The larger the particle diameter d_p of the stationary phase material, the stronger the peak broadening.

λ varies from 1 to 8 for particle size in the range of 50-600 μm.

Π Make a list of the factors which play an important role in axial dispersion.

You should have included:

- particle size;
- shape of particle;
- particle size distribution;
- ordering of particles.

Longitudinal diffusion

The factor B in Equation 7.7 takes the longitudinal diffusion of the component in the mobile phase into account as follows:

$$B = 2\, D_m/\tau \qquad \text{(E - 7.9)}$$

where:

D_m = diffusion coefficient in the mobile phase $m^2\ s^{-1}$

τ = obstruction factor

The diffusion in the mobile phase will be obstructed since solid material is present in the column. The extent to obstruction by the solid material is expressed in the quantity of τ.

The diffusion coefficient D_m in the mobile phase can be predicted using the Stokes Einstein equation:

$$D_m = \frac{KT}{6\,\pi\,\eta r_m} \qquad \text{(E - 7.10)}$$

Where:

K = Bolzmann's constant ($1.38 \cdot 10^{-23}$ J K^{-1})

T = absolute temperature (K)

η = viscosity of the mobile phase (Pas)

r_m = molecular (Stokes) radius m

Π A typical example is provided by myoglobin which has a molecular mass of 17,000 D and a Stokes radius of 2.06 nm. If the viscosity of water at 25°C in 1 mPas, what is the diffusion coefficient of myoglobin in water at 25°C?

Your calculation should show that:

$$D_m = \frac{1.38 \cdot 10^{-23} \cdot 298}{6\,\pi \cdot 10^{-3} \cdot 2.06 \cdot 10^{-9}} = 1.05 \cdot 10^{-10}\ m^2\ s^{-1}$$

This is a typical value for diffusion of a protein in aqueous solutions. For gases the diffusion coefficient is orders of magnitudes higher $D_{gas} \sim 10^{-4} - 10^{-5}\ m^2\ s^{-1}$.

The next table gives the molecular mass, Stokes radius and diffusivity data of some common proteins. (You might like to check the *D* values listed using Stokes Einstein equation - Equation 7.10.

Protein	M (D)	r_m (10^{-9} m)	$\boldsymbol{D}$ (10^{-11} m^2 s^{-1})
Lysozyme	14 400	1.82	11.8
Ovalbumin	45 000	2.94	7.3
Albumin	67 000	3.61	5.94
IgG	156 000	5.37	4.0
Catalase	232 000	5.23	4.1
Myosin	570 000	21.46	1.0

Table 7.2 Molecular mass, Stokes radius and diffusion coefficient for some commonly produced proteins in biotechnology.

From the figures given above for the diffusion coefficient of proteins and gases it may be concluded that for gas separation (chromatography) peak broadening is much more pronounced than for separation of large molecules as proteins.

SAQ 7.2

A protein has molecular mass of 34,000D and its diffusion coefficient in water at 25°C is $0.525 . 10^{-10}$ m^2 s^{-1}. If the viscosity of water at 25°C is 1 mPas and Bolzmann's constant = $1.38 . 10^{-23}$ JK^{-1}, what is the molecular radius (Stokes radius) of the protein?

Resistance to mass transfer

Any stationary phase material exerts a resistance to mass transfer. It takes time to reach equilibrium. The delay between actual contact time (ie flow velocity) and time to reach equilibrium causes peak broadening. Two phenomena contributes to mass transfer resistance:

- transfer from mobile phase through the boundary layer to the stationary layer;
- transfer into the stationary (solid) phase.

It can be derived that the coefficient for C in Equation 7.7 can be written:

$$C = \underbrace{\left(\frac{K}{K+1}\right)^2 \frac{d_p^{\ 2}}{D_m}}_{\text{mass transport in solid}} + \underbrace{\frac{K}{(K+1)^2} \frac{\delta^2}{D_s}}_{\text{mass transfer in boundary layer}} \qquad \text{(E - 7.11)}$$

capacity factor

where:

$K = \dfrac{t_r - t_e}{t_e}$ or $\dfrac{V_e - V_o}{V_o}$ and is called the capacity factor.

t_r = retention time of peak (s)

t_e = mean residence time of carrier or elution medium (s)

V_e = volume of effluent collected from time of injection at the top of the column until it emerges at the maximum concentration at the bottom of the column (m^3)

V_o = interstitial or outer volume of the column (mobile phase volume) (m^3)

D_m = diffusion coefficient in the medium ($m^2\ s^{-1}$)

d_p = particle size of medium material (m)

δ = boundary layer thickness (m)

D_s = diffusion coefficient in particle (solid) ($m^2\ s^{-1}$)

In fact C represents the extent of equilibrium that has been reached.

Substitution of Equations 7.8, 7.9 and 7.11 into 7.7 yields:

$$H = 2\lambda d_p + \frac{2D_m}{\tau v_s} + \left\{ \left(\frac{K}{K+1} \right)^2 \frac{d_p{}^2}{D_m} + \frac{K}{(K+1)^2} \frac{\delta^2}{D_s} \right\} v_s \qquad \text{(E - 7.12)}$$

We do not expect you to remember this equation although you should know how it has been derived and what it means. You should also be able to use it.

Essentially this is a written out version of the Van Deemter equation ($H = A + \frac{B}{v_s} + Cv_s$). H, therefore, reflects the axial dispersion (A) longitudinal dispersion (B) resistance to mass transfer (C) and the superficial velocity (flow rate v_s).

Equation 7.12 holds for liquid (HPLC) or gas (GLC) as a carrier medium in chromatography. It is not usually applied to adsorption.

Figure 7.13 shows a plot of the Van Deemter equation.

According to this equation the axial dispersion (represented by curve A is independent of the flow rate. The dispersion due to incomplete mass transfer is a linear function of the flow rate and is represented by curve C.

The longitudinal dispersion in inversely proportional to the flow rate and is represented by curve B.

Curve D represents the resulting profile of H against v_s.

optimum flow rate is at minimum plate height

As can be seen from this Figure there is an optimum flow rate at which the plate height is minimal. Although the flow rate has no effect on the selectivity, it does affect the resolution via the efficiency, ie plate height.

In particular, the slow mass transfer of the macromolecules leads to peak broadening and poor resolution. The only way to overcome this problem is to reduce the flow rate to some degree. The minimum value for the linear flow rate is about 2-3 cm h^{-1} for large molecules and 100-200 cm h^{-1} for small molecules.

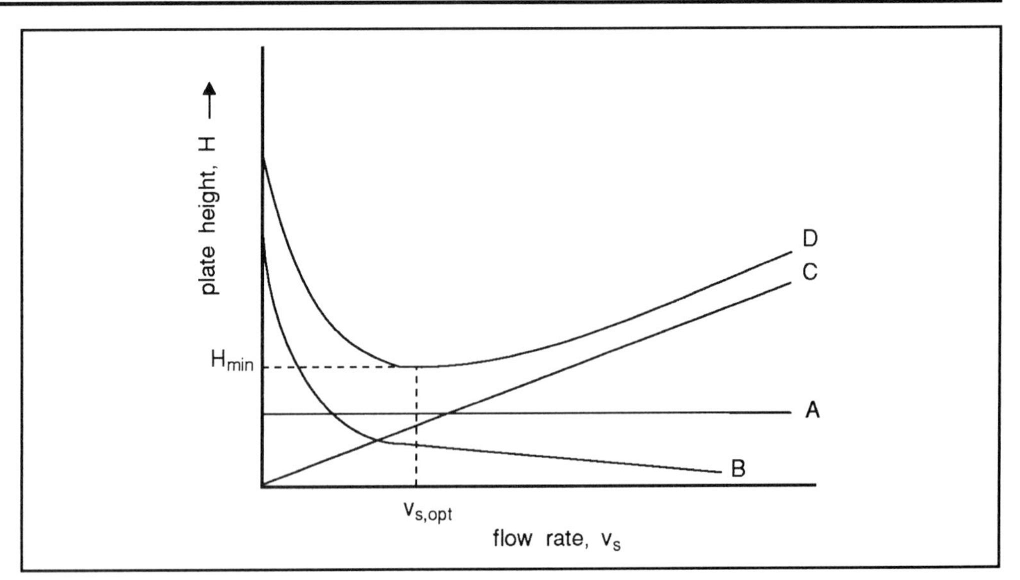

Figure 7.13 Dependence of plate height on flow rate according to Van Deetmter (see text for details).

Π Calculate the axial dispersion (A), longitudinal dispersion (B) and resistance to mass transfer (C) for a protein separation using the following data: $\lambda = 1$; $k = 2$; $d_p = 500\ \mu m$; $D_s = 5.10^{-11}\ m^2\ s^{-1}$; $D_m = 10^{-10}\ m^2\ s^{-1}$; $\delta = 10^{-5}\ m$; $\tau = 10$.

Here is our solution:

Substitution of these values into Equation 7.8, 7.9 and 7.11 gives:

$A = 10^{-3}$ s (from Equation 7.8)

$B = 2.10^{-11}$ s (from Equation 7.9)

$C = 1000$ s (from Equation 7.11)

Π Now re-examine Equation 7.7 (Van Deemter equation) and decide if axial dispersion, longitudinal dispersion or resistance to mass transfer makes the major contribution to the efficiency of this protein separation.

Since $H = A + \frac{B}{v_s} + Cv_s$ (Van Deemter equation, Equation 7.7) and we have calculated

$A = 10^{-3}$ s (axial dispersion)

$B = 2 \times 10^{-11}$ s (longitudinal dispersion)

$C = 1000$ s (resistance to mass transfer)

then we can conclude that A and B contributes very little to H. We can, in effect, write $H = 1000\ v_s$.

Since the plate height (H) is a measure of efficiency, we can conclude that resistance to mass transfer makes the major contribution.

Π If the liquid velocity is 1 mm s^{-1}, what is the number of theoretical plates in column used to separate protein as described in the previous in text activity if the column is 2 m long?

From the above in text activity, H = 1000 v_s.

Thus:

H = 1 m since v_s = 1 mm s^{-1} = 10^{-3} m s^{-1}

But the number of theoretical plates $N = \frac{L}{H}$ where L is the length of the column.

Thus in this case $N = \frac{2}{1} = 2$

This is a low number and would give a very broad peak. Remember the greater the number of plates the sharper the peak (see Figure 7.11).

Π How can we modify the column to give a greater plate number so that we would produce a sharper peak?

There are several options. Essentially what we need to do is to reduce H. If you examine Equation 7.12 you will see the factors which influence H. We have already seen that axial dispersion ($A = 2\lambda d_p$) has little influence on H. Longitudinal dispersion ($B = 2\,D_m/g$) has even less influence on H. C (resistance to mass transfer) is the major influence on H.

Since $\dot{C} = \left(\frac{K}{K+1}\right)^2 \frac{d_p{}^2}{D_m} + \frac{K}{(K+1)^2}\frac{\delta^2}{D_s}$ (see Equation 7.11)

then reducing the diameter of the particle, will greatly reduce C.

Thus in the in text activity we did early, the particle diameter used was 500 μm and we calculated H = 1000 v_s.

If smaller particles (of 10 μm diameter) had been used then H = 0.4 v_s.

Thus, if we used a 2 m long column and a flow rate of 1 mm s^{-1}, then the number of plates $(N) = \frac{2}{0.4 \times 10^{-3}} = 5 \times 10^3$ plates. This will give a much sharper peak.

Thus the difference in H is 1000 v_s compared to 0.4 v_s for columns using all the same parameters except that in the former case large particles (d_p = 500 μm) were used and in the second case small particles (d_p = 10 μm) were used.

Π What superficial velocity (v_s) can be used to achieve the same plate height the column containing small (d_p = 10 μm) particles as that for the column containing large particles (d_p = 500 μm) and with a superficial velocity (v_s) of 1 mm s^{-1}.

The plate height for the column containing large particles

$H = 1000\ v_s$ (see earlier in text activity).

Thus H = 1m.

The plate height for the column containing small particles

$H = 0.4\ v_s$

Thus if H = 1m, then $v_s = 2.5$ m s^{-1}

In other words we can achieve the same value of H using much faster flow rates using the smaller particles. Thus the column separating power in terms of throughput is much higher for small particles than for big particles. However, the resistance to flow ie the pressure drop across (over?) the column, will increase proportional to the square of the particle size ratio, thus:

$$\Delta p_2 = \Delta p_1 \left(\frac{d_{p1}}{d_{p2}}\right)^2 = \left(\frac{500}{10}\right)^2 = 2500$$

If the pressure drop in case of the 500 μm beads was 0.1 bar ($\Delta_{p1} = 10^4\ N\ m^{-2}$) we obtain now 250 bar for the 10 μm beads which can only be achieved on very small scale columns (HPLC). Perhaps you can understand now why HPLC can be equally used to describe high performance liquid chromatography or high pressure liquid chromatography.

Minimum plate height determination

Figure 7.13 shows a minimum in the plate height (H_{min}) at a certain optimum flow velocity $v_{s,opt}$.

This minimum height can be derived by differentiating Equation 7.7 with respect to v_s and putting:

$$\frac{dH}{dv_s} = 0$$

$$\text{Thus } -\frac{B}{v_s^{\ 2}} + C = 0$$

$$\text{or } v_{s,opt} = \sqrt{B/C} \qquad \text{(E - 7.13)}$$

Substitution of $v_{s,opt}$ into Equation 7.7 gives H_{min}:

$$H_{min} = A + \frac{B}{\sqrt{B/C}} + C\sqrt{B/C}$$

rearranging gives:

$$H_{min} = A + 2\sqrt{BC} \qquad \text{(E - 7.14)}$$

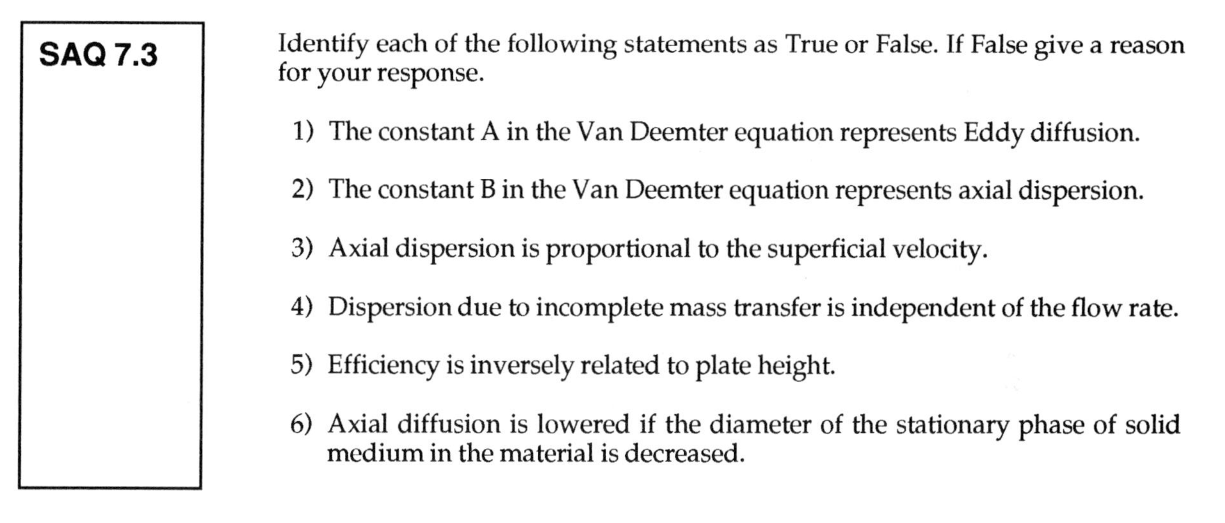

SAQ 7.3

Identify each of the following statements as True or False. If False give a reason for your response.

1) The constant A in the Van Deemter equation represents Eddy diffusion.

2) The constant B in the Van Deemter equation represents axial dispersion.

3) Axial dispersion is proportional to the superficial velocity.

4) Dispersion due to incomplete mass transfer is independent of the flow rate.

5) Efficiency is inversely related to plate height.

6) Axial diffusion is lowered if the diameter of the stationary phase of solid medium in the material is decreased.

Using the relationships described in Equation 7.13 and Equation 7.14 we can determine the optimum flow rate of a chromatograph column. Prove this to yourself by completing SAQ 7.4.

SAQ 7.4

From an experiment we obtain the following values for the Van Deemter equation.

$A = 0.2 \cdot 10^{-3}$ s

$B = 0.82 \cdot 10^{-5}$ s

$C = 0.70 \cdot 10^{-1}$ s

Make a graph of the Van Deemter plot and determine $v_{s,opt}$ and H_{min} graphically and analytically for this system.

7.4.5 Resolution

The performance of a sorption column depends on the efficiency which can be expressed as the plate number and the resolution. Efficiency does not imply a proper separation. Resolution is the ability of a column to separate peaks.

In the case of broad peaks, overlapping can occur even at high plate numbers.

A measure for the result of both effects is the resolution factor R, which is defined as follows for equal quantities of two components:

$$R = \frac{V_{e1} - V_{e2}}{\frac{1}{2}(w_1 + w_2)} = \frac{\Delta t_r}{2(\sigma_1 + \sigma_2)} \qquad \text{(E - 7.15)}$$

where:

V_{e1} = elution volume for component 1

V_{e2} = elution volume for component 2

w_1 = width of peak at base for component 1

w_2 = width of peak at base for component 2

σ_1 = standard deviation for component 1

σ_2 = standard deviation for component 2

Δt_r = distance between maxima of the two peaks (represented by V_{e1} and V_{e2}).

$\frac{1}{2}(w_1 + w_2)$ is the arithmetic mean of the widths of the elution peaks at the base line of a Gaussian curve (see also Figure 7.14).

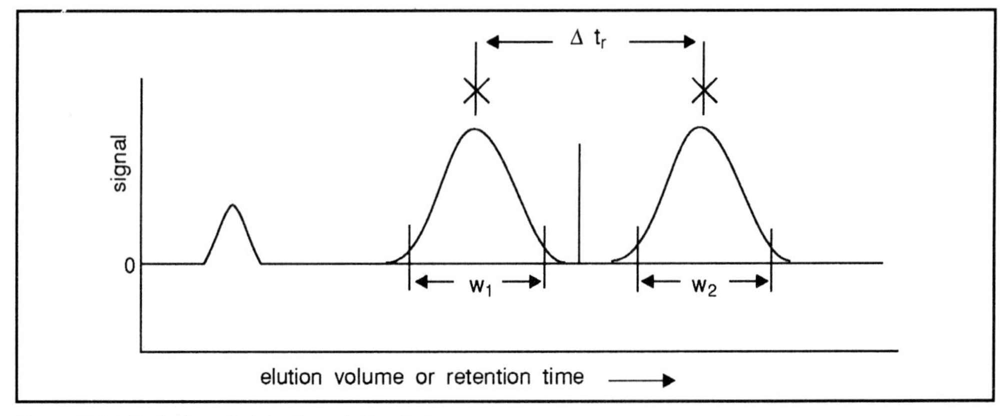

Figure 7.14 Definition sketch of resolution factor.

R = 1 means that the distance of the maxima of two Gaussian elution curves is 4 σ for the case of identical concentration of the two components. The width (w) of the elution peaks is also 4 σ. This means a 98% separation as 2% of peak areas overlap. If R = 1.5 the separation is nearly complete (99%). The distance between the peaks is now 6σ while their widths remain unchanged 4σ.

These values can be calculated using a table for areas under standard normal curves.

7.5 Practical aspects of sorbtion processes

7.5.1 Principles of adsorption and ion exchange

The basic principle of physical adsorption is shown in Figure 7.15.

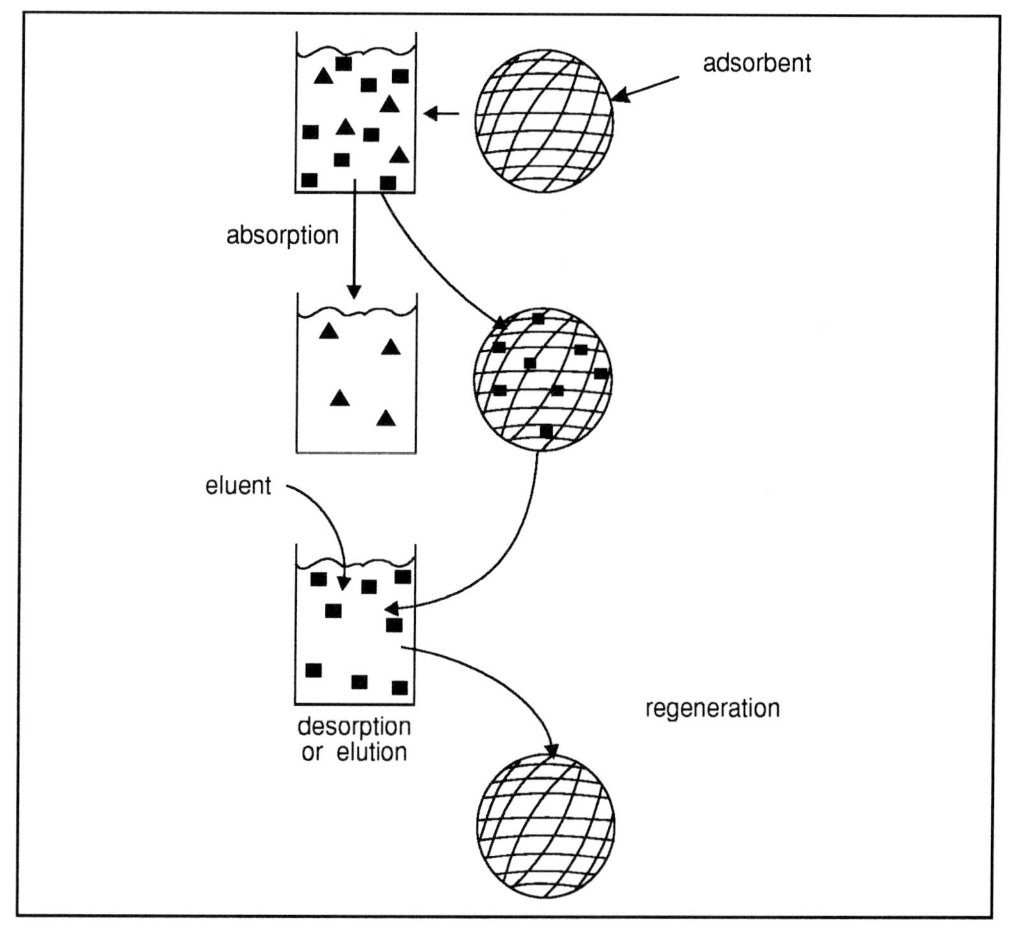

Figure 7.15 Physical adsorption. n product to be removed; s impurities.

A solid porous material (ad)sorbent is added to a solution for example proteins or antibiotics (sorbate). After a certain contact time the loaded (ad)sorbent can be removed from the solution (see Figure 7.15). Then the (ad)sorbent can be eluated to remove the sorbate from the surface. Elution is generally accomplished using aqueous solutions with high molar strength of salts or with aqueous mixtures with organic solvents.

There are two different adsorption phenomena used in biotechnology:

- physical;
- biospecific.

affinity adsorption has high specificity

Physical adsorption is not very specific. To achieve a higher specificity, affinity adsorption can be applied. Here, the binding is accomplished by attaching specific molecules (ligands) to the solid surface, which recognise the molecule to be absorbed. By 'recognise' we mean bind very specifically and tightly. Examples of ligands are Protein A, which interacts with many immunoglobulins, and monoclonal antibodies for biospecific adsorption of proteins (see Figure 7.16). Another examples is to use the substrate (or substrate-like molecules) of an enzyme linked to the immobile support.

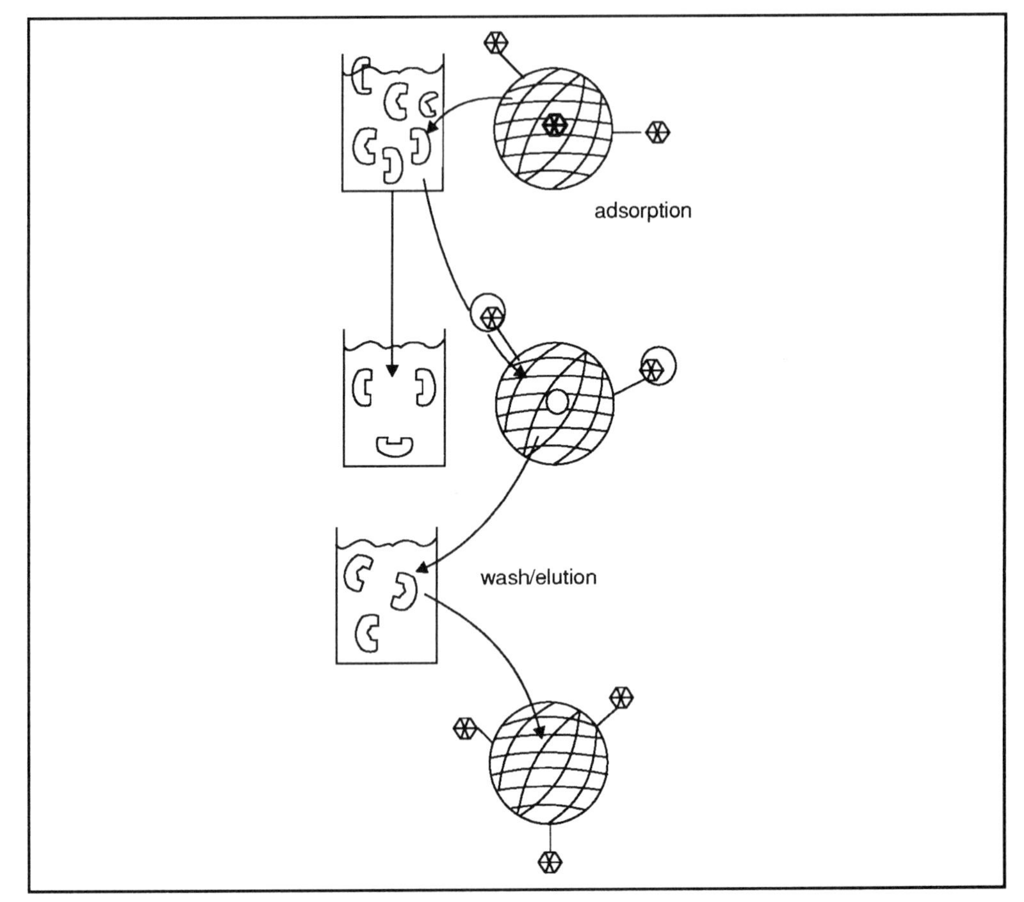

Figure 7.16 Principle of biospecific adsorption.

Passage of a mixture of proteins down the column, will lead to the enzyme binding to the substrate on the column. This type of process is often called affinity chromatography when used in chromatographic format.

In ion exchange the overall charge of a molecule is exploited to achieve 'adsorption'.

Ion exchange resins (matrix) can be charged positively or negatively by using a suitable functional group. Positive charged resins are called cation-exchangers and negative charged resins are called an ion exchangers.

The selection of the most suitable functional group depends on the charge on the product and its pH stability.

If we consider a protein the net positive or negative charge depends on the pH of the (buffer) solution and the iso-electric point (IEP) of the protein. Figure 7.17 shows a simplified filtration curve of a protein.

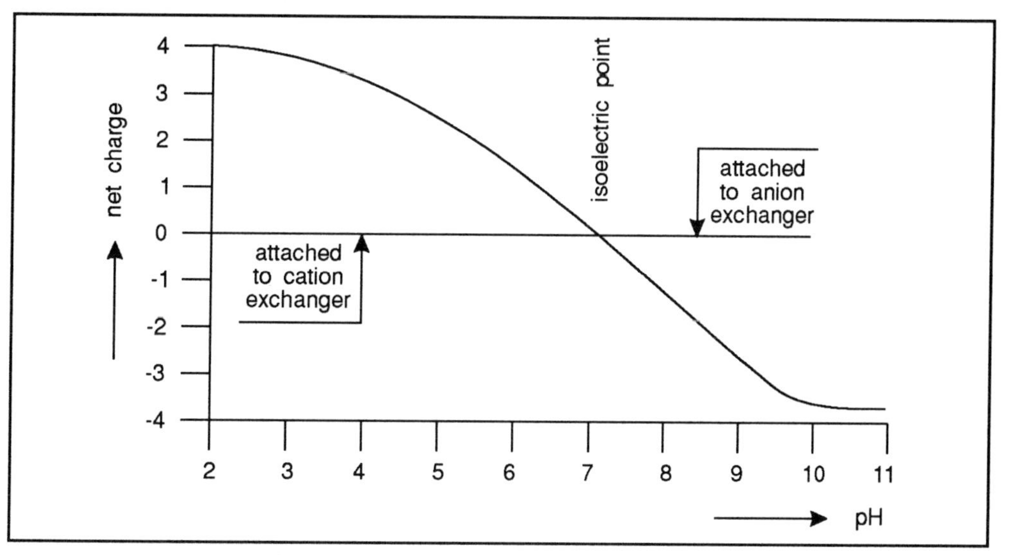

Figure 7.17 Simplified filtration curve of a protein.

At the pH < IEP the net charge is positive so an anion-exchanger may be used. At pH values of > IEP a cation-exchanger can be used. Adsorption will be performed at a suitable pH and at relatively low salt concentrations. The desorption can be performed in various ways depending on the type of adsorbent and the component involved. To release the component from the adsorbent, changes in pH, salt gradient or mixtures of solvents can be applied.

A general scheme is given in Figure 7.18

The ion exchange resin is added to the (protein) solution. After equilibrium has been reached the resin is removed from the solution and treated with an eluent (gradient ions). After elution the resin have to be washed and regenerated with suitable counter ions (to the protein) after washing again the resin is ready for reuse.

Equilibrium in ion exchange is the same as for physical adsorption and is for proteins generally expressed by a modified Langmuir isotherm.

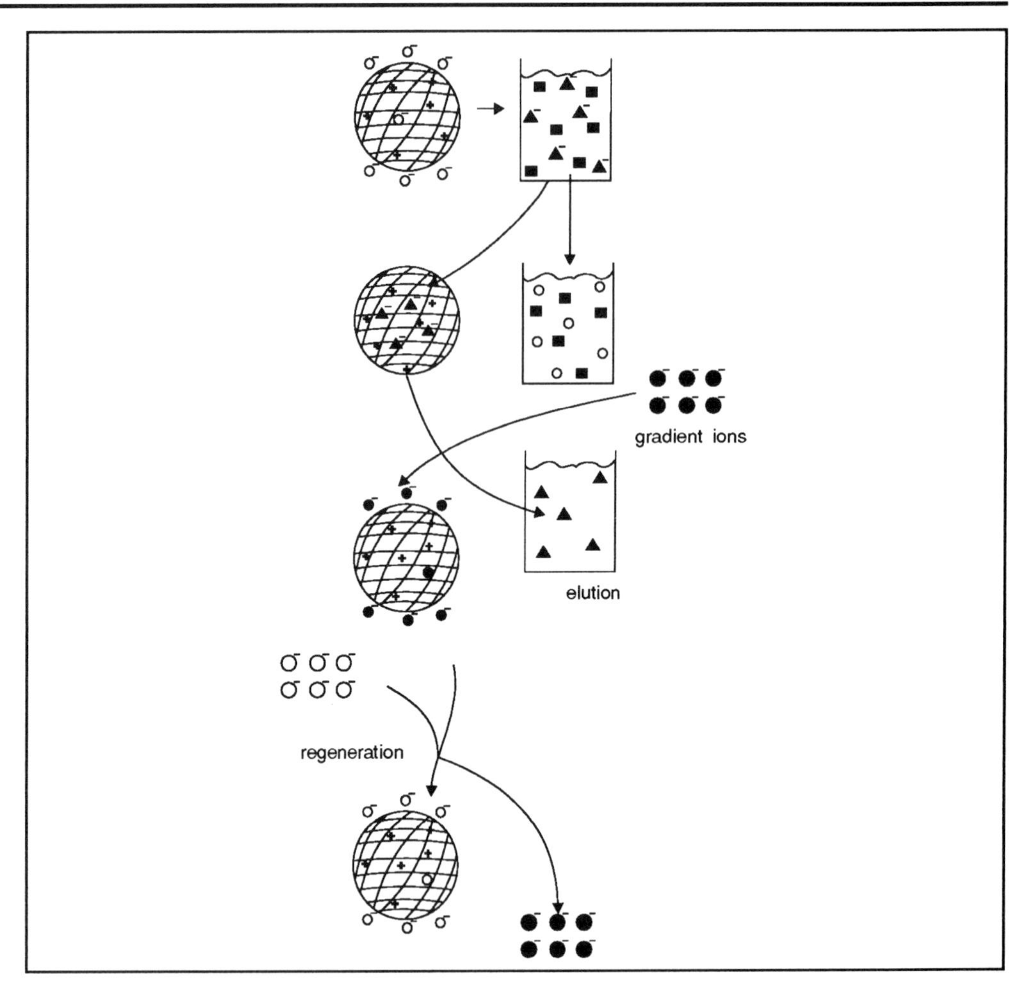

Figure 7.18 Ion exchange: ▲ ions of substances to be separated; ● gradient ions; ❍ counter ions (see text for discussion).

7.5.2 Design of absorption/ion exchange processes

Stirred tank adsorption

Consider a stirred tank which contain L_m kg liquid. S_m kg adsorbent is than added and intensively mixed until equilibrium is reached. The liquid contains a concentration of C_o material to be adsorbed (sorbate). From a mass balance of the adsorbed material, it follows that:

$$q_e = \frac{L_m}{S_m}\,(C_o - C_e) \qquad \text{(E - 7.16)}$$

where:

S_m = mass of absorbent kg

L_m = mass of liquid kg

q_e = solid loading at equilibrium (kg sorbate kg^{-1} sorbent)

C_o = initial concentration in the liquid (kg kg^{-1} liquid)

C_e = equilibrium concentration in the liquid (kg kg^{-1} liquid)

Note: Here concentration on a mass basis are used. In adsorbtion of enzymes, the mass of enzyme present in the broth is generally unknown. The activity is a better measure and is mostly expressed as units ml^{-1} or per mg. q_e can be expressed as units g^{-1} or units ml^{-1}. Many different units can be used here as long as they are consistent.

Equation 7.16 represents a straight line as depicted in Figure 7.19 with a slope of L_m/S_m. We consider now that two different types of adsorbents.

If the capacity of the adsorbent is high (high q_e - value at low concentration) the process is attractive. This is shown in Figure 7.19 (solid line).

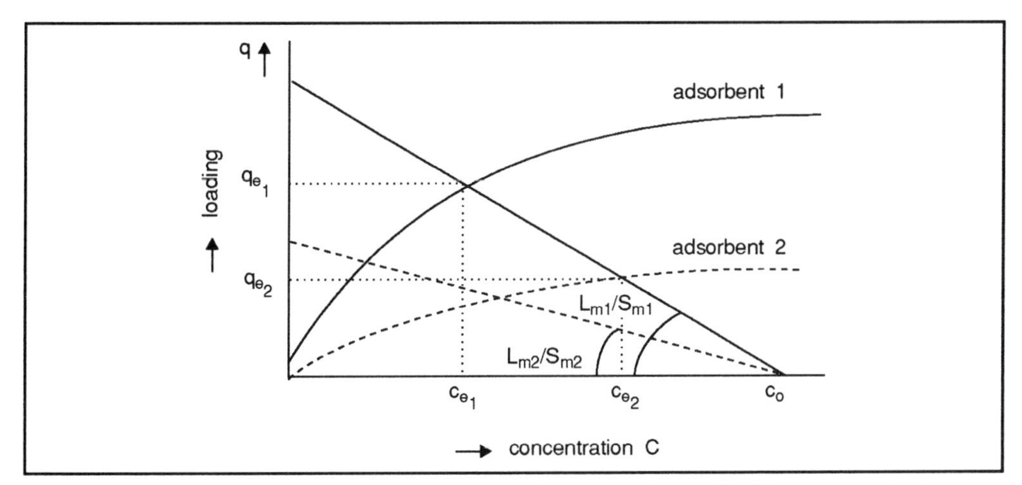

Figure 7.19 Stirred tank adsorption, equilibrium curves for two adsorbents with different capacities. L_m = mass of liquid; S_m = mass of adsorbent; L_{m2} = mass of liquid 2; S_{m2} = mass of adsorbent 2.

adsorbent with high capacity is desirable

The dotted line (adsorbent 2) shows a less attractive process because for the same L_m/S_m ratio, the solid loading (capacity) of the absorbent is lower at all equilibrium concentrations.

Π Would the L_m/S_m ratio have to be increased or decreased to obtain the same equilibrium concentration for adsorbent 2 as for absorbent 1 (Figure 7.19)?

The L_m/S_m ratio would have to be decreased ie more absorbent 2 than absorbent 1 is needed.

Very low concentrations can only be reached at very low L_m/S_m ratios ie large amounts of sorbent will be necessary.

To avoid this problem the adsorption process can be repeated several times. The concentration at the end of the first step (C_{e1}) is the initial concentration of the second step and so on. Figure 7.20 shows the result for a constant L_m/S_m ratio.

Thus if we start with an initial concentration C_o and a particular L_m/S_m ratio, then at equilibrium, the concentration will be C_{e1}. We now remove the adsorbent, and add fresh adsorbent to give the same L_m/S_m ratio. At equilibrium, the concentration will have declined to c_{e2}. In Figure 7.20 we have shown this cycle of events repeated for 5 times. The slopes of the lines labelled 1-5 in Figure 7.20 are given by L_m/S_m. If we change this ratio for any of the cycles, then the slope of the line will be changed. For example, if we use relatively more adsorbent (ie L_m/S_m is smaller), then the slope of the line will be shallower and C_e will be lower. It might be a helpful to draw such an example onto Figure 7.20 to remind yourself that there in practice. it is not necessary to keep L_m/S_m constant for each cycle.

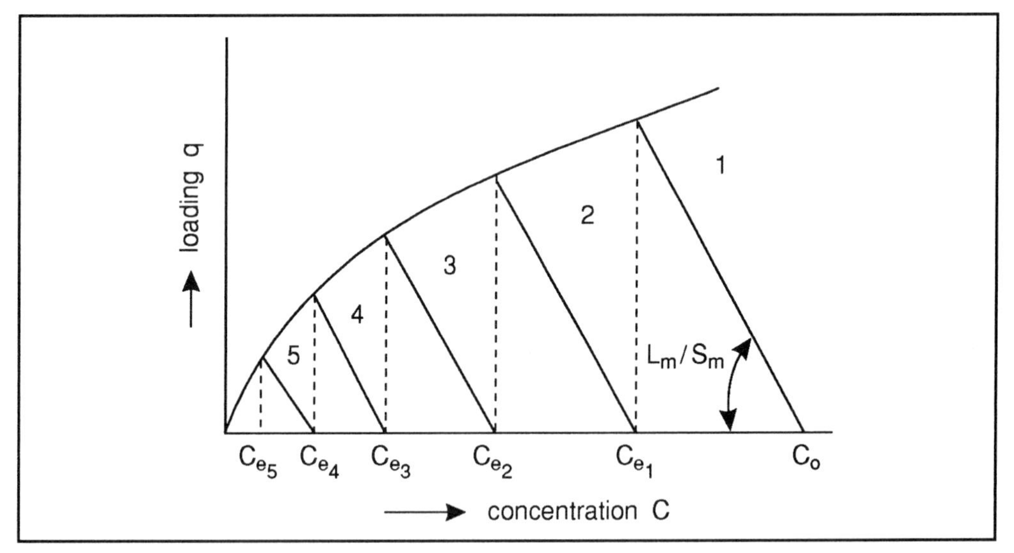

Figure 7.20 Multiple adsorption with constant L_m/S_m ratio.

This way of processing is very labour intensive because of the addition, settling and removal of the adsorbent has to be done several times. Batch extraction, therefore, can only be efficient if:

- the adsorbent has a high in capacity;
- the process in not too large;
- the added value of the product is high.

If a large number of equilibrium steps is necessary, a packed bed adsorption process is a much better option.

Packed bed adsorption

no mass transfer

The design of a packed bed, inclusive of mass transfer resistance and non-linear isotherms, is rather complicated. Here we will restrict ourselves to a system with no mass transfer resistance. From the equation for the propagation velocity Equation 7.3.

The time lapse (t) before breakthrough of initial concentration (C_o) is given by:

$$t = \frac{L}{v_s\,\varepsilon}\left(\varepsilon + (1-\varepsilon)\frac{\Delta\bar{q}}{\Delta C}\right) \qquad \text{(E - 7.3b)}$$

where:

L = column length (m);

v_s = superficial velocity (ms^{-1});

ε = void fraction;

$\frac{\Delta\bar{q}}{\Delta C}$ = slope of the isotherm;

L/v_s = is the residence time of the liquid in an empty column;

$(1-\varepsilon)$ = corrects for the presence of adsorbent.

In designing a packed bed adsorption process it would be essential to be able to work out how much liquid can be passed through the adsorption bed before breakthrough occurs. Let us do the calculations ourselves.

If the superficial velocity of the liquid is v_s and the cross sectional area = A, then in time t we will have treated $t \,.\, v_s \,.\, A$ m^3 of liquid feed (L_v).

We also know that the net volume of adsorbent in the column is $L \,.\, A \,.\, (1-\varepsilon)$.

Let us suppose, for example, $\varepsilon = 0.5$ and that $\frac{\Delta\bar{q}}{\Delta C} = 2$. then according to Equation 7.3b.

$t = 3\,L/v_s$

But we have already calculated that the total volume of liquid feed treated in time t is $t \,.\, v_s \,.\, A$ m^3. Substituting in $t = \frac{3L}{v_s}$, the total volume of liquid treated in 3LA.

Thus the ratio of liquid feed to solid adsorbent (ie L_v:S_v where L_v = volume of liquid feed and S_v is volume of adsorbent) is $\frac{t\,v_s\,A}{L\,.\,A\,.\,(1-\varepsilon)} = \frac{3\frac{L}{v_s}v_s\,A}{L\,.\,A\,.\,(1-\varepsilon)} = \frac{3}{1-\varepsilon} = 6$

In other words we can treat a liquid volume 6 times that of the volume of the solid before we get complete breakthrough. In practice this ratio will be lower because of front spreading.

Obviously the ratio of liquid feed that can be treated to the solid adsorption used will depend upon the void volume (ε) and on the adsorption characteristics of the solid (since this will influence $\frac{\Delta\bar{q}}{\Delta C}$ which is in fact an equilibrium constant (k) for this absorption).

SAQ 7.5

Complete the following statements using the appropriate words from the selection provided below.

1) L / v_s is the [ ] of the liquid in an [ ] column.

2) High [ ] of absorbent per unit volume is indicated by a [ ] q_c value.

3) In a multiple stage adsorption process, a high L_m/S_m ratio will [ ] the number of cycles needed to achieve the same final concentration

Word selection: empty, full, capacity, high, low, lower, higher, front, residence time, superficial velocity, increase.

SAQ 7.6

You are asked to design a packed bed adsorption process based on the following data.

$t = 360$ s

$L = 0.5$ m

$A = 0.3$ m^2

$\varepsilon = 0.6$

$v_s = 8.8 \ . \ 10^{-3}$ ms^{-1}

1) Determine the value of $\frac{\Delta \bar{q}}{\Delta C}$

2) How much feed has been treated during this period (360s)?

3) Determine the ratio of the liquid feed volume (L_v) to solid adsorbent volume (S_v).

4) Explain how you could double the volume of feed treated in unit time.

7.5.3 Chromatography

Various chromatographic techniques are available for the purification of biological products. The main ones are listed in Table 7.3.

Method	separation criterion	important parameters
Gel permeation	particle size	column length
Ion exchange chromatography	charge	pH, ionic strength
Hydrophobic interaction	hydrophobicity	polarity, ionic strength
Reversed phase chromatography	hydrophobicity	
Affinty chromatography	biospecific interaction	ligand, eluent

Table 7.3 Overview of major chromatographic techniques.

Table 7.4 shows the resolution and capacity of the various chromatographic techniques that may be employed.

Chromatographic technique	Resolution	Capacity
ion exchange	low/moderate	very high
hydrophobic interaction	high	high
affinity	very high	high
dyes		
lectins		
antibodies		
metal ion		
HPLC (reversed phase)	very high	high
Gel permeation	moderate	moderate

Table 7.4 Methods for large scale protein chromatography.

As an example we will examine gel permeation (gel filtration) processes in more detail.

7.5.4 Gel permeation

Gel permeation is a form of partition chromatography used for separating molecules of different sizes. It has been described by several other terms including gel filtration, gel exclusion and molecular sieve chromatography.

The basic principle of gel permeation chromatography is that molecules are partitioned between a solvent and a stationary phase. This stationary phase (matrix) which can be in a gel or a bead form has a defined porosity. The separation process will be explained using Figure 7.21.

A mixture of molecules (illustrated in Figure 7.21 with three different sizes) enters the column at the top (pulse feed). The largest molecules are not able to penetrate into the gel matrix, the middle size molecules penetrate partly and the smallest molecules penetrate to a large extent into the gel. The result is that the smallest molecule are delayed in their passage down the column compared to the larger molecules. The large molecules thus leave the column first followed by the smaller molecules in the order of their sizes. This behaviour has been illustrated in Figure 7.8.

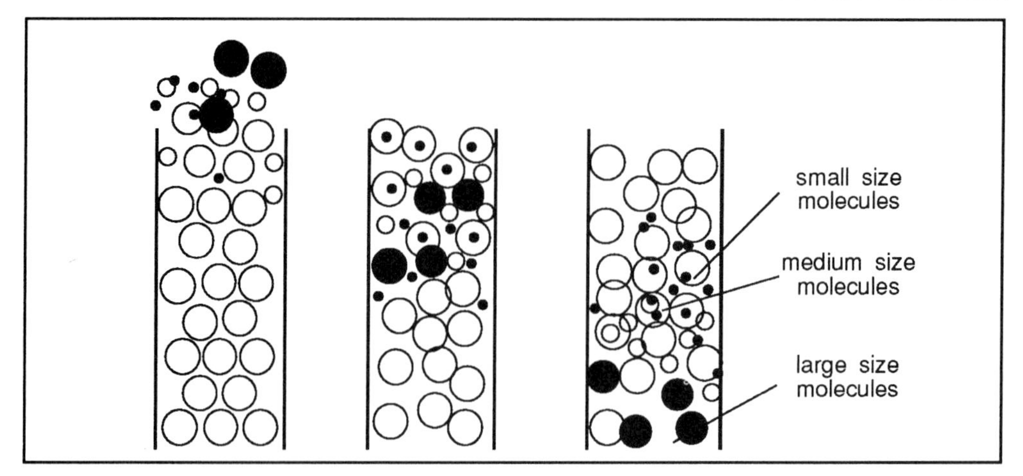

Figure 7.21 Basic principle of gel permeation chromatography (see text for a description).

The elution peaks can be described generally by a Gaussian distribution function. The outlet concentration of a compound can be mathematically expressed as follows:

$$C_t = C_{max} \exp\left\{-\frac{(t-t_r)^2}{2\sigma^2}\right\} \qquad \text{(E - 7.17)}$$

where:

C_f = concentration at the discharge outlet kg m^{-3};

C_{max} = concentration at the top of the peak kg m^{-3};

t = elapsed time s;

t_r = retention time s

σ = standard deviation (s).

The characteristics of this type of curve have already been shown in Figure 7.10.

The plate number can be calculated using Equation 7.6.

Finally we may rewrite Equation 7.15 for the resolution in case of gel permeation as follows:

$$R = \frac{\alpha - 1}{4\alpha} \cdot \frac{K}{1 + K} \sqrt{N} \qquad \text{(E - 7.18)}$$

where:

α = relative retention or selectivity;

K = distribution coefficient, or less correctly capacity factor or retention factor;

N = number of theoretical plates.

The selectivity is defined as follows:

$$\alpha = \frac{V_{e2}}{V_{e1}} \qquad \text{(E - 7.19)}$$

where:

V_{e1} and V_{e2} are the elution volumes for two components (1 and 2).

Π If we quadruple the number of theoretical plates in a gel permeation system, by how much do we increase the resolution?

Since the resolution $R = \frac{\alpha - 1}{4\alpha} \frac{K}{1 + K} \sqrt{N}$ (see Equation 7.18)

If we increase N to x 4 then R will be increased by a factor of $\sqrt{4} = 2$. Thus R will be doubled. (We have assumed that α and K have remained constant).

Π If the relative retention of two components is 5 and the capacity factor of the system is 0.8 what will be the resolution of the two components if the theoretical plate number is 100?

Since $R = \frac{\alpha - 1}{4\alpha} \frac{K}{1 + K} \sqrt{N}$ then

$$R = \frac{4}{20} \cdot \frac{0.8}{1.8} \cdot \sqrt{100} = 0.88$$

This is a rather low value.

Π Make a list of the ways we improve the resolution of the system described above.

You may gave suggested many different ways. Our solution is as follows. Re-examine Equation 7.18. You will see that R is influenced by α, K and N. In principle if we can change α, N and/or K, then we will influence the resolution (R).

We evaluate the selectivity (α) first. Figure 7.22 shows a plot of $\frac{\alpha - 1}{\alpha}$ as function of α. Increasing α-values increase the resolution.

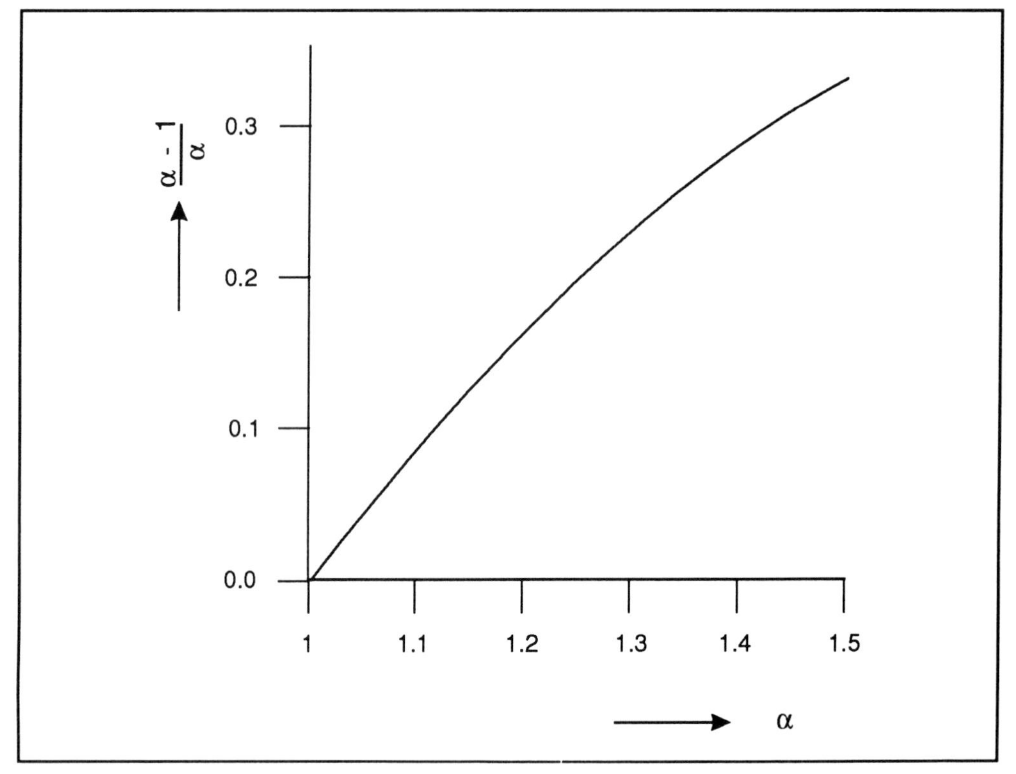

Figure 7.22 Plot of $\frac{\alpha - 1}{\alpha}$ as function of α.

The same holds for the capacity factor plotting $\frac{K}{K+1}$ as function of K gives Figure 7.23.

The largest effect can be found if $1 < K < 5$, shadowed area in Figure 7.23. Finally we will evaluate the effect of the number of plates on the relative retention. For example if R = 1 and K = 4 then Equation 7.18 gives:

$$\left(\frac{\alpha}{\alpha - 1}\right)^2 = \frac{N}{25}$$

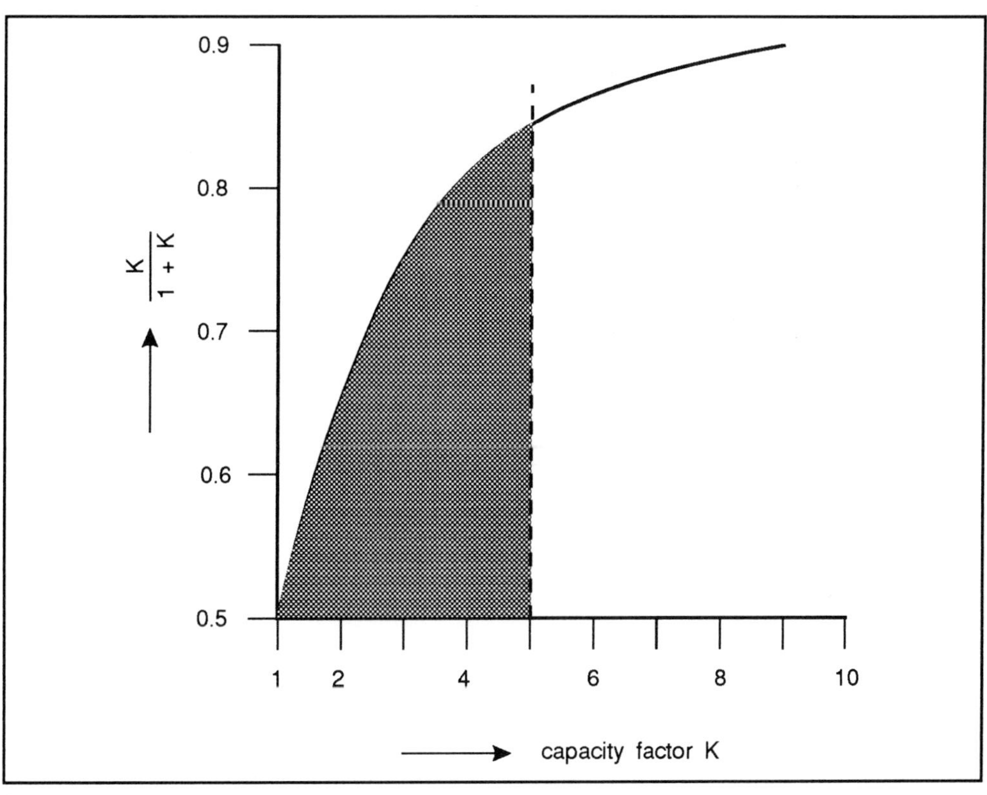

Figure 7.23 Plot of $\frac{K}{K+1}$ as function of K.

A plot of α as function of N yields Figure 7.24. A decrease in α has to be compensated by a tremendous increase of H to obtain a constant resolution.

From Figure 7.24 and from Equation 7.18 it should be self evident that if the selectivity (α) of the system is high then it is easier to achieve a high resolution.

7.5.5 Guide lines for the scale up of gel permeation and chromatographic processes

Guide lines for gel permeation or chromatographic process in general differs from scale up of adsorption processes and are relatively simple. We offer the following advice.

First optimise the purification scheme on laboratory scale.

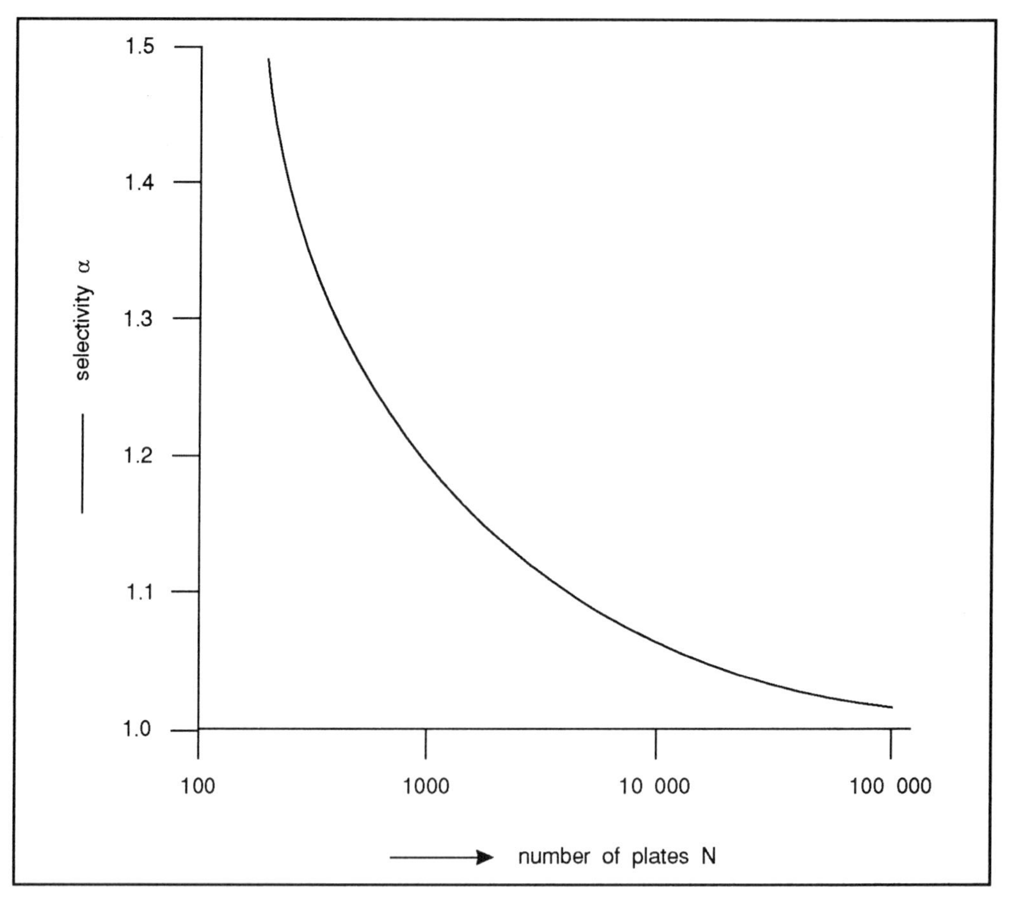

Figure 7.24 Relationship between selectivity (α) and the number of plates (N) needed to achieve the same resolution.

Then on scale up:

maintain

- bed height;
- linear flow;
- sample concentration.

increase

- sample load;
- volumetric flow rate;
- column diameter.

One of the restrictions in increasing the linear flow rate and subsequently the volumetric flow rate is the ability of gel and beads to deform. At increasing linear flow rate the pressure drop across the bed (column length) rises exponentially as shown in Figure 7.25.

The column diameter is also an important parameter. Increasing the column diameter affects the linear flow rate (at constant pressure drop) as can be seen in the figure.

This effect is due to a decrease in the support given by the wall to the medium at increasing column diameter.

When choosing the most suitable gel matrix for a particular application, several factors

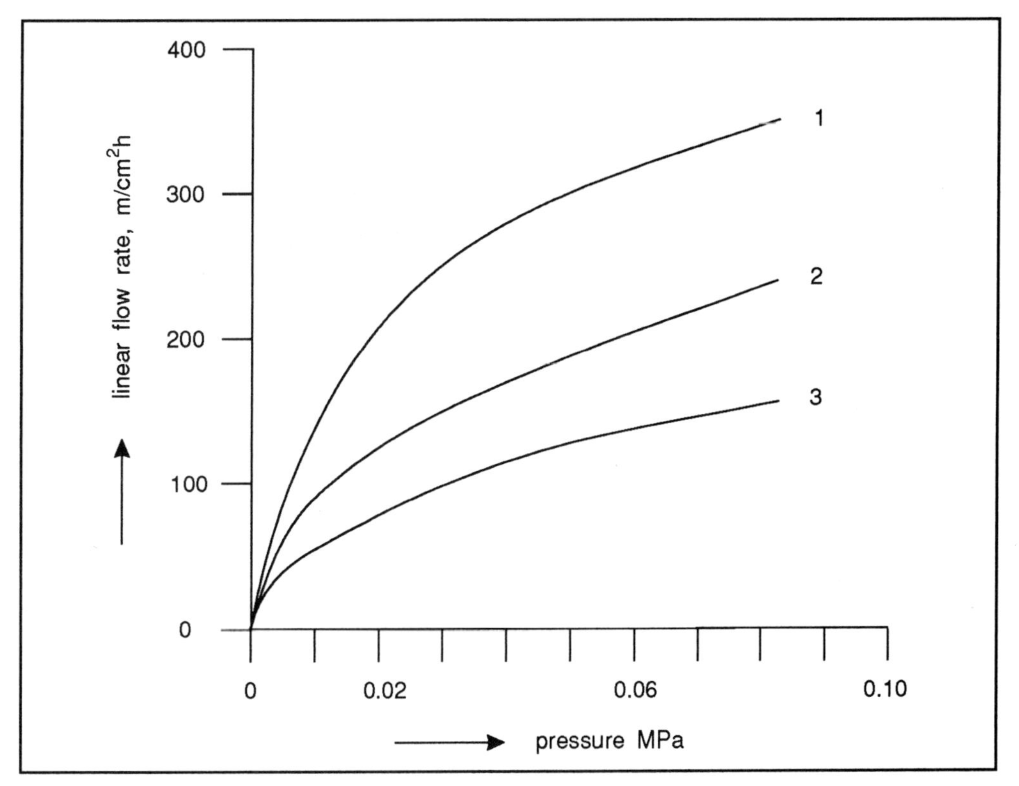

Figure 7.25 Compressibility of gel matrices curve 1: 5.0 cm diameter; curve 2: 25 cm; curve 3: 80 cm.

should be considered:

mode of operation

- desalting;
- fractionation.

resolution

- smaller beads usually give better resolution.

stability

- temperature;
- pH;
- organic solvent.

There is, therefore, quite a considerable amount to consider in choosing a gel matrix. Our advice is that you should consult the supplier before starting experiments. The experience of suppliers is very valuable.

Concluding comment

Despite the many relationships we have described in this chapter, sorption is more an art than a science. Experience plays an important role. Nevertheless, the discussion presented in this chapter will have alerted you to the important parameters in sorption processes. Undoubtedly this area of bioprocess technology is of great importance in the purification of desired products from bioreaction mixtures. You should anticipate significant further advances in the understanding and application of sorption processes in the future.

SAQ 7.7

1) Which of the following has least effect on the resolution achieved in a gel permeation process.

   a) doubling N;

   b) doubling α.

2) If the theoretical plate number is increased from 100 to 1000, by how much will the resolution be increased?

3) In scale up of a gel permeation, which of the following parameters used in the optimisation of the process in a laboratory scale should be

   a) increased;

   b) kept the same.

bed height, column diameter, linear flow, volumetric flow rate, sample load, sample concentration.

Summary and objectives

Adsorption and chromatography are the main steps in the purification of many bioproducts. Adsorption of small molecules is well described and has been applied on a large scale for many years, mainly in the recovery and purification of antibiotics. Adsorption of proteins is applied on a much smaller scale, but will be more common in the near future. Packed bed adsorption is predominantly applied in industry, although in some cases small scale stirred tank operations are preferred. A wide variety of adsorbents are available for nearly every application and based on the adsorption isotherm, the column behaviour can be predicted.

The large scale application of chromatography is limited, the limitations however moving to intermediate scale operations. There is no satisfactory mathematical description for the chromatography of proteins and the results largely depend on experience.

Now that you have completed this chapter you should be able to:

- predict the behaviour of breakthrough curves for the various types of adsorption isotherms;
- calculate breakthrough times from supplied data;
- mathematically describe various types of adsorption curves;
- explain how various factors could influence the design of stirred tank and packed bed adsorption processes and use relationships to calculate the ratios of liquid feed treatment to absorbent used;
- determine the number of plates and the resolution factor from supplied data;
- use the Van Deemter equation to explain how various factors influence peak height and to optimise flow rate of a chromatographic process;
- list the parameters which should be considered in the selection of chromatographic media and in the scale up of chromatographic purification.

Formulation and application

Formulation and application

8.1 Introduction

importance of formulation

Product formulation is an important subject in the production of biotechnological and pharmaceutical products. It is the final and probably most essential step between production and application. Whatever the magnificent power of, say, penicillin may be, it is of no use if the penicillin is destroyed before it can exert its power. Although penicillin has an obvious prophylactic effect, straight-forward per-oral administration make little sense, since it is very unstable at low pH (see Chapter 7). A pH of 1.0 in the stomach, and a residence time of more than 30 minutes (at a temperature of 37°C), will destroy all activity. Therefore, the penicillin has to be formulated, (for example encapsulated) to prevent degradation of the penicillin in the stomach and to allow its release in the small intestine, where the pH is 5-7. This so called 'enteric coating' of the capsule prevents it from dissolving at low pH, but it will do so at higher pHs. (It must be noted that the stability at low pH of modern penicillin derivatives has been improved considerably). This type of formulation technique is well established in the pharmaceutical industry. Elsewhere it is rather less common. In biotechnology, formulation is becoming increasingly important. Immobilisation of enzymes and cells and encapsulation of detergent enzymes are significant examples.

Outside of the pharmaceutical field, there is not much general literature about formulation because general knowledge on the subject is still under-developed and individual applications are often too specific to have general appeal. Fortunately, an extensive patent literature is available. Possessing a patent with a proper formulation of a given perparation, can be a trump-card in the effort to obtain a significant part of the market. It is, therefore, amongst the patent literature that we must search for details of formulation practice.

As mentioned before, the formulation may be so specific that even the required production equipment is not available on the market and must also be developed. In the first part of this chapter, examples will be given of formulation and formulation processes to provide you with some insight into the range of formulation processes that are being developed. Particular emphasis is placed on formulations involving proteins (enzymes). The reader should note that discussion of formulations is an integral part of the case studies included in the BIOTOL text relating to specific business sectors (eg Biotechnological Innovations in Healthcare; Biotechnological Innovations in Food Processing; Biotechnological Innovations in Chemical Synthesis).

important questions to be answered by formulation

An important aspect of formulation strategy is to address the questions 'Will the product meet the requirements of the market/customer?' and 'Can we improve the product by careful formulation?' Important in answering these questions is the need to carry out research into how the customer wishes to use the product. The requirements for formulation is therefore largely driven by the perceived needs of the customer rather than the needs of biotechnologists. The research into the needs of the customer and the consequences it has on the final formulation of the product are discussed in the second part of the chapter. We call the research into the requirements of the market concerning a particular product, applications research.

8.2 Formulation of bakers' yeast

The production of bakers' yeast (*Saccharomyces cervisiae*), which is one of the classic biotechnological products, is depicted in Figure 8.1. The broth obtained from aerobic fermentation is concentrated in a centrifuge. The concentrated yeast, which is still fluid, is subsequently cooled in a storage tank and filtered on a rotary vacuum filter. Then, the dry matter content is about 35%. In a strand press, the yeast, with a consistency similar to putty, is pressed into a rectangular strand, which is further cut into regular pieces with a knife.

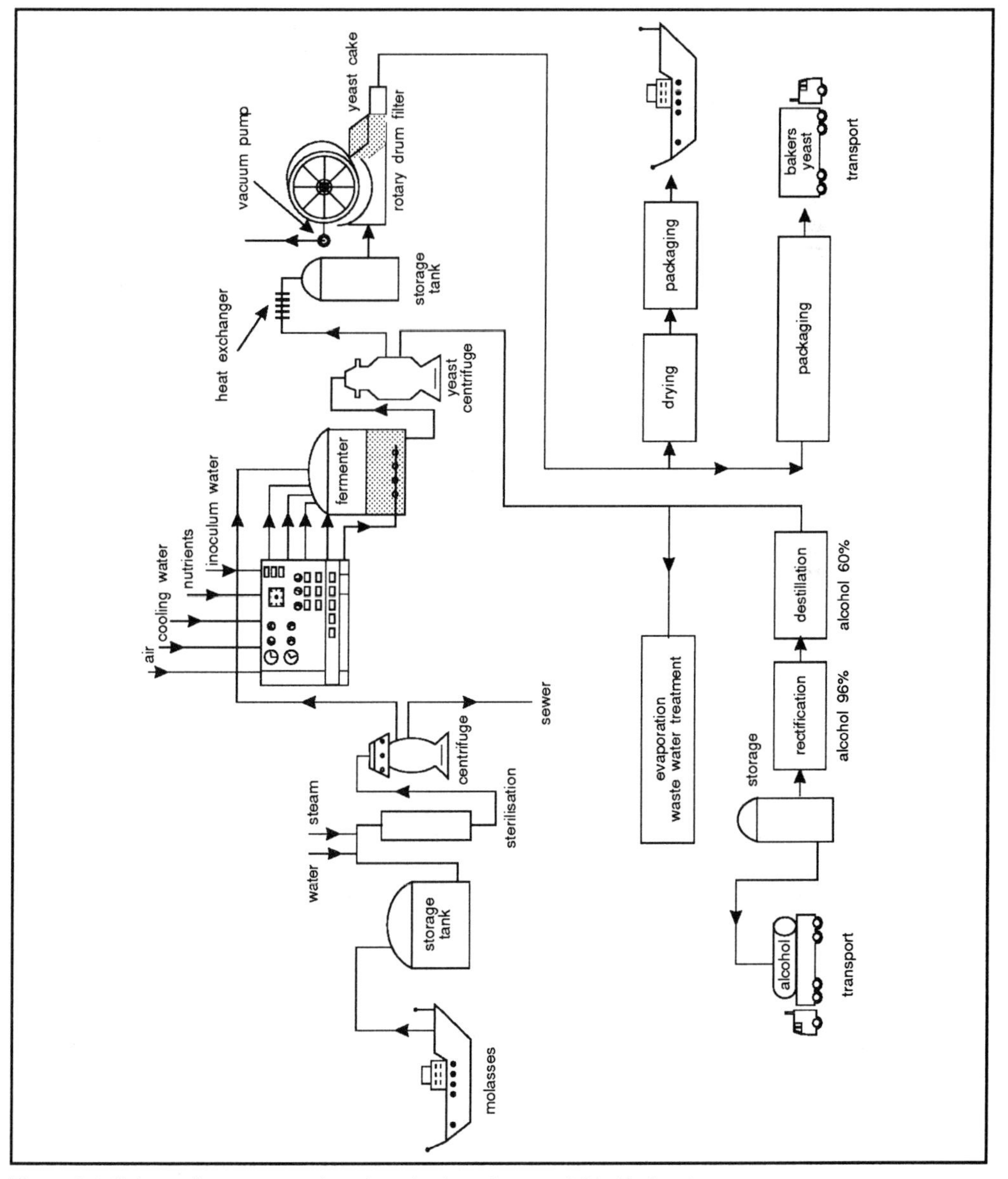

Figure 8.1 Schematic representation of production of wet and dried bakers' yeast.

Π Is this putty-like product stable? If not what is the consequence for the production and sale of the product?

limited shelf life

Your knowledge about biological systems should have led you to the conclusion that the product would not be stable. Yeast, like many other cells, in a damp environment (the putty) without a supply of nutrients would begin to undergo autolysis (self-hydrolysis) resulting in the death (inactivation) of the yeast. This process, even though slowed by refridgeration, still continues. Thus the wet yeast will loose its quality within a few days, even at low temperatures. For this reason, it is unsuitable for export. Therefore, fresh yeast has to be produced locally. Problems will arise as soon as the transport distances become too large, or in tropical areas, where the required low temperature cannot be guaranteed. Hence, the formulation problem is to enhance the stability.

dried and 'instant' yeast

First a dry yeast and later on, an 'instant' yeast were developed. For example, after vacuum filtration, the yeast is extruded into long spaghetti-like filaments. These are further broken and dried in a fluidised bed. Special yeast strains have been developed such that they can first be dehydrated without loss of biological activity. Activity is restored by soaking in water just before use.

Π In drying the yeast in a fluidised bed, what temperature is used?

We are not going to give a specific temperature as different processes use slightly different temperatures. They all, however, have on thing in common, they use a high enough temperature to enhance the rate of drying but not so high that the yeast is unactivated as a result of protein denaturation. A good target, therefore, for genetic engineers is to produce strains of yeast that are more tolerant to the dehydration process. Typically air temperatures are initially >100°C which drops during drying. The bed temperature is usually 40-60°C.

8.3 Formulation of enzymes

immobilisation as a form of formulation

As you will already be aware, enzymes are biologically active catalysts which can induce very specific reactions under gentle conditions. They are large complex organic molecules with a molecular weight of 5,000 to more that 2×10^6 Daltons. They are used in important industrial processes such as the conversion of starch to glucose syrups (α-amylase and amyloglucosidase) and the production of fructose syrups from glucose (glucose-isomerase). Some enzymes, such as α-amylase, are rather cheap and are therefore used only once. It is not worthwhile to separate the α-amylase from the glucose syrup for re-use. With glucose-isomerase the situation is quite different. This intracellular enzyme is expensive to produce and cannot be applied economically without re-use. A solution to this is the fixation of the enzymes on an insoluble carrier. This process is called 'immobilisation' and can be considered as a particular form of formulation. It allows continuous processing in a packed bed or fluidised bed reactor. No enzyme is left in the solution, and contamination and toxicity problems are less likely to occur. The main immobilisation principles are shown in Figure 8.2. We do not intend to examine the actual process of immobilisation in detail here but to briefly summarise the various strategies. For further details, the reader is referred to the BIOTOL text 'Technological Applications of Enzymes'.

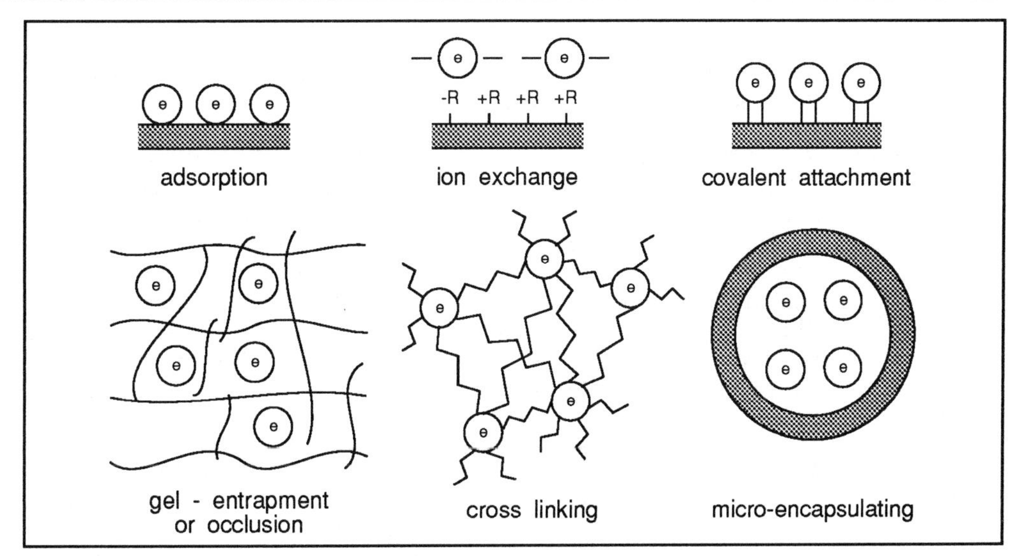

Figure 8.2 Schematic representation of several techniques for immobilising enzymes.

Π Which of the processes illustrated above depend upon a) physical, b) chemical interaction.

Your answer should be:

a) adsorption, ion exchange, get entrapment, micro-encapsulation;

b) covalent attachment and cross-linking.

In immobilisation by adsorption or electrostatic interaction, the enzyme is adsorbed and interacts electrostatically with a large surface area carrier material, eg alumina, clay, carbon, ion exchange resins, cellulose or glass. Van der Waals forces are important in adsorption.

In immobilisation by covalent attachment, the functional groups of a number of amino acids are attached to the surface of a chemically modified support material. Only functional groups of enzymes are taken into account which can interact with other molecules. The important ones are:

- ε-amino groups of lysine;
- α-amino groups of N-terminal acids;
- β and δ-carboxyl groups or aspartic and glutamic acids;
- α-carboxylic groups of C-terminal amino acids;
- phenol groups of tyrosine;
- imidasol groups of histidine.

Note these amino acids have ionisable side chains and are also important in binding enzymes to ionised supports. The charge on these amino acids is, of course, dependent upon pH. pH is very important in this type of immobilisation.

common support materials

For the common covalent binding of enzymes, the first four groups are the most interesting in practice. Synthetic polymers, cellulose, agarose and porous glass are frequently used as support materials and frequently chemical agents are used to link the enzyme to the support.

enzyme entrapment

In immobilisation by occlusion or entrapment, the enzyme molecule is captured in a polymer matrix. Polyacrylamide, starch, gelatin, carrageenan and alginate have been used for the polymer matrix. To stabilise the gels, crosslinking with bi- or polyfunctional agents may be necessary. Immobilisation by direct crosslinking is an alternative. In immobilisation by micro-encapsulation, the enzyme molecules are enclosed in polymeric capsules of 10-250 μm diameter.

Table 8.1 shows some current industrial applications of immobilised biocatalysts.

Biocatalyst	Method of immobilisation	Application
L-amino-acylase	ionic binding to anion exchangers	L-amino acids from synthetically made mixtures of D- and L-amino acids
Glucose-isomerase	Co-crosslinking with gelatin or immobilisation of dead cells	Fructose containing syrups (HFCS) from glucose
Penicillin acylase	Carrier binding or immobilisation of dead cells	6-Amino penicillanic acid for the oroduction of semisynthetic penicillins
L-aspartase	Entrapment of dead *E. coli* cells	L-Aspartic acid from fumaric acid
Lactose	Entrapment into cellulose acetate fibres	Lactose hydrolysis in milk

Table 8.1 Some current industrial applications of immobilised biocatalysts.

One of the most successful immobilised product is glucose isomerase. The two main processes used were developed by NOVO and Gist-brocades respectively and will be briefly described here. Both processes use a whole cell immobilisation technique, ie the enzyme is not liberated from the cells. This is possible because the substrate to be converted (glucose) can easily diffuse into the inactivated cells.

8.3.1 Glucose isomerase formulation

Glucose isomerase

importance of glucose isomerase

The enzyme glucose isomerase is a key enzynme in the production of HFCS (High Fructose Corn Syrup). In the early days, the isomerisation of glucose into fructose was performed chemically under alkaline conditions. This route was not very effective, led to, for instance, psicose formation and in addition to objectionable colour formation. The enzymatic process was developed relatively soon after the discovery of a thermostable glucose isomerase (GI). Note, the process of glucose isomerase immobilisation is reviewed in detail in the BIOTOL text 'Biotechnological Innovations in Food Processing'.

The NOVO GI process

As source of the glucose-isomerase, NOVO uses a special thermophilic mutant of the strain *Bacillus coagulans* in a continous culture. The harvested cells are concentrated by centrifugation prior to homogenisation, then, crosslinked with glutaraldehyde, diluted and flocculated with a cationic flucculant to give a clear water phase. The mixture is filtered and the moist, crosslinked aggregate is extruded by means of an axial extruder. Finally, the particle are dried in a fluid-bed dryer and sieved (see Figure 8.3). The product obtained is put on the market under the trade name Sweetzym^R. Recently, NOVO introduced a new immobolised glucose isomerase produced by a selected strain of *Streptomyces murinus*. The major advantages of the new immobilised enzyme are the significant higher productivity and lower by-product formation.

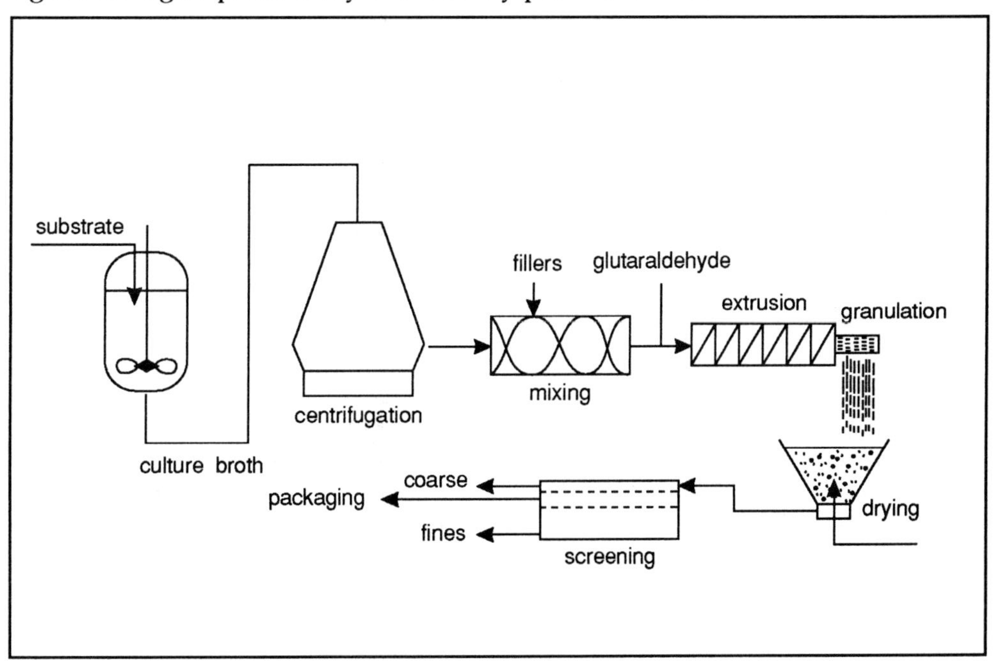

Figure 8.3 Immobilisation method for glucose-isomerase according to NOVO.

The Gist-brocades process

Gist-process (GB) has developed a process using a gel-entrapment or occlusion method. The source of the intracellular enzyme is a strain of *Actinoplanes missouriensis*. The fermentation is carried out under aerobic conditions and at neutral pH. The process, developed by GB, is depicted in Figure 8.4. The enzyme containing cells are mixed with gelatin at a temperature of about 40°C. The final gelatin concentration is about 8% (w/w). The mixture is then prilled in a column containing a cold, water and a immiscible organic solvent like butylacetate. You may not have met the term 'prill' before. It is a term which describes the solidification of droplets as they pass through an immisible solvent. Thus, while falling through the column, the enzyme containing droplets will solidify. The spherical particles are collected at the bottom of the column, dehydrated, crosslinked with glutaraldehyde, and washed. Finally, the product is classified and preserved in propyleneglycol. The product is known under the trade name Maxazym GI-Immob^R.

Note that Gist-brocades have also developed a more heat stable enzyme; not by strain selection but by protein engineering using site directed mutagenesis. The new product is made by *Streptomyces olivochromagenes*.

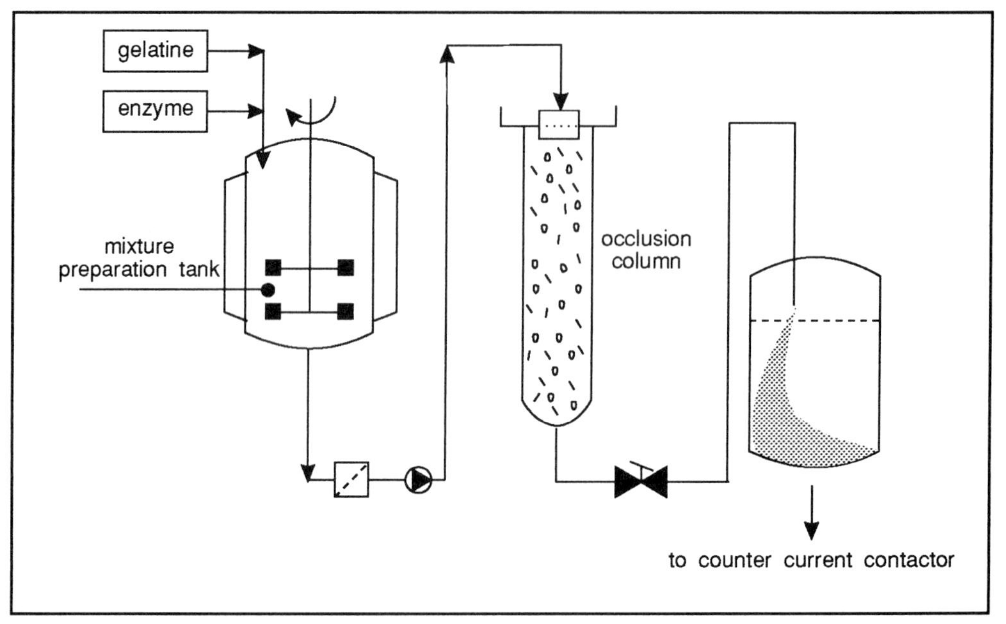

Figure 8.4 Flow sheet of the production unit of immobilised enzymes.

It is interesting to contrast these two processes. We will do so in the form of an SAQ.

SAQ 8.1

In the production of immobilised glucose isomerase put each of the following items under the appropriate company. (Note some may be directly relevant to both or neither).

Items: use of glucose isomerase from *Streptomyces murinus;* use of glucose isomerase from *Actinoplanes missouriensis;* use of glutaraldehyde as a crosslinked agent; dried on a fluidised bed; use of propyleneglycol as a preservative; use of gelatin; use of butylacetate; use of glucose isomerase from *Bacillus coagulans.*

Companies: NOVO; Gist-brocades.

If you cannot answer this perhaps you should re-read Section 9.3.1 and make yourself a summary table.

8.3.2 Formulation of detergent enzymes

serine proteases

problems with original formulation and market needs

Proteolytic enzymes suitable for laundry detergents are without exception serine proteases. Proteases are hydrolytic enzymes that are able to hydrolyze peptide bonds in proteins and therefore remove proteinaceous stains. Because of the high pH (10) of laundry detergents only alkaline proteases, generally produced by *Bacillus* species, are effective. One of the first alkaline proteases applications in Europe was marketed by Kortman & Schulte in the Netherlands. It was incorporated in the detergent B10-40, later known as Biotex. The introduction took place in the middle of the 1960s. At that time, the enzyme was added to the detergent as a dry powder. However, various allergenic and skin reactions seemed to appear with people who worked in the production of the proteolytic enzyme, even to such an extent that Ralph Nader (a well known activist on consumer protection in the USA) took legal action against the detergent industry. Consequently, in the early seventies, the application of proteolytic enzymes in detergents was practically nil in the USA. This problem can be overcome by a proper formulation, which would encapsulate the proteolytic enzyme in order to prevent release of the enzyme dust. At the same time, also other market demands were made, with respect to the flexibility of the particles. These were required to dissolve quickly and be of uniform particle size with an average value of about 0.6 mm with no release of odour and a colour contrasting with that of the detergent used. Moreover, the formulation had to be adapted in such a way that the product would have a good shelf life at 30°C in a high relative humidity environment. Furthermore, the formulation had to offer protection against various aggressive components of the washing powder. Table 8.2 gives a survey of the composition of a typical European detergent. Increasingly more bleachers and bleach acticvators were added (such as TAED: tetra-acetyl-ethylene-diamine) to facilitate bleaching at low temperature. Thus the formulation had to offer protection against this type of aggressive compounds. Only a few manufacturers succeeded in developing a formulation that meets all the requirements. Again, NOVO and GB were, however, successful. Both formulation processes will be described briefly.

surface active compounds	anionics	2 - 10%	
	nonionics	0.5 - 6%	in rare cases 10 - 15%
	soap	1 - 5%	
sequesting agents	polyphosphate	30 - 50%	or zeolite 15 - 18% polyphosphate 20 - 22%
bleach	Na-perborate	20 - 30%	
enzymes (proteases)		0.3 - 0.6%	
cellulose compounds		0.5 - 2%	
optical colorants		3 - 5%	
perfume		0.1 - 0.3%	
Na-sulfate		up to 100%	

Table 8.2 Composition of West-European detergent powders. Content as % (w/w). Note: The term detergent refers to laundry detergents. Detergents used in dish washers are not taken into consideration.

The NOVO encapsulation process

granule process

NOVO has developed several encapsulation processes. One of these, the granule processes, will be discussed briefly (Figure 8.5).

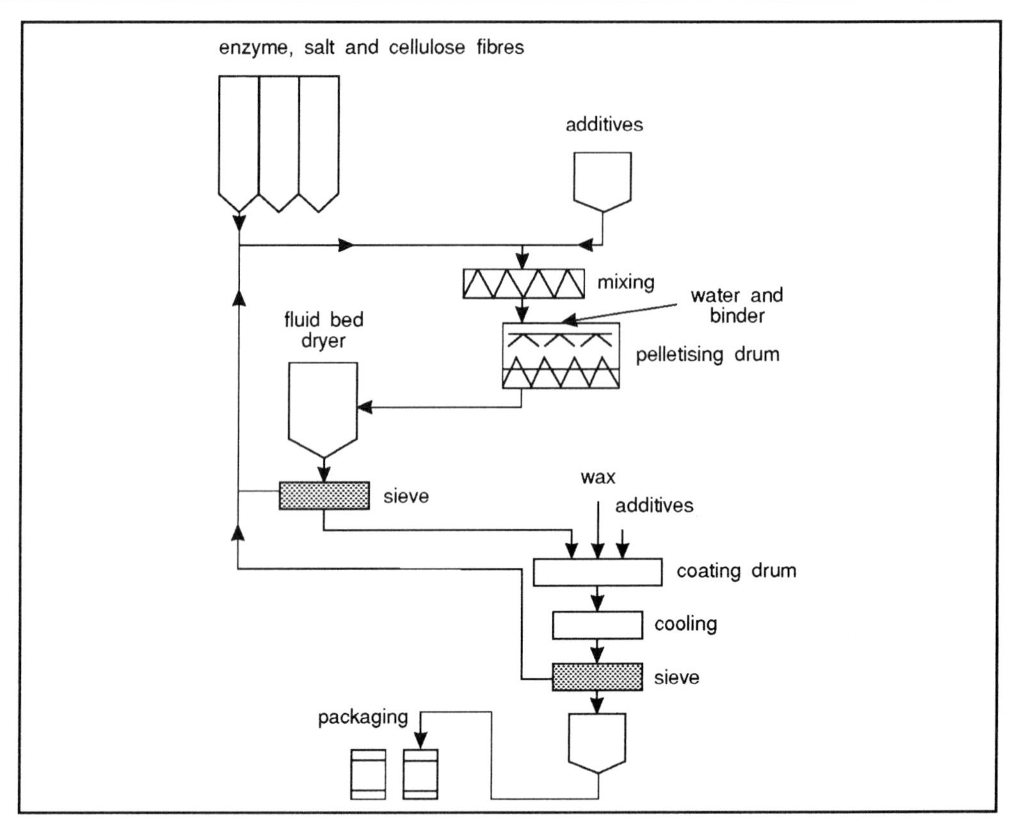

Figure 8.5 Flow sheet for the preparation of a dust-free enzyme product: Granule-T.

NOVO introduced in 1981 a granulate, produced in a pelletising drum (Lodige type mixer), (see Figure 8.5). To prevent sticking on the mixer wall, they added cellulose fibres. Another advantage in using cellulose fibres was that the pellet size did not exceed a certain maximum value, which is, in practice, in the order of the maximum acceptable particle size. This gives some flexibility in the operation of the process. The product made in this way is called Granule-T. Depending on the type of enzyme they used, the product has also been marketed as Savinase^{R} or Alcalase^{R} Granule-T. Savinase refers to an alkaline protease, whereas Alcalase refers to a highly alkaline protease ($pH_{opt} \geq 10$).

The Gist-brocades process

The enzyme is prepared from a conventional bioreactor. The culture is filtered to remove biomass (the enzyme is an exo-enzyme) and checked that the filtrate is free of organisms. The enzyme is precipitated out and collected by filtration. The wet crystals are then dried, sieved and the pure enzyme saved as powder. The GB process for formulating the enzyme is presented in Figure 8.6.

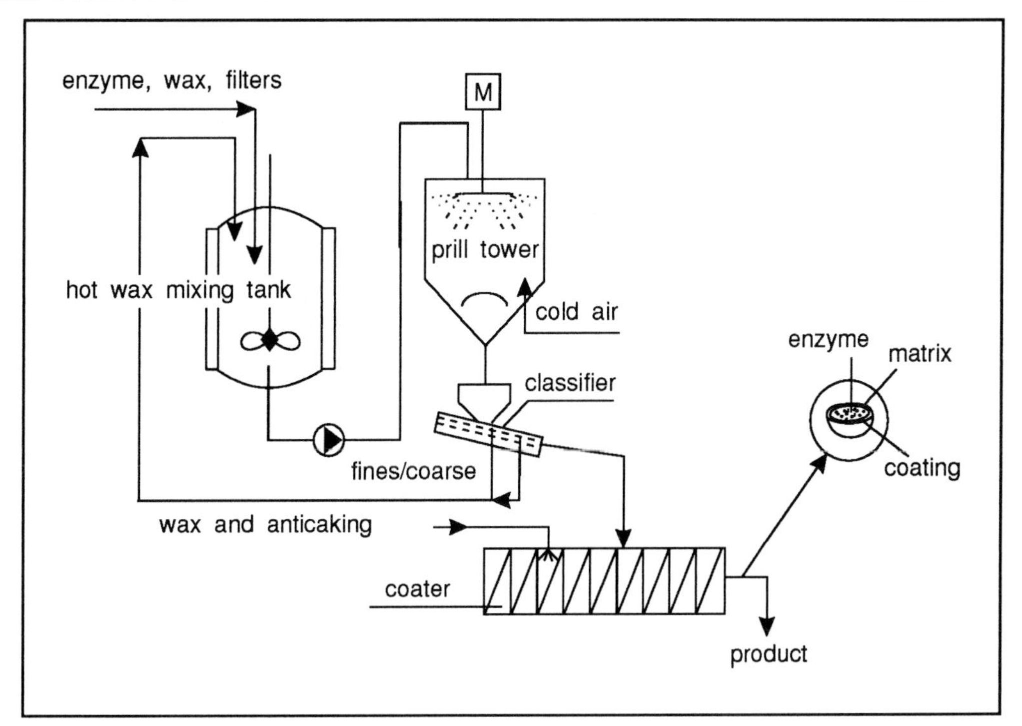

Figure 8.6 Flow sheet of GB process of formulation of dust free detergent enzymes.

The formulation is as follows.

prilling

The enzyme powder is mixed with several additives in a non-ionic melt (etoxylated-C_{18}-fatty alcohol). The mixture is fed to a prill tower where the particles are turned into spherical form on a rotary prill wheel. The drops solidify during their fall. The coarse and fine fractions are remelted, the proper fraction is coated to prevent dust formation during handling. The product is put on the market under the names Maxatase^R and Maxacal^R. Maxacal refers to a highly alkaline protease ($pH_{opt} \geq 10$).

Again it is interesting to draw up a table contrasting the two companies approaches. You will notice that Gist-brocades use prilling for both its detergent enzyme and for glucose isomerase.

Π What will happen to the spherical particles of C_{18}-fatty alcohol containing the protease when it is put into water along with the other detergent components?

You should, of course, realise that the surface active components would emulsify the fatty alcohol, thereby realising the enzyme from the particles. The particles would, in effect, disintegrate.

8.4 Formulation of pharmaceutical products

The formulation of pharmaceutical products is best known for drugs for human consumption. However, also products in the diagnostic field such as colour strips to demonstrate sugar, proteins, and hormones (predictors) are significant examples of formulation.

In Section 8.1 we mentioned the formulation of penicillin. As a further example we will briefly discuss the so-called 'effertvescent tablet'. The analgesic acetylsalicylic acid ('aspirin') does not dissolve easily in the gastric juice when it is still contained in the tablet. Only when the tablet is first vigorously stirred in a sufficient amount of water, does the activity improve. This problem is tackled by the effervescent tablet which contains not only acetylsalicylic acid and additives but also other acids, such as citric acid, and sodium bicarbonate. The two latter react, when they are dissolved in water, according to the equation:

$$\text{citrate COOH} + HCO_3^- \text{--->} \text{citrate COO}^- + H_2O + CO_2$$

The disintegration of the tablet starts by the penetration of water into the pores of the tablet and is then strongly enhanced by the production of CO_2 in the pores, which 'blow up' remaining particle remnants. Furthermore, the better soluble sodium acetylsalicylate is produced. The analgesic is now finely dispersed in water and can even dissolve completely. An additional advantage is that in this way only few particles of salicylic acid are brought into contact with the stomach-lining.

We will not extend the discussion of healthcare products here. Several examples including diagnostics, vaccines and proteinacious hormones used as therapeutics are dexcribed in the BIOTOL text 'Biotechnological Innovations in Healthcare'. We now move on to the other important aspect of formulation - namely application research.

8.5 Application research

importance of application research

In addition to formulation, application research is also of importance in the development and marketing of bioactive preparations. A formulation will be developed, based on a general assessment of the demands of the market. In application research, it must be verified whether the product meets the requirements or not in particular cases of application and the formulation will have to be adjusted in the light of the demands of particular customers. For healthcare and food products, not only has the product to do the job it was designed to do effectively but it must also be safe. To market such products, market authorisation has to be gained from the relevant authorities. To gain market authorisation, details of the final formulation must be included. Thus considerable effort must be expended in application research means that the effect of its different properties on its activity in particular industrial processes must be examined. This would include average particle size (distribution), pH-optimum, temperature dependence of the activity, effect of substrate concentration and the effect of co-factors on the product.

For detergent enzymes, of course, the enzymes should contribute to the washing result. Thus, elaborate washing experiments are necessary. The product has to dissolve quickly, must possess a long shelf life and detergent stability. It must be active at low temperatures in the presence of bleachers and bleach activators at a high pH (10-11). The preparation should not be allowed to release dust, nor have a odour and undesirable colour. It should, of course, not be toxic. Therefore, elaborate research is necessary, which is usually carried our in co-operation with the 'soapers'.

For both products we will give here a brief explanation of what application research means.

8.5.1 Application research on immobilised GI

After immobilisation, a lot of additional questions have to be answered in the application of an immobilised enzyme.

Π Make a list of as many factors you can that you think should be examined to ensure the product meets market expectation.

We believe you will have generated quite a list. There are, indeed, many factors to consider. Here we will discuss some of the major ones.

It is important to know the life time of the immobilised enzyme as function of the process conditions such as substrate concentration, pH and temperature.

Because of its application in packed bed columns the product must have sufficient mechanical strength to withstand the force induced at high linear flow velocities (5-10m h^{-1}). Furthermore, the immobilised enzyme must maintain its physical stength at temperatures of 60-65°C for process life-times of as long as possible. (In practice this enzyme may have a process life-time of about 2000 hours).

The kinetics of the equilibrium reaction between glucose and fructose can be described by Michaelis Menten kinetics. According to this equation the maximum specific rate of fructose formation and the (pseudo) Michaelia Menten constant is experimentally determined. It can be shown that under the conditions prevailing in a fixed bed the Michaelis Menten equation can be simplified to:

$$V_f = K (C_s - C_s^*) \qquad \text{(E - 8.1)}$$

Where V_f = rate of fructose formed (mol m^{-3} s^{-1}); C_s = glucose concentration (mol m^{-3}); C_s^* = equilibrium concentration of glucose (mol m^{-3}); K = (pseudo) first order rate constant

We do not propose to carry out the derivation here as this is covered elsewhere in the BIOTOL series. But, because the rate constant is temperature dependent, this relationship can be evaluated by an Arrhenius-type equation. Thus we might anticipate the higher the temperature the faster the reaction.

importance of enzyme inactivation

We must remember, however, that there is a second temperature dependent process which is significant to the performance of enzyme catalysed reactions namely the de-activation of enzyme. This decay process can be described with first order kinetics in which the pseudo first order rate constant will decrease with time. Also here an Arrhenius-type relationship holds because the de-activation strongly depends on the temperature. All the mentioned parameters have to be determined carefully. The next step is the mathematical description of the productivity P_t. The productivity of immobilised glucose isomerase is defined here as volume (m^3) of glucose converted into fructose of a 45% w/w solution per m^3 column volume. The result of this modelling is expressed in the next equation:

$$P_t = \frac{6\ D\,AH\ (1-\varepsilon)\ C^*}{K_d\ R_p^2} \left\{ \ln \frac{\sinh \phi_o}{\phi_o} - \ln \frac{\sinh \phi_t}{\phi_t} \right\} \qquad (E-8.2)$$

The terms are defined below.

Equation 8.2 is plotted in Figure 8.7.

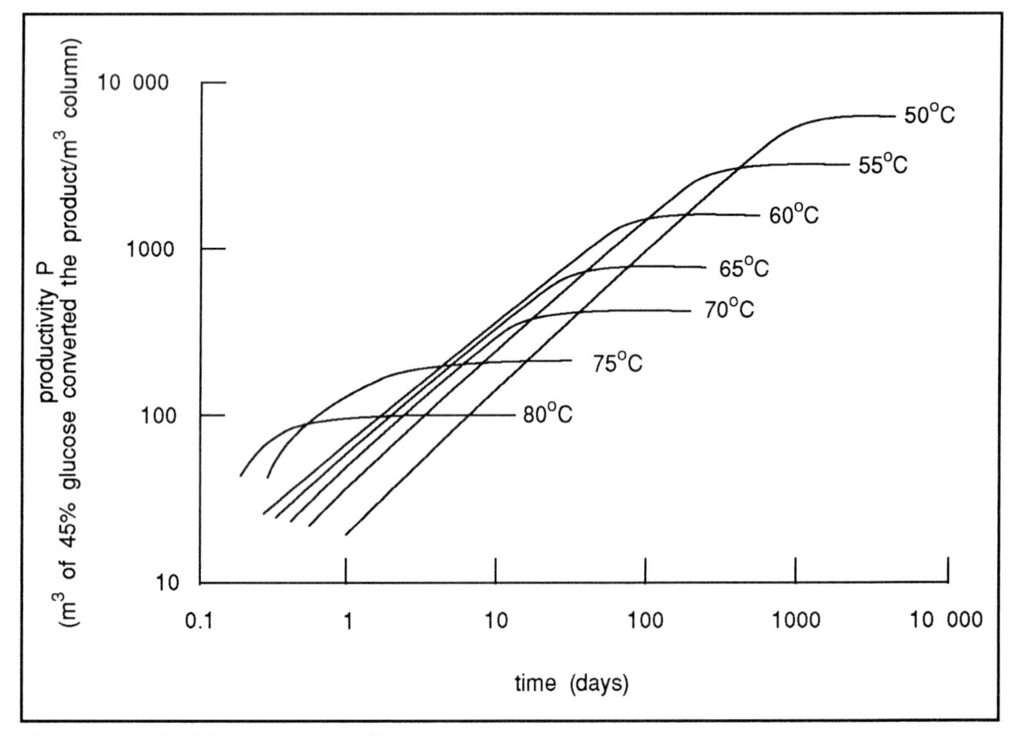

Figure 8.7 Productivity of commercial GI as a function of operating time over a range of temperatures.

From this equation it will be clear that the productivity per m^3 column volume will in inversely proportional to R_p^2. A decrease in particle size gives a rise in productivity. However, the pressure drop in a packed bed increases also inversely proportional to R_p^2:

$$\Delta p\ \alpha\ \frac{H\ \eta\ v_s}{R_p^2} \qquad (E-8.3)$$

(Where Δp = pressure drop; H is the column height, η is the dynamic viscosity, v_s = superficial flow velocity).

Thus if we use very large particle size, this will have an effect on the pressure drop.

These last two equations may look rather horrific but they have been derived in the BIOTOL text 'Bioreactor Design and Product Yield'. We choose not to go through the derivation again here but let us see if we can make sense of them.

P_t is the productivity of the system at time t

D is the diffusion coefficient of glucose into the particle

A is the area of the column (m^2)

H is the height of the column (m)

C^* is dimensionless concentrations $\left(\frac{C_{So} - C_s^*}{C_{St} - C_s^*}\right)$

K_d is the first order rate constant (s^{-1}).

R_p is the radius of the particles containing the enzyme (m).

ϕ is a dimensionless number called Thiele Modulus ($\phi = R_p \sqrt{K/D}$ where K is the pseudo first order rate constant).

ϕ is effectively a measure of the extent of diffusion limitation of glucose in the particle.

Sinh is a mathematical term, it is a hyperbolic function.

Sinh $x = \frac{1}{2}(e^{+x} - e^{-x})$. Thus $\sinh \phi = \frac{1}{2}(e^{\phi} - e^{-\phi})$.

Note that ϕ_o equals mutual Thiele modulus; ϕ_t is Thiele modulus at time t.

Π Now let us do a little practical exercise. For each of the facors listed below, see if you believe that if each is increased, P_t is increased or if the reverse is true. We will give you a start by doing the first few.

We would generally agree that increased rates of diffusion would increase productivity because at high diffusion rates the substrate is brought to the enzyme more quickly.

Thus we could expect $P_t \propto D$.

Likewise a bigger column (larger A and H values) would also be expected to increase product formation.

Thus:

$P_t \propto AH$

On the other hand increased rates of enzyme denaturation (ie high K_d values) would result in lower production.

Thus $P_t \alpha \frac{1}{K_d}$.

Would the size of the particle (R_P) and the substrate concentration C^* effect the rate and amount of product formation. If so, in what way?

You should have concluded that large particles are likely to limit the access of substrate to the enzyme. It has a smaller surface to volume ratio, thus $P_t \alpha \frac{1}{R_p^2}$. Higher substrate concentrations are likely to increase the amount converted.

Thus $P_t \alpha C^*$

Π Now add all these relationships together.

You should get $P_t \alpha \frac{D\ A\ H\ C^*}{K_d\ R_p}$

Now if you look back to Equation 8.2 you should begin to understand the nature of the relationship described. The remaining part of the equation taken in to account the rate of the reaction (K) which is part of the Thiele module ϕ.

Thus although from this rather limited approach we cannot derive Equation 8.2 exactly, we can understand that it makes sense.

We should remember, however, that there should be a balance between the productivity on the one hand and the allowable pressure drop on the other hand.

Because of the fact that immobilised particles of Gb-glucose isomerase are somewhat compressible, the relation to the superficial flow velocity in the column and the compressibility must be investigated.

The superficial flow velocity as function of the pressure drop across the bed is plotted in Figure 8.8. As can be seen an increasing pressure drop causes a non-linear increase of the flow velocity caused by bed compression.

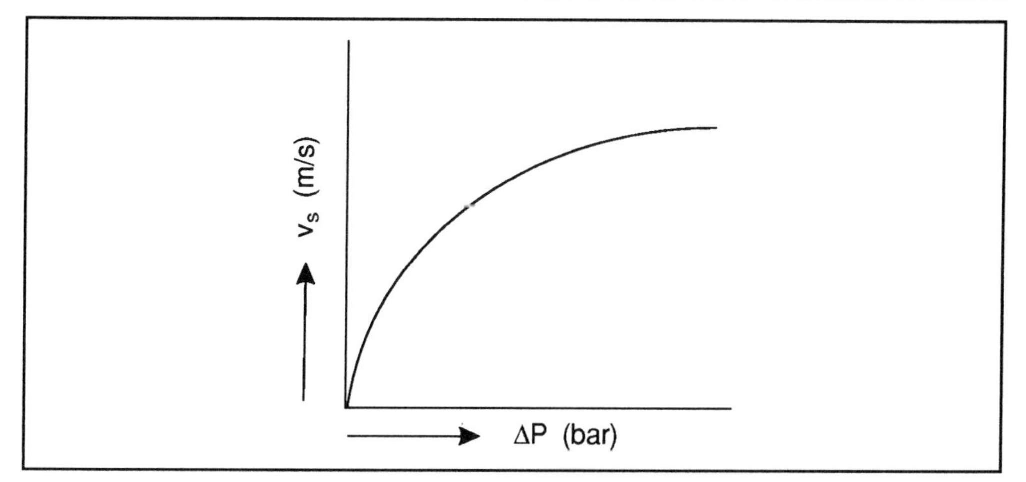

Figure 8.8 Pressure drop as function of the superficial flow velocity.

It can also be derived from Figure 8.7 that the lower the temperature the longer the life time of the enzyme.

To prevent infection (contamination) of the syrup a temperature of 58-60°C has to be maintained.

You should have realised from this discussion that by choosing an immobilised formulation of the enzyme a lot of additional questions have to be answered. Intensive laboratory and pilot plant studies are necessary to answer them. An outstanding example of application research!

SAQ 8.2

Answer true or false to the following statements (give reasons for your choice).

1) Gisr-brocades formulation of detergent protease involves embedding the enzyme in C_{18}-fatty alcohols by a process of prilling.

2) In principle the solubility of the analgesic acetylsaticylic acid could be improved simply by adding bicarbinate as a component of the dry tablet.

3) It is always better to use a high temperature when using an immobilised enzyme system as this increases the productivity of the system.

4) It is always best to use the smallest particles that can be produced of an immobilised enzyme to be used in a packed column.

8.5.2 Application research on detergent enzymes

It should be self evident that the main part of application research with this product is aimed at the contribution of the enzyme to the washing activity. To this purpose the composition of the laundry detergents and the washing process must be studied. The composition of the detergent is already described in Section 8.4 (Table 8.1). Over the past few years, the washing process has been changed considerably, in particular the washing temperature. To an increasing extent the washing process occurs at lower temperatures. In this situation, however, the bleach components are less effective. This has led to the introduction of bleach activators such as TAED.

The degree of stain removal is determined by the percentage of remission of light that is reflected from a dirty patch (stained with proteins and artificial dirt and subsequently washed with a detergent and a protease). Several artificially stained patches are available to test both the washing process and the enzyme. Figure 8.9 shows the result of such a test.

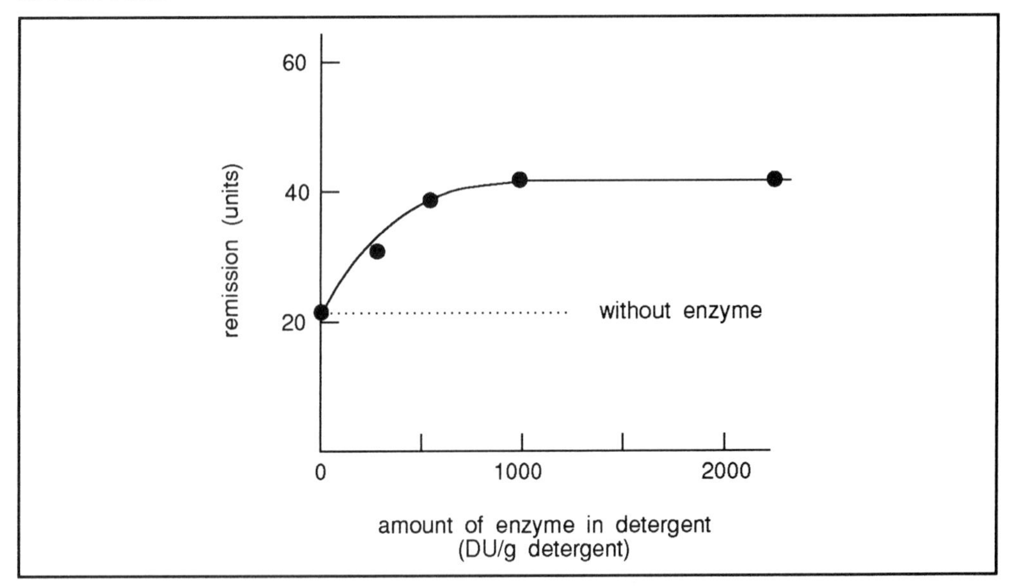

Figure 8.9 Washing test for proteases. DU = Delft units.

Also many washing tests are carried out with 'real' laundry to examine the effect of the enzyme. However, in the latter case the objective measuring methods have been replaced by subjective ones, which makes the application research much more difficult.

Summary and objectives

In this chapter we have introduced the concepts of formulation and applications research. We did not attempt to give an exhaustive list of examples but choose to focus on some specific examples. Thus in formulation we choose a whole cell system (yeast) and two types of enzymes. Some comparison were made between the stategies adopted by two major producers of industrial enzymes. To explain the importance of applications research, we again choose selected enzymes to show how the decision to market these in an immobilised format demands extensive applications research.

Now that you have completed this chapter you should be able to:

- explain, using suitabe examples, the importance of formulation especially relating to safety, efficacy and shelf-life;
- describe the stages in the production of a 'formulated' product especially in relation to encapsulation and other forms of immobilisation;
- explain what is meant by applications research and describe the objectives of applications research.

Process integration and alternative separation processes

Process integration and alternative separation processes

9.1 Process integration

Designing efficient downstream processes is more than sequencing a large number of unit operations. Development of cost-effective product recovery is essential in high volume/low value product markets. More and more attention will have to be given to hazard assessment, risk quantification and quality assurance. Recombinant DNA based production technologies require strategies for the validation of processes and production under containment conditions.

The design of downstream biotechnology processes is no longer seen as being a sequence of unit operations of low resolution. Separation is largely based on the (bio)specific properties of the product. An outstanding example is affinity chromatography using antibodies. Only one molecular species will be bound.

yield and the number of process steps

Reduction of the number of steps also has a substantial impact with respect to the loss of product. In Figure 9.1 the overall yield is plotted as function of the number of process steps. The yield of each step is given as a parameter in this figure. So a process with 8 steps with an average step yield of 0.85 which is rather high gives an overall yield of 27%. This means that 73% has been lost. Reduction to 4 steps almost doubles the yield (!) and considerably reduces the cost price.

Two ways of process integration are coming on stream now:

- whole broth treatment;
- *in situ* recovery.

In this chapter, we will explain both of these approaches to process integration and examine a range of separation techniques that have recently been introduced and which are particular applicable to bioproduct recovery.

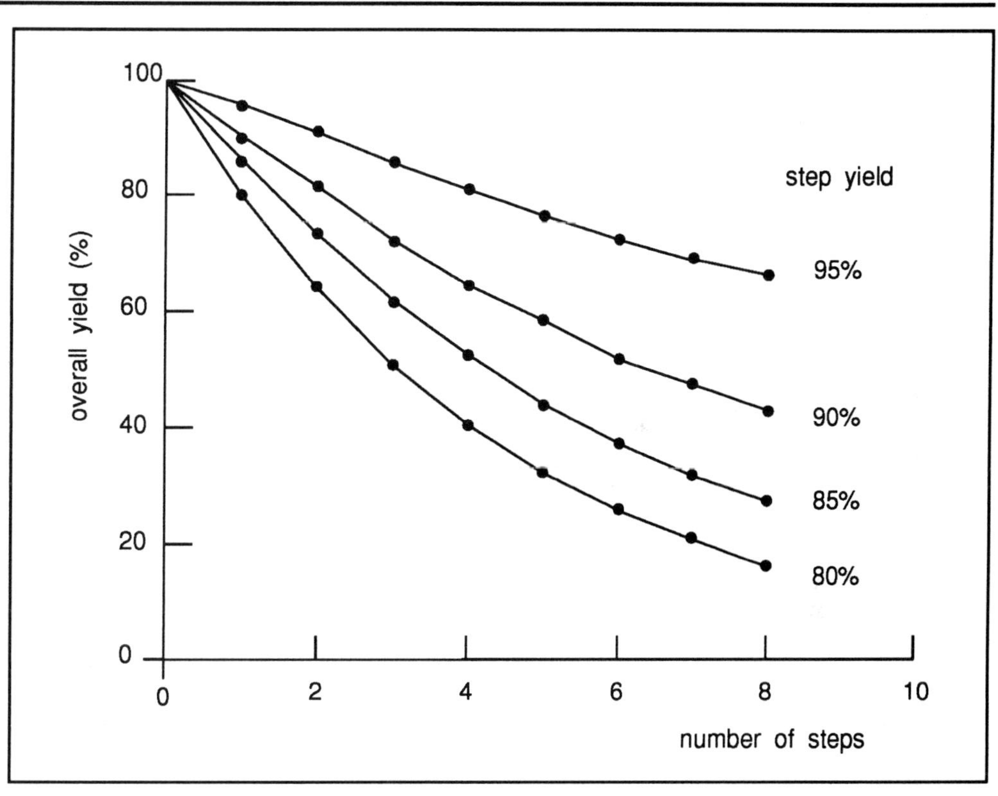

Figure 9.1 Overall yield as function of the number of process steps.

9.1.1 Whole broth treatment

modified decanter

In whole broth treatment the expensive solid/liquid separation can be avoided by direct extraction with solvents or adsorbents. This can be achieved by modifying a standard decanter to allow the introduction of separate phases (solvents) at appropriate point and also modifying the scroll. The scroll now forms a cylindrical channel about 3 m long which is used for counter current extraction (see Figure 9.2). Each end of the bowl is used for phase separation. A separating disc is fitted in the bowl at the point where the cylindrical section changes to conical and holds the required interface.

The heavy phase flows along the conical section together with transported solids and discharges at atmospheric pressure.The light phase is collected in a chamber at the opposite end, from which it is removed by an in-built centripetal pump.

The use of direct extraction is on the increase especially in the production of antibiotics such as penicillin and erythromycin. It is also used in the extraction of vitamins, alkaloids and steroids.

Currently, direct adsorption has limited application because of operational problems with fluidised bed columns in the presence of feed solid material.

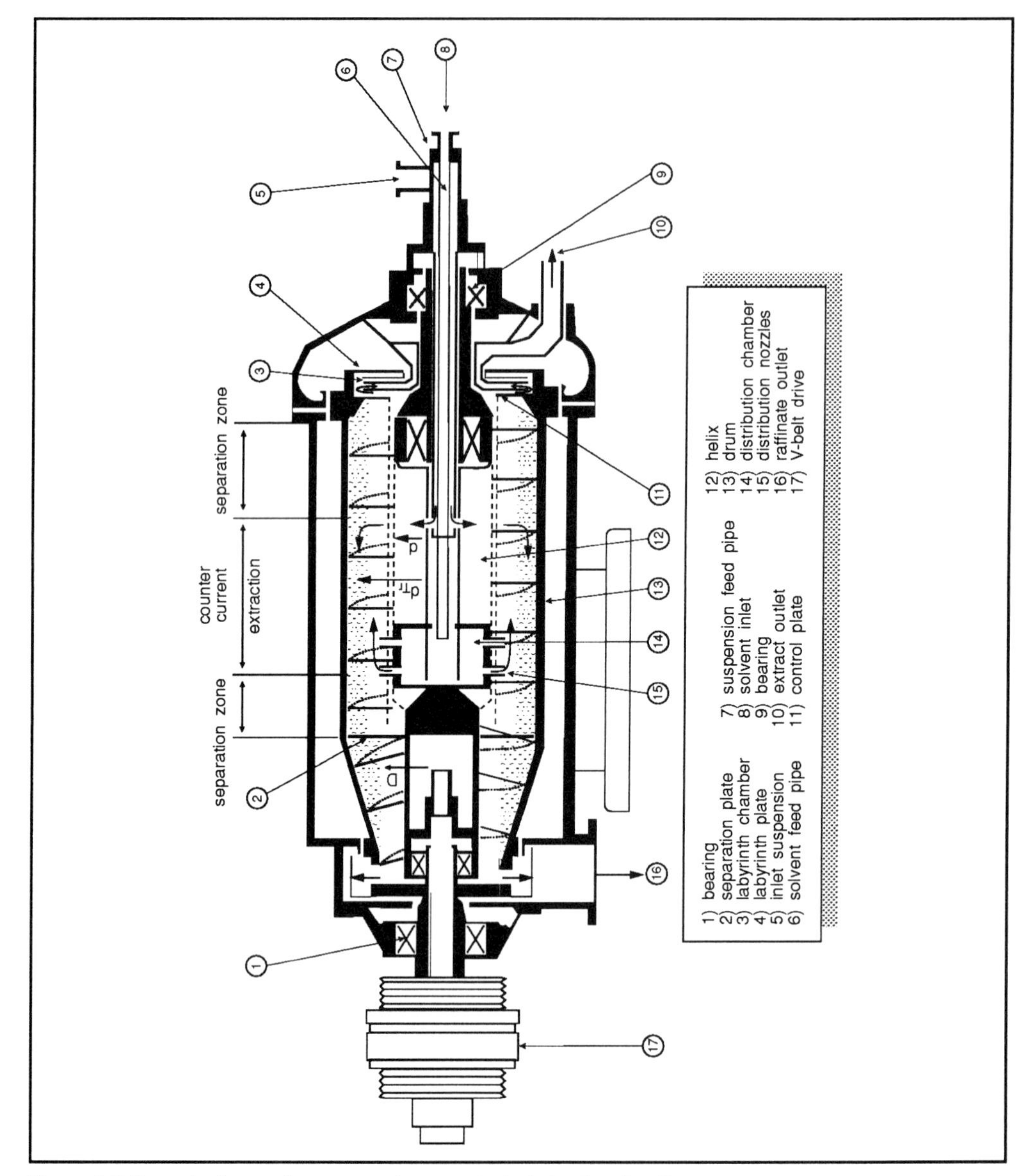

Figure 9.2 Decanter for continuous countercurrent whole broth extraction.

9.1.2 *In situ* recovery

In situ recovery is being applied on a small scale.

It refers to a procedure in which the product is immediately removed from the fermenter. In this way the overall performance of a bioreactor may be improved considerably.

In situ recovery can enhance the productivity in particular in the case of:

- limited solubility of the product in the broth;
- product inhibition;
- degradation and autodegradation of the product.

The general concept of *in situ* product recovery is given in Figure 9.3. Basically there are three modes of operation:

- direct product removal from the broth;
- the broth is circulated outside the fermenter; the product is then removed by putting a selective agent in the cycle stream. Examples of selective agents are adsorbents, membranes and solvents. Membranes and solvent extraction are considered to have the greatest potentials in combination with immobilised cell reactors;
- to prevent direct contact between cells and separating agent the cells may be removed first and the cell free broth can then be treated further.

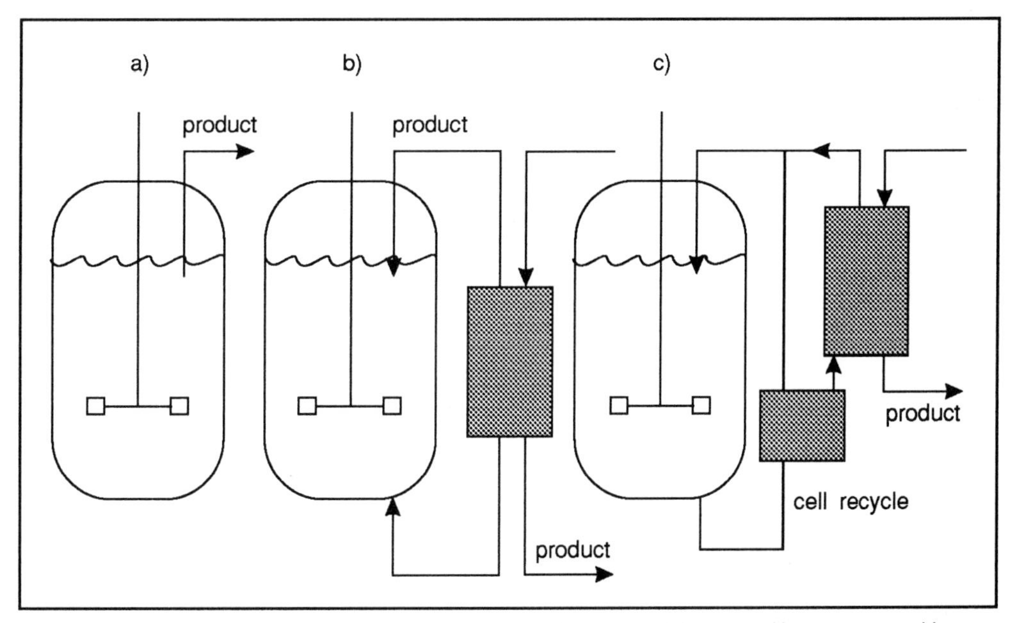

Figure 9.3 Configurations for *in situ* recovery, a: Direct product removal, b: external loop, c: external loop with cell recycling (see text for further details).

An integrated process for ethanol produced by yeasts has already been commercialised. In this process the yeast cells are separated from the whole broth by a centrifuge. The slurry is recycled to the fermenter. The supernatant is distilled immediately after centrifugation. Since it contains fermentation nutrients, the bottom product of the distillation column is also recycled to the fermenter (Figure 9.4).

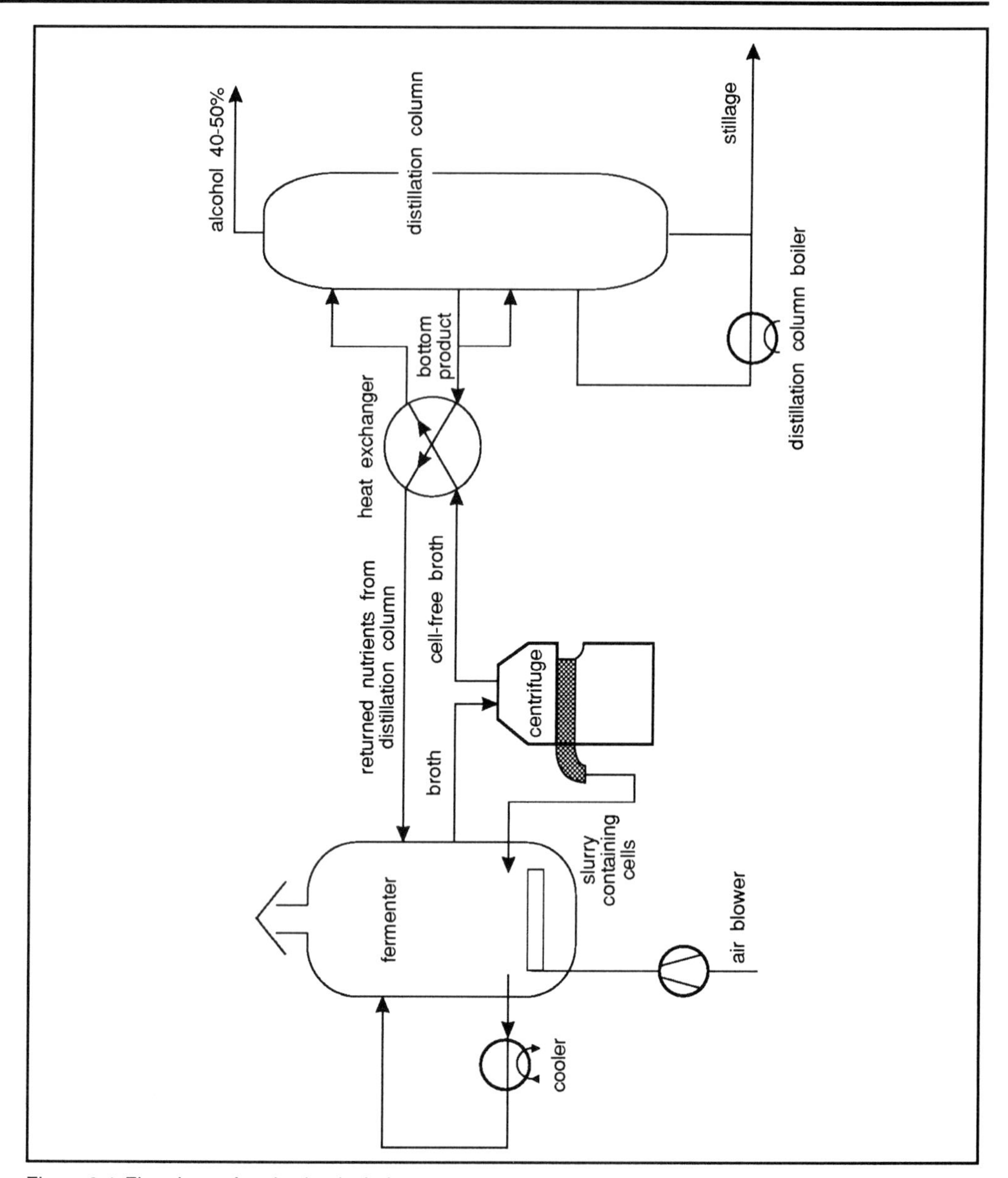

Figure 9.4 Flowsheet of an *in situ* alcohol recovery process plant.

9.2 Alternative separation methods

In recent years many alternative separation methods have been developed, in particular for protein purification.

Several of these developments are listed in Table 9.1.

Separation methods
Electrokinetic processes
electrodialysis
electrophoresis
Membranes
affinity
liquid
chromatography
pervaportion
pertraction
Extraction
supercritical
reversed micelles
affinity
rDNA for recovery
protein fusion

Table 9.1 Alternative separation processes.

Here we will discuss electrophoresis, pervaporation, supercritical and reversed micelles extraction and finally rDNA for recovery.

9.2.1 Electrophoresis

electrophoretic mobility

Almost all biomolecules and cells contain charged groups. Therefore it is possible to separate these materials using methods which make use of an electrical field. Charged molecules will migrate under the influence of an electrical field with a velocity according to their charge, size, and the field strength applied. Because of the differences in migration velocity, also called electrophoretic mobility, for different biomolecules they can be separated.

The most simple method for separation of proteins is electrophoresis. A simple electrophoresis device consists of two electrodes to apply a field across a thin chamber (see Figure 9.5).

The chamber is filled with a gel. A gel is used instead of a liquid to suppress mixing. If we now place a charged molecule say a protein between the cathode and the anode, the molecule will move through the gel depending on its net charge towards the anode or cathode. If the pH of the buffer solution in the gel is lower than its iso-electric point (IEP = pH at which the molecule carries not net charge) the protein will be charged positively and move to the anode.

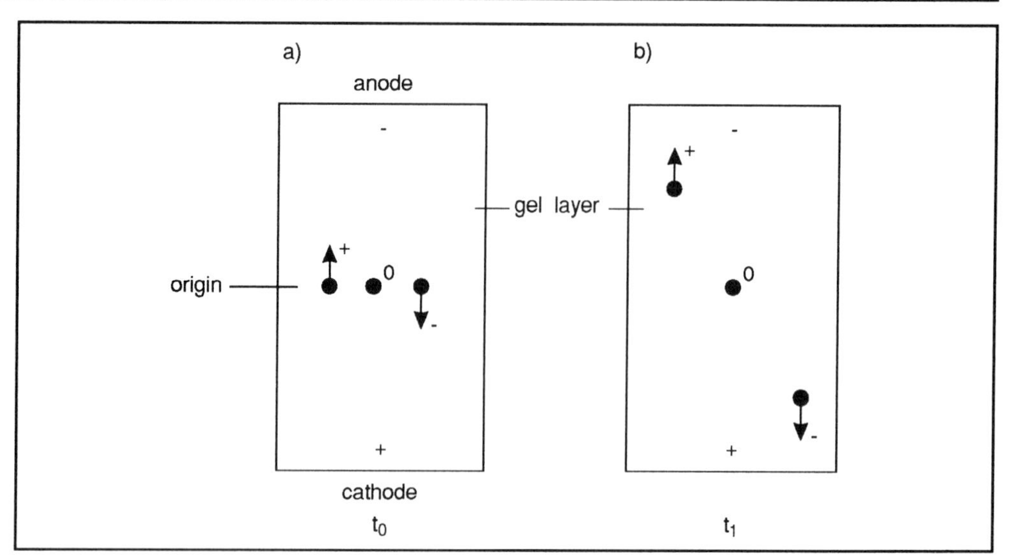

Figure 9.5 Basic principle of electrophoresis. a) In an electric field, positively charged ions begin to more towards the negative electrode (anode). Negatively charged ions move towards the cathode. Molecules carrying no net charge do not migrate. b) After some time (t_1), the three types of molecules are well separated.

Π See if you can list the factors which will influence the rate at which a molecule will migrate in the gel.

We anticipate you will have included the following:

- the potential difference between the anode and cathode. The greater this difference, the faster a charge particle will move, more strictly it is the gradient in potential difference that is important;
- the size of the particle. We might anticipate that smaller particles will move faster than large particles;
- the charge on the particle. We might anticipate that particles with a high net charge will move faster than a molecule with a low net charge.

Essentially, therefore, if we establish a potential difference between two electrodes, the rate at which components will move between these electrodes is influenced by the ratio of their charge and mass (or size). Thus, highly charged, small molecule will migrate much faster than large molecules carrying a small charge.

Mathematically we can calculate the movement of charged particles in an electric field from:

$$v_{epm} = \frac{-eZ}{6\pi r_m \eta} \qquad \text{(E - 9.1)}$$

where:

v_{epm} = electrophoretic mobility ($m^2 V^{-1} s^{-1}$)

e = charge of an electron ($1.6 * 10^{-19}$C)

Z = net charge of protein

r_m = Stokes radius (m)

η = dynamic viscosity (Nsm^{-2})

Note that the electrophoretic mobility defines the mobility ($m s^{-1}$) in a standard electric field ($V m^{-1}$). Note also that the net charge (Z) on a protein (and on many other biomolecules and on cells) is dependent upon the pH.

Typical mobility values are:

low molecular weight ions	$(4 - 9) * 10^{-8} m^2 V^{-1} s^{-1}$
proteins	$(0.1 - 1.0) * 10^{-8} m^2 V^{-1} s^{-1}$
cells	$(0.1 - 4.0) * 10^{-8} m^2 V^{-1} s^{-1}$

SAQ 9.1

1) Calculate the electrophoretic mobility (v_{emp}) of β-lactoglobulin using the following data.

   The Stokes radius of β-lactoglobulin is about 20×10^{-10} m and the viscosity of the system is 1.0 mPas. At the pH at which the electrophoresis is run lactoglobulin carries a net charge of +5.

2) Given the same set of conditions as described in 1), how fast will β-lactoglobulin migrate in an electric field in which the voltage drop is 10 V cm^{-1}?

3) Will the value of electrophoretic mobility (v_{epm}) increase/decrease/remain the same if the pH of the system is charged such that the pH is closer to the isoelectric point of β-lactoglobulin?

uses of electrophoresis

Electrophoresis is now in wide spread use as an analytical tool for characterising bio-molecules but its application even on preparative scale in very limited. One of the major problems is the heat generation in the gel due to the passage of electrical current through the gel. This heat generation causes many molecules to denature and become inactive. Several semi-continuous systems have been developed but until now no system have been commercialised successfully.

Philpot apparatus

One example of the more modern adaptions is the electrophoresis unit of Philpot (Figure 9.6). His apparatus consist of an annular space across which an electric field is applied. The outer electrode rotates while the inner electrode is stationary. The feed enters at the bottom (together with a carrier fluid) and flows upwards. The separation takes place in the radial direction. At the top the mixture is collected different fractions (about 30). The throughput of this apparatus is typically in the order of 100 g h^{-1}.

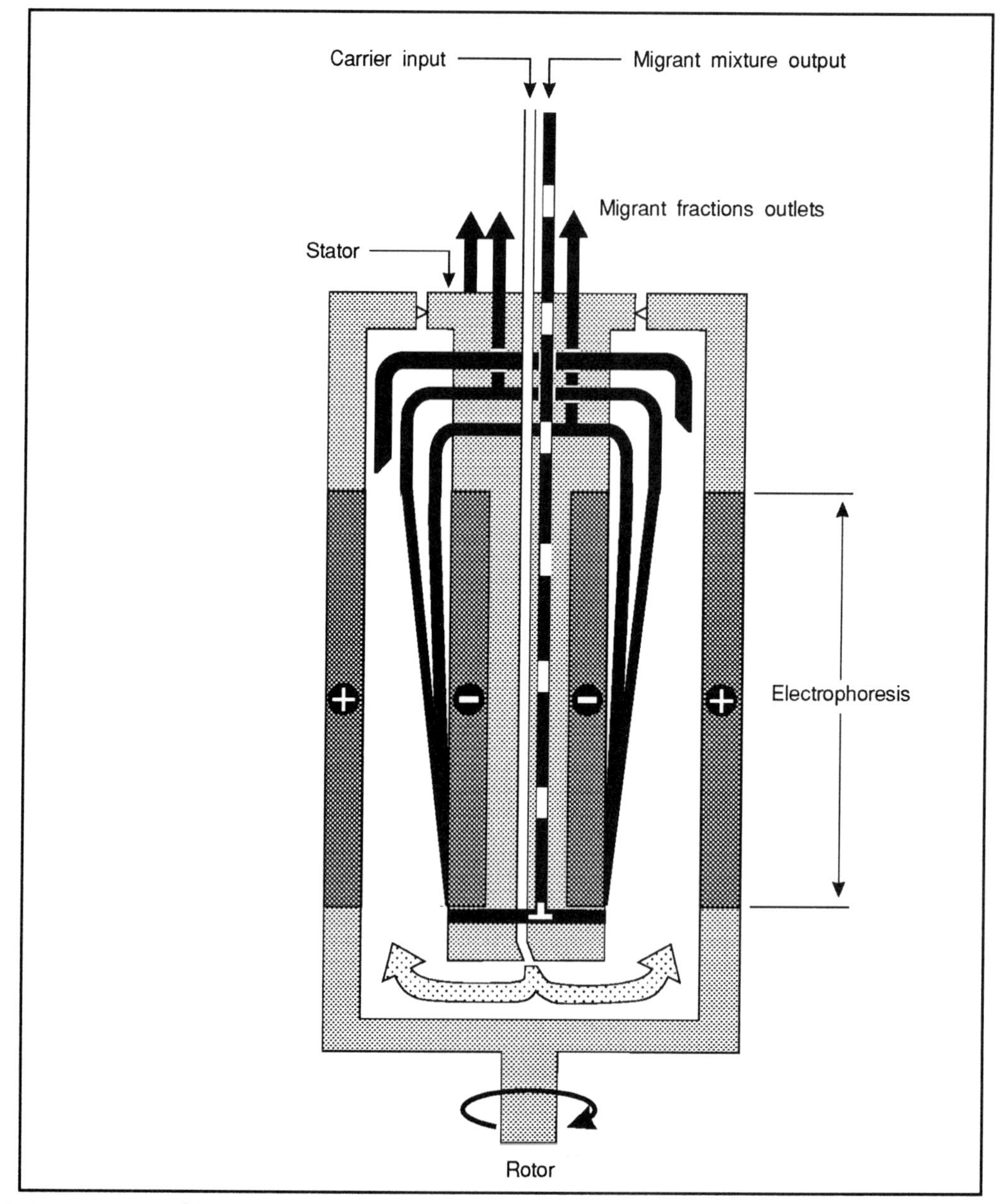

Figure 9.6 Philpot electrophoresis unit.

9.2.2 Pervaporation

Pervaporation is a membrane process in which a liquid mixture is in direct contact with the membrane (upstream side) and where the permeated product is removed as a vapour on the other side (downstream side) by applying a very low pressure.

This can be achieved either by creating a vacuum or by employing a carrier gas. Pervaporation is the only membrane process where a phase transition occurs going from upstream to downstream side. Figure 9.7 shows a scheme for pervaporation.

Pervaporation is developing but its commercialisation is rather slow. The three main reasons for this are:

- the energy consumption is relatively high compared to other membrane processes because a phase transition occurs so the heat of vapourisation has to be supplied;
- rather low permeation rates (fluxes) and/or insufficient selectivities;
- process design is difficult because of a temperature drop across the membrane.

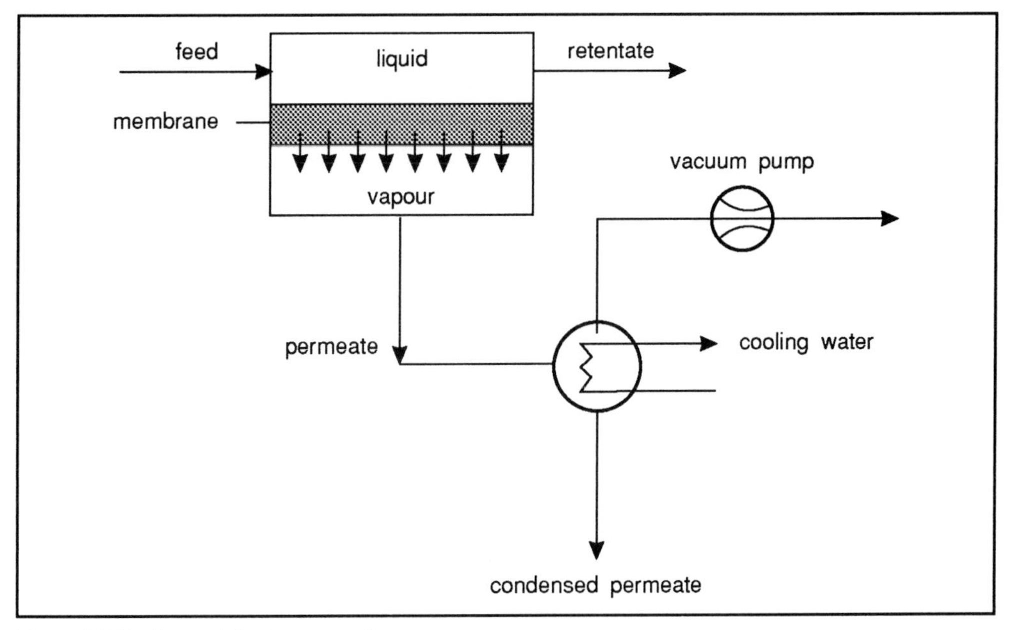

Figure 9.7 Basic principle of pervaportion.

Pervaporation can be used to separate organic liquids which are difficult to separate by distillation such as azeotropic mixtures and mixtures with close boiling points. Some examples are listed in Table 9.2.

The performance of a pervaporation process depend heavily on two major factors: the permeate flux and the selectivity. We will examine each of these in turn.

Permeate flux

factors influencing permeate flux

The permeation rate or mass transfer is governed by the diffusion of the component to be separated through the membrane and the membrane thickness. The diffusion coefficient in its turn depends on the solubility of that component in the membrane material. Pervaporation membranes can be considered as homogeneous swollen polymers. The diffusion coefficient therefore depends on the membrane material as well as the extent to which the material is swollen. The driving force is the partial vapour pressure difference across the membrane for the component to be separated. This is described in Equation 9.2.

water/methanol
water/ethanol
water/butanol
water/acetone
water/pyridine
hexane/chloroform
benzene/ethanol
toluene/hexane
benzene/toluene
ethanol/chloroform
ethanol/acetone

Table 9.2 Examples of liquid mixtures which may be separated by pervaporation.

$$\phi_m'' = \frac{D\,K\,C}{d_m} \left(1 - \frac{p_{2,s}}{p_{1,s}}\right) \qquad \text{(E - 9.2)}$$

where:

ϕ_m'' = mass flux of permeate (kg m^{-2} s^{-1})

K = solubility constant

D = diffusion coefficient (m^2 s^{-1})

C = concentration in membrane (kg m^{-3})

$p_{2,s}$ = partial pressure permeate side (Nm^{-2})

$p_{1,s}$ = partial pressure retentate side (Nm^{-2})

d_m = membrane thickness (m)

It should be noted here that this equation is simplified by assuming an ideal behaviour of vapours and liquids.

As can be seen from the equation the flux increases at increasing *D* and K; higher concentrations in the feed and using thin membranes. Note that an increase in temperature reduces the particle pressure ratio. The pressure on the retentate side of the membrane has no effect on the mass flux. The process can be applied if one of the components is sufficient volatile. It can be applied at all feed compositions and the osmotic pressure does not have to be taken into account.

Selectivity

The selectivity for a liquid mixture is expressed by the selectivity factor which can be defined by:

$$\sigma = \frac{(C_A/C_b)\ \text{perm}}{(C_A/C_B)\ \text{feed}} \qquad \text{(E - 9.3)}$$

where:

C_A and C_B = concentration of components A and B in the permeate and feed

The selectivity of a membrane depends on the type of material used, its swelling behaviour and the nature of the component to be separated.

Application in biotechnology

A promising application of pervaporation is the selective removal of solvents (ethanol, acetone butanol etc) during fermentation to prevent inhibition during the fermentation process as described in the section concerning *in situ* recovery.

pervaporation in alcohol purification

Another example is the purification of ethanol after fermentation to obtain pure (> 99%) alcohol. The alcohol is removed from the fermenter and concentrated by pervaporation up to 25%. The second step is the removal of water by distillation. The reason for this is a pure economical one: distillation is cheaper in the concentration range of 20-80% Outside of this range pervaporation is used. Thus in preparing pure alcohol two pervaporation stages and a single distillation stage are used (see Figure 9.8).

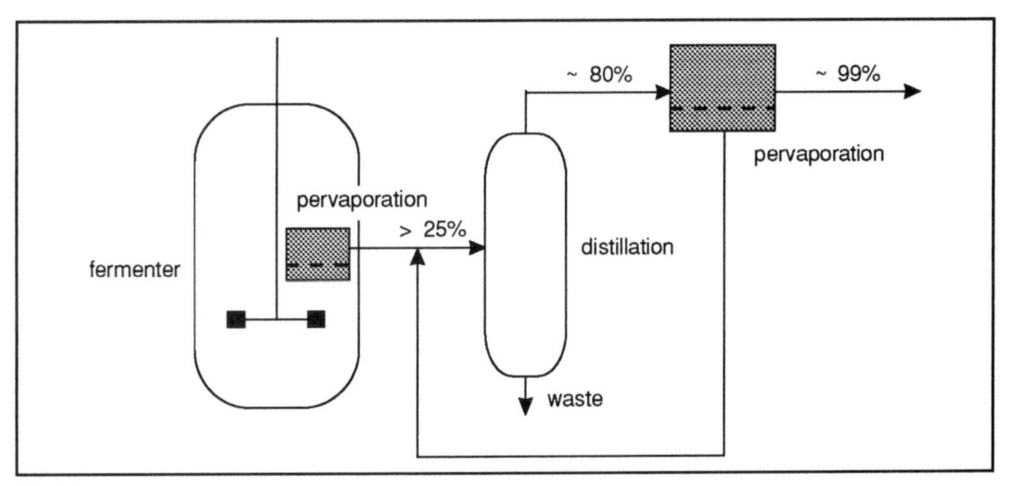

Figure 9.8 Combination of distillation and pervaporation for the purification of alcohol.

It is well established that the vapour of a boiling mixture of water-ethanol is richer in the more volatile component than the liquid. By removal of this vapour and condensing it to a liquid, the liquid will have a higher concentration of ethanol than the original feed. This process can be repeated several times in order to concentrate the ethanol. Separation by distillation is based on differences in relative volatility of the components to be separated. For ethanol-water (of atmospheric pressure), however, the relative volatilities vary with concentration and at a concentration of 96% (w/w) the relative volatilities equal 1. This point is called the 'azeotropic point'. Higher concentrations than the azeotrope are not possible in distillation with distillation. (In fact the azeotropic composition is a function of the pressure, lowering the pressure shifts the azeotrope towards 100% , for ethanol).

The vapour-liquid diagram is shown in Figure 9.9.

In distillation it is common to plot the concentration of the most volatile component in the liquid phase as function of its concentration in the vapour phase. For pervaporation processes it is common to do the other way around as can be seen in Figure 9.9 because of the fact that only water is permeated through the membrane.

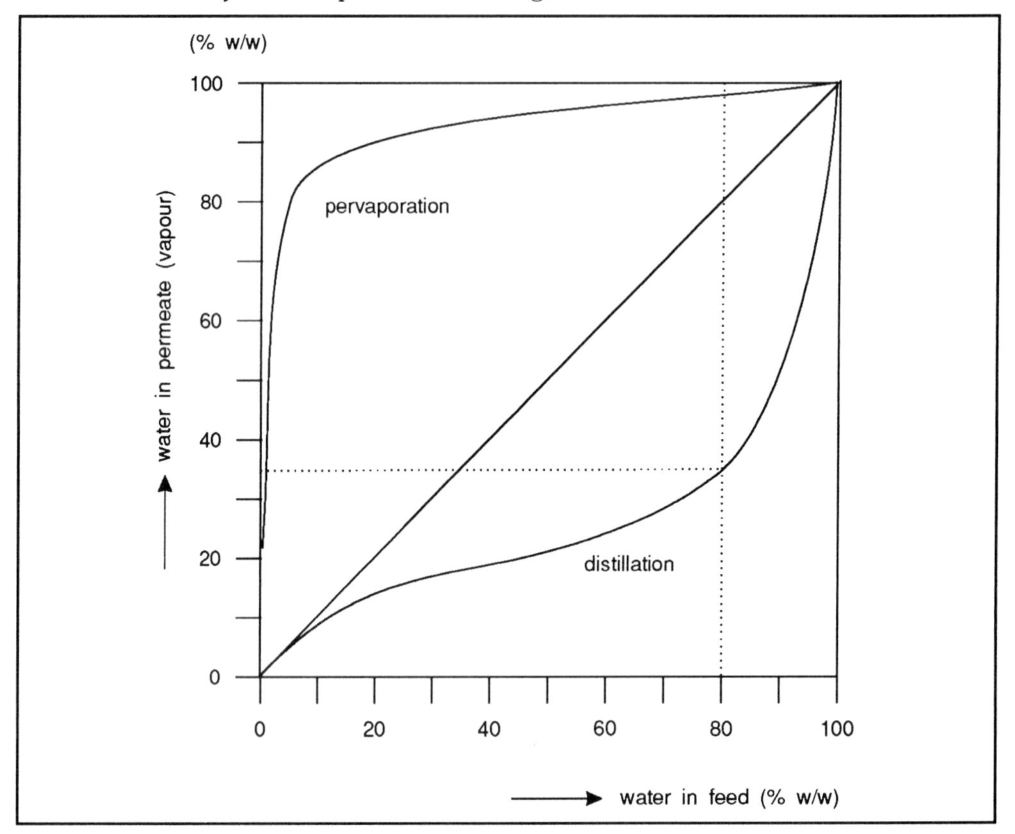

Figure 9.9 Vapour liquid diagram for water-ethanol (see text for an explanation).

The curve for distillation is based on thermodynamic equilibrium for a liquid and a vapour for the binary mixture water-ethanol. The curve for pervaporation, however, is based on differences in solubility and diffusivity and it thus dependent on the membrane material!

Returning to the figure it shows that 80% water in the feed correspond with about 35% water in the vapour whereas in the case of pervaporation about 95% of water is in the permeate (vapour). The result is that the feed is more concentrated in by pervaporation.

The selectivity can now be calculated also using the date Figure 9.9. X_A is now the fraction water in the permeate and Y_A the fraction water in the feed:

Since:

$$\sigma = \frac{(C_A/C_B)\ \text{perm}}{(C_A/C_B)\ \text{feed}} \quad \text{(see Equation 9.3)}$$

then:

$$\sigma = \frac{0.95/0.05}{0.8/0.20} = 4.75$$

SAQ 9.2

Calculate the selectivity for the distillation process using the data from Figure 9.9 and compare this selectivity with that of the pervaporation process.

Another approach is the selective removal of ethanol from a feed. This is frequently done in case of *in situ* recovery. The choice of the membrane material is very important here. Hydrophobic membranes retain water and may preferentially permeate ethanol, while in the case of very hydrophillic membranes water preferentially permeates.

The mass fluxes as already mentioned are very low compared to ultrafiltration membranes. Fluxes of 0.03-2 kg m^2 h^{-1} are reported. The selectivities however are very high and ranges from 12500 to 20! As long as the fluxes are so low with such membranes distillation will be more attractive if the separation can be achieved by distillation.

9.2.3 Supercritical extraction

During the past 20 years supercritical extraction (SCE) has developed from a laboratory curiosity to large scale commercial processes. It is, however, employed in a limited way in biotechnology and then often at a pilot plant scale for the extraction of thermally labile non-volatile products.

How does supercritical extraction works?

principles of SCE

SCE refers to a process in which a supercritical fluid (highly compressed gas) is used as a solvent. A supercritical fluid is a gas that has been pressurised and heated beyond its critical temperature and pressure.

Figure 9.10 shows a typical diagram for a pure solvent plotted on a relative scale. Horizontally the reduce density ρ/ρ_{crit} is plotted against reduced pressure. p/p_{crit} for different reduced temperatures T_R (= T/T_{crit}). This diagram is based on carbon dioxide which is frequently used as an extractant.

The temperature and density range for extraction purposes is indicated by different hatching. In these areas the physical properties of carbon dioxide behaves in between those of liquids and gases as can be seen from the data in Table 9.3.

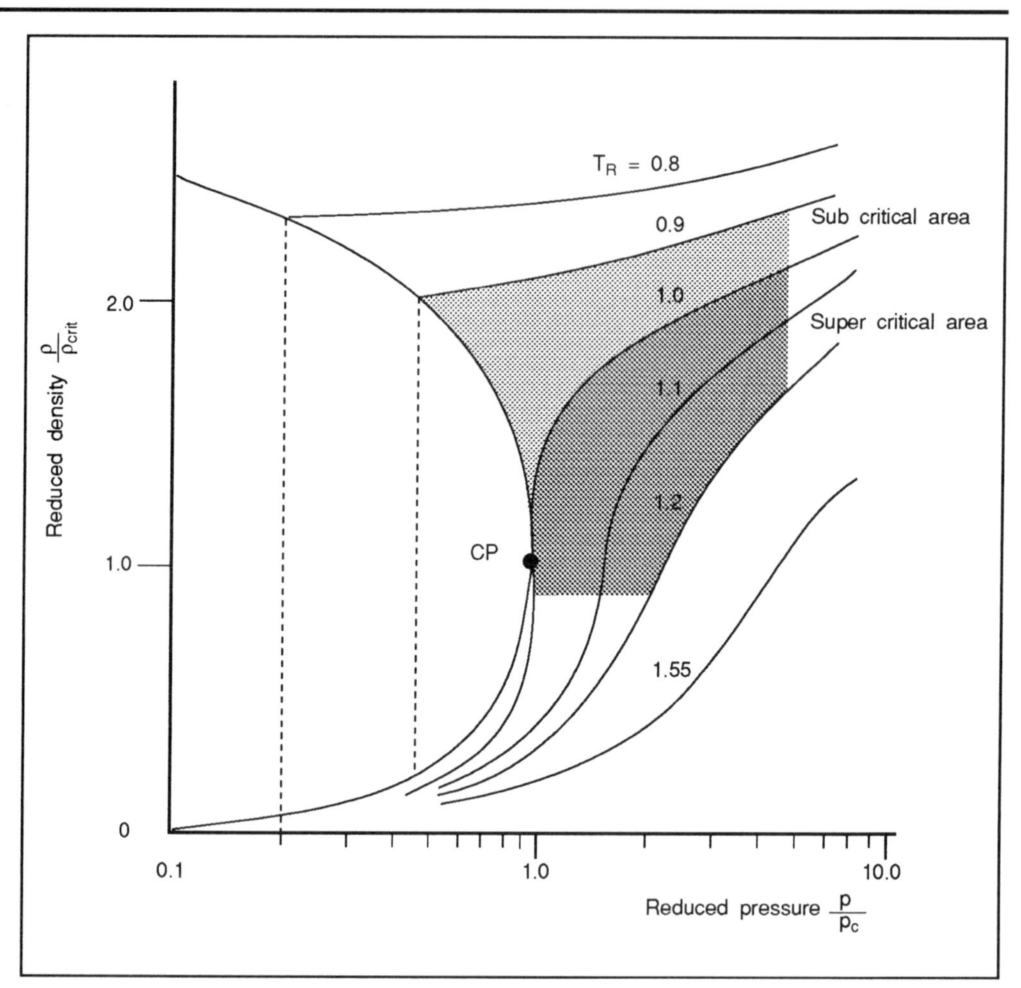

Figure 9.10 Data of supercritical dioxide. CP = critical point.

	Density kg m^{-3}	Viscosity mPas	Diffusivity m^2 s
Gas (1bar, 20°C)	0.6-2.0	0.01 0.03	3.10^{-5}
Sup Crit Solvent	200-500	0.01-0.03	7.10^{-7}
Liquid (20°C)	600-1600	0.2-0.3	10^{-9}

Table 9.3 Comparison of some physical properties of carbon dioxide under different conditions.

The viscosity is relatively low and makes supercritical fluids easier to handle than liquids.Because the diffusivity is much higher and the viscosity much lower than for liquids the mass transfer process proceed much faster than in ordinary liquids. In Figure 9.11 a process flow diagram is depicted of a simple supercritical extraction unit.

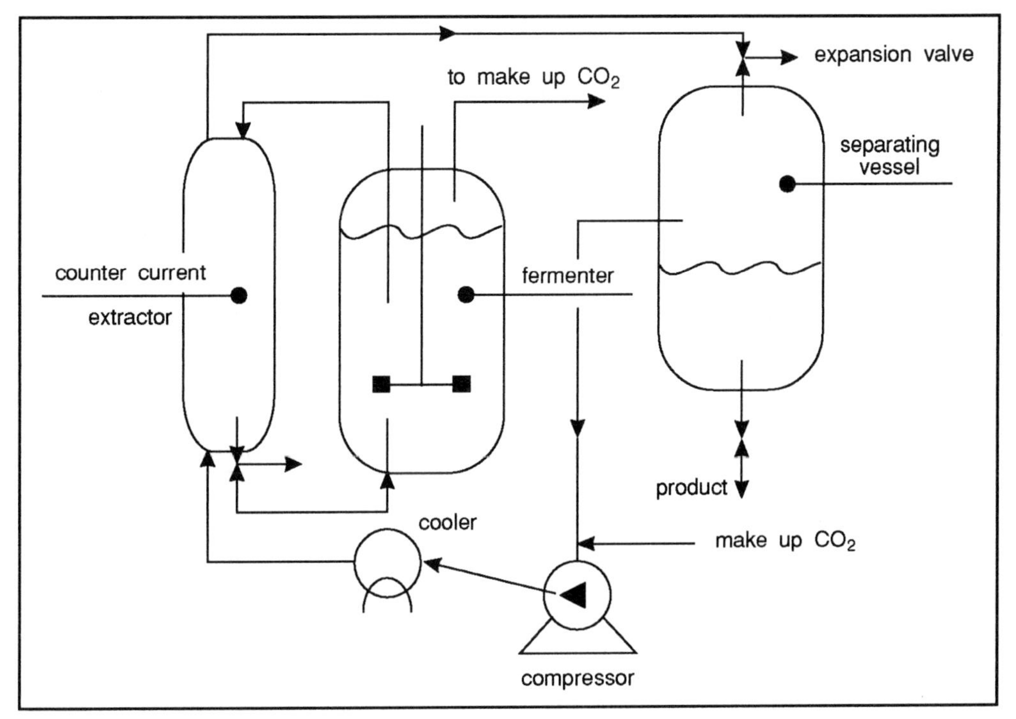

Figure 9.11 Flow scheme of a simple SCE system.

The carbon dioxide is compressed to the desired pressure (200 bar or more) and than cooled down to 40-80°C. The supercritical fluid is then pumped into an extraction column which contain solids (or liquid) to be extracted. The loaded SC-fluid is then transferred via an expansion valve to a separator.

After lowering the pressure (say to 80 bar) the extracted material precipitates and can be removed out of the separator. The solvent that remains in the separator has a low pressure and is free of extracted solutes. It is then re-pressurised and recycled to the extraction column.

The solute solubility depends strongly on the pressure applied. The higher the pressure the higher the solubility at a given temperature.

To enhance the dissolving power, gas mixtures and/or co-solvents (entrainers) can be used.

Π Make a list of the advantages of using supercritical fluid extraction over organic solvent extraction (think of CO_2 extraction as an example).

The list we anticipate you might write is:

- no residue is left in the product;
- non-toxic solvent;
- mild conditions (temperature);
- rapid extraction.

Potentially a wide range of solvents can be used.

There are also disadvantages:

- CO_2 has a relatively high triple point pressure (5.3 bar);
- critical pressure for CO_2 is rather high (73 bar) (propane 50 bar);
- dissolving power is lower than for conventional solvents;
- it does not extract hydrophillic substances to an appreciable extent;
- economical distribution coefficients can only be obtained at molecular masses below 1000 D.

Despite these disadvantages supercritical extraction is used on a large scale for the extraction of hops and de-caffeinisation of coffee beans. As has already been mentioned the application in biotechnology is limited mainly because we deal with hydrophillic compounds. Undoubtedly, it will find a growing application in the future. Table 9-4 gives an overview of possible products suited for SC-CO_2 and N_2 - extraction.

Products
Alkanes
Nicotines
Jojoba oils
Triacylyclycerols from plants
Choloesterol
Stigmasterol
Ergosterol
Pregnanadiol
Cortisone
Progesterone
Andosterone
Amino acids
Proteins

Table 9.4 Products suitable for SCE.

9.2.4 Reversed micelles extraction

Reversed micellar solutions are an attractive solvent system for use in the liquid-liquid extraction of biological products because products recovered from such a system do not lose their native structure and function.

A reversed micelle is a droplet, on a nanometre (nm) scale, of an aqueous solution stabilised in an apolar environment by the presence of surfactant at the interface.

Surfactants

Surfactants are molecules that posses both hydrophilic and hydrophobic parts as shown in Figure 9.12.

Due to their structure, such amphiphiles (molecules processing both hydrophobic and hydrophilic property exhibit abnormal behaviour in solution e.g. formation of aggregates.

In water aggregates are oriented in such a way that the hydrophilic moieties are in contact with the solvent and the hydrophobic are turned away from it.

The reverse orientation is encountered in apolar solutions as can be seen in Figure 9.12.

Micelles are most well known and extensively studied of the different types of aggregates.

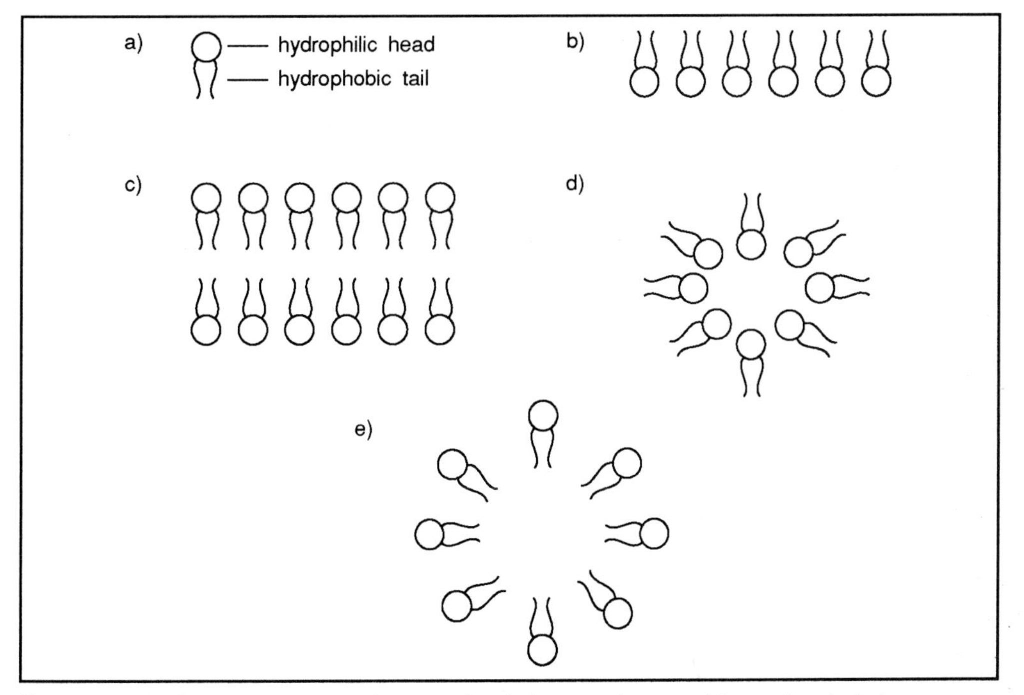

Figure 9.12 Micellar structures a = surfactant molecule, b = monolayer, c = bilayer, d = micelle in aqueous solution, e=micelle in a polar solvent (reversed micelle).

Surfactants can be classified according to the nature of their hydrophilic parts into:

- anionic;
- cationic;
- nonionic.

The most frequently studied surfactants are:

- trioctylmethylammonium chloride (TOMAC) which is cationic;
- sodium di-2-ethylhexyl sulphosuccinate (AOT) which is anionic;
- nonylphenolpentaethoxylate, which is nonionic;
- dodecyltrimethylammonium (DTAB), which is also cationic.

Micelles

The term micelle was first used in the early 1900's to describe the aggregation of amphiphilic electrolytes in aqueous solutions. The size and shape of the micelles vary significantly with the type of surfactant-solvent system but an order of magnitude is 1-10 nm. They are also a function of pressure, temperature, ionic strength and surfactant and solvent concentrations.

critical micelle concentration

A phenomenological observation for most surfactant systems is the existence of a critical micelle concentration (CMC). It is the minimum concentration of surfactant necessary for the formation of micelles. The CMC is characteristic of the system and depends on the parameters just mentioned and the molecular structure of the surfactant. CMC-values vary from 0.1-1.0 m mol l^{-1} in water or in non polar solvents

In fact (reversed) micelles are one of the various possible association structures of microdomains which compose micro-emulsions as can be seen in Figure 9.13.

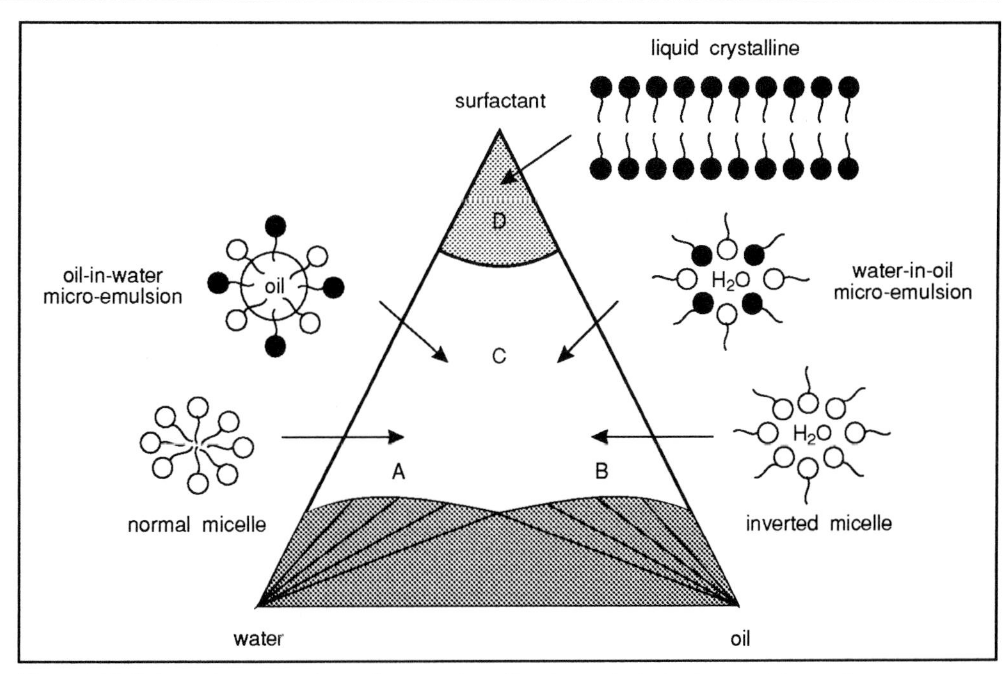

Figure 9.13 Schematic ternary phase diagram of an oil-water-surfactant microemulsion system.

Protein solubilisation

Most proteins are only sparingly soluble in apolar solvents, however, many proteins have been solubilised in reversed micellar solutions of a polar solvent without denaturation. It has also been demonstrated that these proteins can be transferred from the micellar phase into the aqueous phase.The enhanced solubility in micelles can be explained as follows.

A reversed micelle can be envisaged as a nanometre scale droplet of an aqueous solution stabilised in an apolar environment by the presence of surfactant on the interface. In this water-shell model the protein is enclosed in the water pool and is separated from the inner surface of the reversed micelle by a layer of water as can be seen in Figure 9.14.

Using reverse micellar phase as a liquid membrane batch and continuous extraction from and stripping to aqueous solutions can be achieved using mixer-settlers and column extraction units.

limitations of reversed micelles

The mass transfer rates however are low and the reuse of the chemicals is difficult and sometimes not possible, which makes the process expensive. Also the chemistry of the process is, as yet, not well understood. A lot has to be done before it can be applied on a commercial scale.

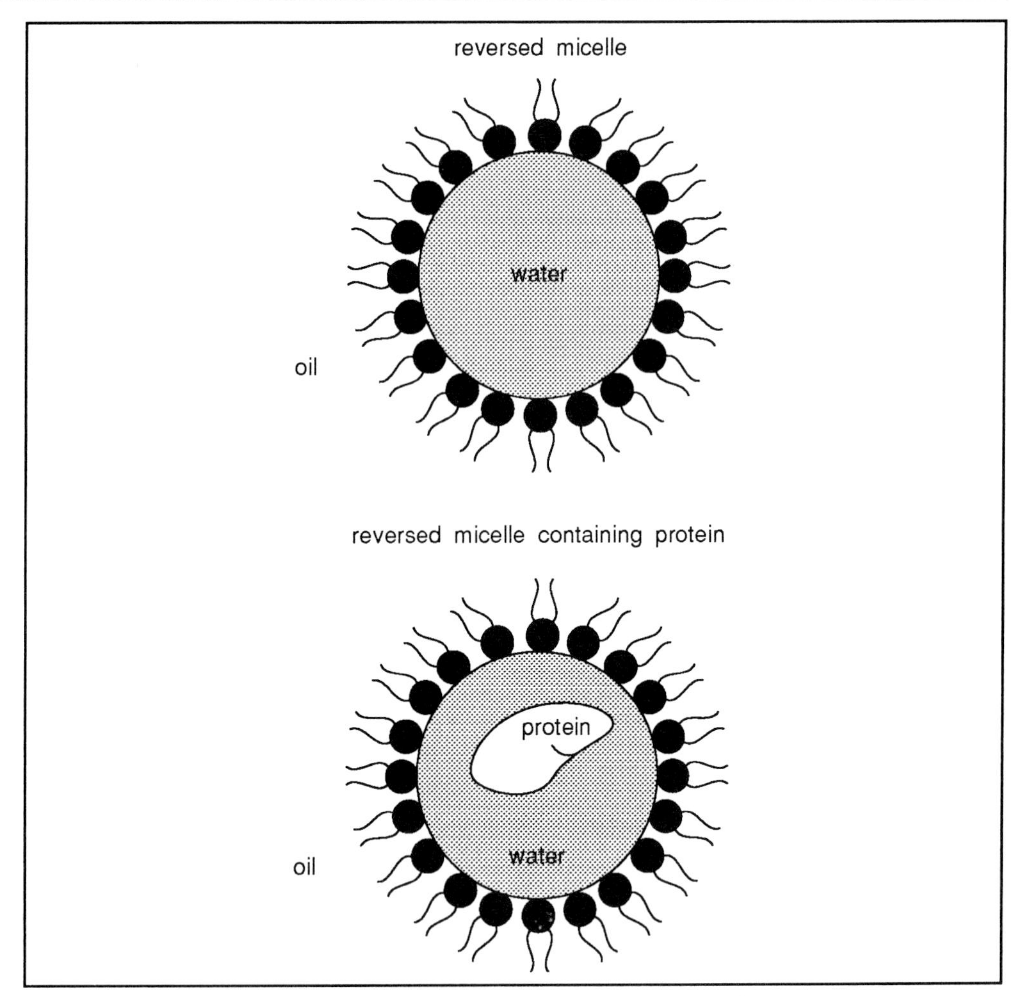

Figure 9.14 Schematic representation of a protein enclosed in a reversed micelle.

9.2.5 rDNA for recovery

An elegant method utilising protein engineering to improve and simplify purification is protein fusion. Using genetic engineering technique it is possible to fuse two genes which will produce a hybrid (chimeric) protein (made up of the two proteins coded for by the original genes).

Fusion of a target protein with a protein with favourable separation characteristics is the basis for this technique.

A well known example of a fusion protein is β-galactosidase. This protein can be applied to enhance the distribution coefficient in the aqueous two phase extraction of Staphylococcus protein A (SpA). β-galactosidase as such has a partition coefficient of about 100 in an aqueous two phase system (PEG 4000/potassium phosphate) whereas the partitioning coefficient of SpA is only 0.2. The fused product shows a partitioning coefficient of 3.0.

The same technique can be used for the separation of very similar proteins. Problems associated with this technique will occur when the fusion protein must be split from the target protein and subsequently when the splitting enzymes are to be separated from the mixture.

Despite these problems, it is believed that protein engineering may contribute to the simplification of downstream processes.

A possible process flow sheet is given in Figure 9.15.

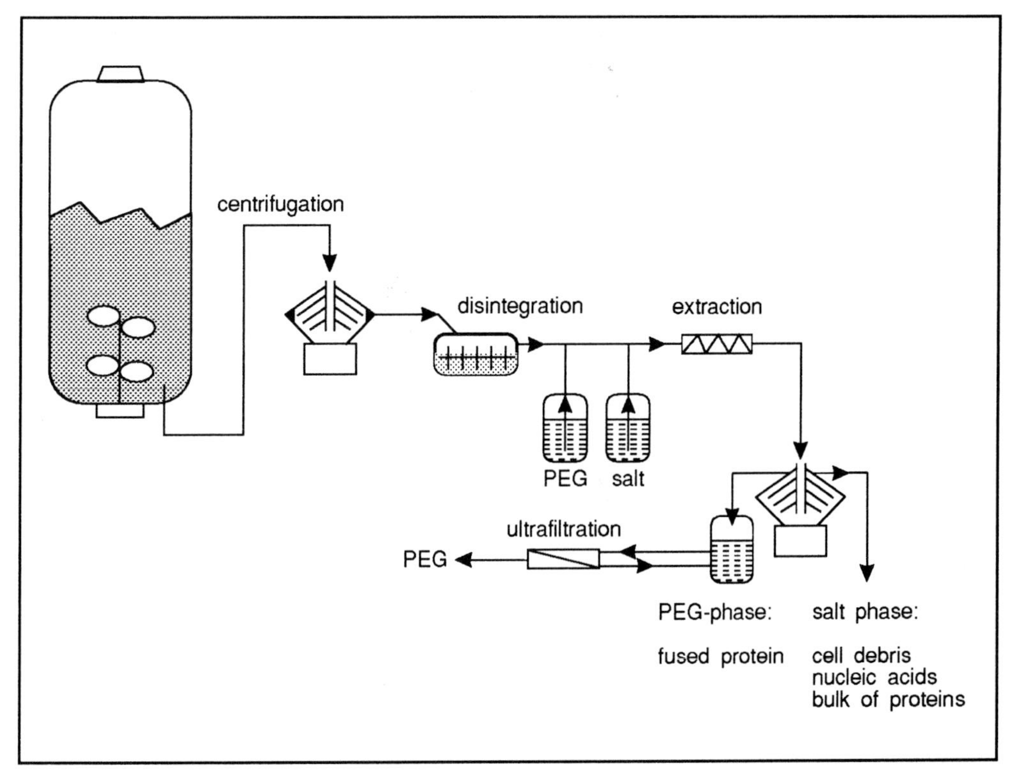

Figure 9.15 Process flow sheet for continuous separation of fused proteins in an aqueous two phase system.

SAQ 9.3

1) Examine Figure 9.15 carefully. At the bottom is the fused protein. What further steps would be needed before the desired protein can be produced.

2) We wish to prepare large quantities of a protein (protein A) which has a n almost equal mixture of hydrophobic and hydrophylic amino acids. We carry out some genetic engineering and add to the gene which codes for protein A some genetic information which codes for a sequence of 40 aspartic acid residues to be added to the protein. In what ways might this new gene product be easier to purify?

Summary and objectives

In order to simplify the recovery and purification of biomolecules the number of process steps should be minimal. This has a positive effect on the overall yield. Integration can be done in several ways. In this, *in situ* recovery is the tool with the highest potential for the moment.

A large number of alternative separation methods are coming on stream, some of which have already been commercialised on a small scale. The most promising are:

- supercritical extraction;
- reversed micelle extraction;
- pervaporation and perstraction;
- electrophoresis on a preparative scale.

rDNA also may offer potential solutions to separation problems in the near future for products with a high added value.

Now that you have completed this chapter you should be able to:

- describe the importance of process integration and the whole broth treatment and *in situ* recovery strategies for product recovery;
- describe a variety of novel extraction procedures;
- list advantages and disadvantages of a variety of novel extraction procedures;
- calculate electrophoretic mobility and selectivity from supplied data;
- explain the potential of using recombinant DNA technology (genetic engineering) to help product recovery.

Responses to SAQs

Responses to Chapter 1 SAQs

1.1

1) Total turnover = (selling price) . (total market volume)

   For enzyme A = 25 . 10^6 US \$ y^{-1}

   For enzyme B = 2 . 10^8 US \$ y^{-1}

2) Annual net profit = added value x total market volume

   For enzyme A = 6.10^6 US \$ y^{-1}

   For enzyme B = 2.10^7 US \$ y^{-1}

3) By reference to Figure 1.1 we can see from the selling price that enzyme A is a diagnostic enzyme and enzyme B is a bulk enzyme.

4) Enzyme B is likely to have the highest concentration in the starting material because it is a bulk market sector enzyme (see Figure 1.1).

1.2

The following characteristics are more commonly associated with classical biotechnology than with modern biotechnology.

1) Unit recovery operations have low resolution
3) Purification factor is low
4) Recovery has large number of unit operations
6) Process is well described mathematically

1.3

The correct pairings are:

Drying with formulation

Sedimentation with separation

Cell disruption with pre-treatment

Precipitation with concentration and purification

Chromatography with purification

Membranes with concentration

Sterilisation with pre-treatment

Centrifugation with solid/liquid separation

Tabletting with formulation

You can also check your response by referring to Figure 1.3.

1.4 You can also check your response by reference to Figure 1.4.

Size range	Appropriate unit operations
1) Ionic	c, f, g, h and i
2) Macromolecular	a, c, e, f, g, h, i, j and k
3) Micron particle	a, e, g, j, k and m
4) Fine particle	b, d, e, k, l and m
5) Coarse particle	b, d, and m

Responses to Chapter 2 SAQs

2.1

1) True.

2) False - yeast cell walls are composed of polysaccharides.

3) False - plant cells have a cell wall but animal cells do not.

4) True - it is the rigidity of the murein in the cell wall that gives the cell its shape.

5) True.

2.2

1) For highly dilute suspensions, relative viscosity is best described by the Einstein equation.

2) Shear stress is a linear function of shear rate.

3) The value of K_1, is determined by the particle shape and is 2.5 for spherical particles.

4) For concentrated suspensions, relative viscosity is best described by the Eilers equation.

5) The viscosity is independent of shear rate.

2.3

1) For plots 1 and 2, $\tau = \eta \dot{\gamma}$ holds. You may not have spotted 5. Clearly there is proportionality between τ and $\dot{\gamma}$ in this case as the plot of τ against is a straight line but it does not go through the origin. What this means is that a finite yield stress has to be applied. Thus in this case $\tau = \tau_o + \eta \dot{\gamma}$ - see the answer to 4).

2) n is less than 1 for plots 4 and 6.

3) n is greater than 1 for plot 3.

4) For plots 5 and 6 a finite yield stress has to be applied - see 1).

5) τ_o is finite and n is less than 1 for plot 6.

6) All plots are influenced by the viscosity of the medium.

2.4

1) False. We told you that n was 0.4 to 0.8 for Actinomycites. For liquids to be dilatants, then n has to greater than 1.

2) True.

3) True.

4) True.

5) False - the apparent viscosity increases markedly as the shear rate increases; n > 1, see Figure 2.7 curve 3.

Responses to Chapter 3 SAQs

3.1

To determine the value of K_b we need to consider the initial rate of product release.

From Figure 3.3, for the polyurethane impeller, we can see that the rate of release is linear for the first 10 seconds.

So, from the graph, at t = 10, $\frac{C_r}{C_r^{max}} = 0.35$

The value of C_r^{max} for the polyurethane impeller is 0.084.

So:

$$\frac{C_r}{0.084} = 0.35$$

$$C_r = 0.029$$

Substitution into Equation 3.2 we have:

$$\ln \frac{0.084}{0.084 - 0.029} = K_b.10$$

$$0.431 = K_b.10$$

$$K_b = 0.043 \text{ min}^{-1}$$

A similar approach can be taken for the stainless steel impeller, in this case a K_b value of around 0.019 min^{-1} should be obtained.

3.2

1) Increasing temperature decreases broth viscosity and has a positive effect on cell disruption.

2) $\frac{\Delta p}{c_p}$ equals $\Delta\theta$.

3) Power input is proportional to the homogenisation pressure.

4) $(C_r^{max} - C_r)$ represents the concentration of unreleased product.

5) The size and cell wall thickness and composition of a microorganism influences cell disruption.

3.3

Rearranging Equation 3.6 we obtain:

$$\Delta p^{\beta} = \frac{1}{K_h . N} . \ln\left(\frac{C_r^{max}}{C_r^{max} - C_r}\right)$$

By entering the values given, we obtain:

$$600^{\beta} = \frac{1}{1.65\ 10^{-8}} \cdot \ln\left(\frac{0.063}{0.063 - 0.051}\right) \rightarrow \frac{1}{1.65 \cdot 10^{-8}} \cdot 1.66 = 10^{8}$$

$600^{\beta} = 1 \cdot 10^{8}$. Thus $600^{\beta} = 10^{8} \rightarrow \beta = 2.88$

We asked you however to do this graphically. We have done this in the following way. We can calculate log ΔP^{β} for different values of β. Thus log 600^{β}:

= 2.78 when β = 1; = 5.56 when β = 2; = 8.33 when β = 3; = 11.11 when β = 4

We can now plot log ΔP^{β}.

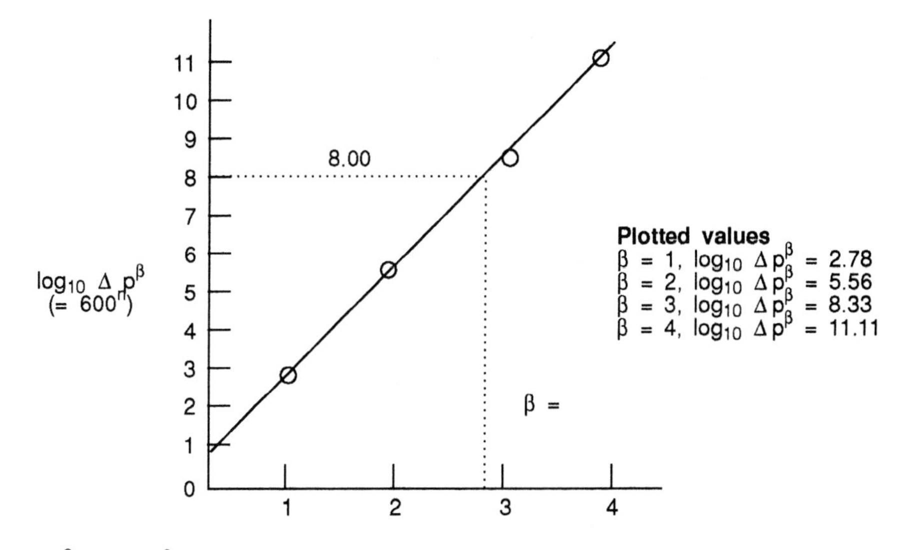

Since $\Delta P^{\beta} = 1 \times 10^{8}$, we obtain the value of n in this expression by taking the log of $1 \cdot 10^{8}$ (ie 8) and reading from the graph. Using this approach, β = 2.88.

3.4

1) False. Temperature rise is only a function of the applied pressure and the broth properties (Equation 3.9), not a function of scale.

2) False. Gluconase is an enzyme that digests the cell walls of yeasts.

3) True

4) False. Cellulase and pectinase digest cell walls of plants.

5) False. Although scale-up of homogenisers is achieved by installation of a bigger plunger pump and discharge valve, the homogenisation pressures can remain constant.

Responses to Chapter 4 SAQs

4.1

1) Membrane filter presses have the advantages of: higher yield; drier cake; easy cake removal.

2) Refer to Table 4.1 to see if you have remembered the various types of filter media and suitable examples of each.

3) Choice of filter medium is influenced by: fouling behaviour/cleaning procedures; chemical resistance; resistance to flow; durability. You may have also included the ability to trap the solids.

4) Types of cakes discharge are: knife (scraper) discharge; string discharge; belt discharge; roller discharge.

5) The three main stages of vacuum filtration are:

   filtering; dewatering; discharge (see Figure 4.3).

4.2

1) For incompressible cakes, the resistance in the cake is assumed to be directly proportional to amount of cake deposited.

2) The mass of cake deposited per unit area is a function of time in batch operation and concentration of solids in the broth.

3) For constant pressure filtration, the slope of a $\frac{t}{V}$ versus V plot is a measure of specific cake resistance and the intercept a measure of the medium (cloth) resistance.

4) False. However, the reverse is true.

5) False. See Figure 4.6.

4.3

1) Multichamber type.

2) Tubular type.

3) Disc type and decanter type.

4) Multichamber type.

5) All types.

6) Tubular type.

7) Decanter type.

8) Decanter type.

9) Decanter type.

4.4

1) Buoyancy force can be determined using Equation 4.12 and requires a, c and j.

2) Gravitational sedimentation rate can be determined using Equation 4.15 and requires a, b, c, h and j.

3) Throughput of sedimentation tank can be determined using Equations 4.18 and requires a, b, c, d, h and j.

4) Centrifugal sedimentation rate can be determined using Equation 4.19 and requires parameters a, b, c, h and k.

5) Gravitational force can be determined using Equation 4.11 and requires parameters a, b and j.

6) Mean residence time of liquid in a continuous sedimentation tank can be determined using Equation 4.16. This requires determination of throughput (see 3) and the volume of the tank must also be known ie e, f and g from the list. So the full list a, b c, d, e, f, g, h and j must be known.

4.5

We begin by calculating the Σ value (see Equation 4.22a and 4.22b).

$$\Sigma_{tube} = \frac{\pi (r_o^2 - r_i^2)}{\ln \frac{r_o}{r_i}} \frac{\omega^2 L}{g}$$

$$r_o = \frac{L}{2. * 10} = \frac{0.75}{2 * 10} = 0.0375 \text{ m}$$

thickness of broth layer 0.1 d_o, so $r_i = \frac{0.9 \,.\, 0.75}{2} = 0.03375$ m

Substitution gives:

$$\Sigma_{tube} = \frac{\pi (0.0375^2 - 0.03375^2)}{\ln \frac{0.0375}{0.03375}} \,.\, \frac{650^2 \,.\, 0.75}{9.8} = 257 \text{ m}^2$$

The sedimentation rate under gravity conditions is given by Equation 4.15.

$$v_g = \frac{d_p^2}{18\,\eta} \Delta\rho\, g$$

$$= \frac{(10 \,.\, 10^{-6})^2 * 50 * 9.8}{18 * 2 * 10^{-3}} = 1.36 * 10^{-6} \text{ (m s}^{-1}\text{)}$$

According to Equation 4.23:

$$\phi_V = v_g \Sigma_{tube}$$

$$= 1.36 \,.\, 10^{-6} * 257 = 3.5 \,.\, 10^{-4} \text{ (m}^3 \text{ s}^{-1}\text{)}$$

$$= 1260 \text{ l h}^{-1}$$

4.6

1), 2), 3) and 4) are correct. 5) is wrong. We can see from Equation 4.29 and 4.30 that increasing the diameter (and therefore radius), number of discs (Z) and angular speed (ω) all increase the volumetric flow rates. Since an increase in length will also increase cone angle (φ - see Figure 4.15), the Equation also show that this will increase the volumetric flow rate. All these parameters can therefore be used to scale-up a disc stack centrifuge.

Particle diameter, however, is a characteristic of the fermentation broth and cannot be used as a scale-up parameter.

Responses to Chapter 5 SAQs

5.1

Statement 1 and a) plate evaporator

Statement 2 and c) natural circulation evaporator

Statement 3 and b) falling film evaporator

Statement 4 and e) backward feed evaporator

Statement 5 and d) forward feed evaporator

5.2

1) Plate evaporators have a small construction volume and are flexible in their use.
2) Falling film evaporators can be used to treat viscous products, heat sensitive solutions, as well as clear, foaming and corrosive products, at low pressures.
3) Natural circulation evaporators are relatively cheap and easy to operate.
4) Multiple-effect evaporators have reduced heat consumption.
5) Forward feed evaporators can be used for heat sensitive products.
6) Backward feed evaporators have improved heat transfer.

5.3

Δh_c = change in enthalpy of concentrate (kJ)

θ_f = temperature of feed (°C)

ψ_s = specific steam consumption (no units), it is simply a ratio

ϕ_v = mass flow rate of vapour (kg s^{-1})

$c_{p,f}$ = specific heat of feed (kJ kg^{-1} $°C^{-1}$) or (kJ kg^{-1} K^{-1})

$\phi_{s,c}$ = mass flow rate of steam as condensate (kg s^{-1}).

5.4

1) True. Compare Equations 5.15 and 5.13.
2) True. See Equation 5.15.
3) False. Overall energy balance is given by Equation 5.20.
4) False. Heat of evaporation is large compared to $(\phi_c \, c_{p,c} \, \theta_c - \phi_f \, c_{p,f} \, \theta_f)$
5) False. Heat losses in the evaporator increase the value of ψ, since the quantity of steam to the evaporator will be increased relative to the quantity of evaporated solution.

6) True - well almost! $c_{p,f}$ and $c_{p,c}$ are usually very similar but not identical. Remember that the feed and the concentration contain different amounts of solute. This influences the specific heats of these two solutions.

5.5

1) The heat transfer rate (ϕ_q) would be increased (see Equation 5.24, A would be increased thus ϕ_q would increase).

2) The heat transfer rate would increase if copper replaced iron as tube material in the evaporator (see Table 5.1).

3) Forced circulation would increase the heat transfer rate in a vertical evaporator (see Table 5.2).

5.6

The correct response to this SAQ can be determined from Figure 5.7. Thus:

	Particle size	Pressure difference (bar)
Ultrafiltration	0.01μm	5
Microfiltration	5μm	0.7
Hyperfiltration	0.8nm (= 0.0008μm)	40

Note that particles of 0.1nm (= 0.0001μm) are really too small to separate even by hyperfiltration. Particles of 0.1mm (= 100μm) are too big even to separate by microfiltration.

Pressure differences of 500 bar are too large to use with membranes. On the other hand pressure differences of 0.01 bar are too small to use with membranes.

5.7

1) In membrane separations the permeated stream contains relatively small components.

2) Membrane selectivity is often expressed in terms of molecular weight cut off.

3) For asymmetrical membranes only a thin top layer determines the selective barrier.

4) A membrane a sharp molecular weight cut off will have a narrow pore size distribution.

5.8

$\alpha = 2.30$.

We know that: concentration factor (α)

$$\alpha = \frac{\phi_{v,f}}{\phi_{v,r}}$$

also, $\psi = \dfrac{\phi_{V,r}\, C_f}{\phi_{v,f}\, C_f}$ - see Equation 5.30

So:

$$\psi = \frac{C_r}{C_f} \cdot \frac{1}{\alpha}$$

From Figure 5.17, at R = 0.5 and Δ = 0.6, the ψ = 0.58.

Using Equation 5.26, $R = 1 - \frac{C_p}{C_r}$

Substituting, $0.5 = 1 - \frac{10}{C_r}$

$$C_r = \frac{10}{0.5} = 20$$

Substituting into $\psi = \frac{C_r}{C_f} \cdot \frac{1}{\alpha}$

$$0.58 = \frac{20}{15} \cdot \frac{1}{\alpha}$$

$$\alpha = \frac{20}{15} \cdot \frac{1}{0.58} = 2.30$$

5.9 $K = 8.3 * 10^{-13}m$, $R_m = 1.2 * 10^{12}m^{-1}$

K can be calculated from $K = \frac{\varepsilon R^2}{8d_m\tau}$ (see Equation 5.43)

$$\text{Thus } K = \frac{0.2\,(10^{-7})^2}{8 * 10^{-4} * 3} = 8.3 * 10^{-13}m$$

R_m is given by $R_m = \frac{1}{K} = 1.2 * 10^{12}m^{-1}$

5.10 $j = 2 * 10^{-5}m\,s^{-1}$ or $72\ l\,m^{-2}\,h^{-1}$

Since $j = \frac{K\Delta p}{\eta}$ (see Equation 5.44) and $Rm = \frac{1}{K}$ (Equation 5.45)

$$\text{then } j = \frac{10^{-13} * 200 * 10^3}{1 * 10^{-3}} = 2.10^{-5}\ m\,s^{-1}$$

Note the units of j from this calculation are in $m\,s^{-1}$ or m/s.

$$\text{But } \frac{m}{s} = \frac{m^3}{m^2s} = \frac{3600\ m^3}{m^2\,h}$$

So the flux in $lm^{-2}h^{-1}$ will be:

$2 \cdot 10^{-5} \cdot 3600 \cdot 1000 = 72 \text{ lm}^{-2} \text{h}^{-1}$

This value is quite typical for an ultrafiltration membrane.

5.11

1) A very small pore radius (r) would reduce membrane flux by reducing the value of the permeability coefficient ($K = \frac{\varepsilon r^2}{8\tau\delta}$). Also, with very small pores the osmotic pressure off-sets the applied transmembrane pressure, this also reduces membrane flux (ie $j = \frac{K}{\eta}(\Delta p - \Delta\pi)$; Equation 5.46).

2) During concentration polarization the flux is limited by the balance of convective flow to the membrane surface of the protein and its counter diffusion to the bulk as a result of the concentration difference.

3) An increase in solute concentration at the wall indicates concentration polarization. Here, the flux becomes independent of the transmembrane pressure but dependent on the concentration gradient ie the higher the wall concentration relative to that of the bulk liquid, the greater the membrane flux ($j = K \text{ Ln} \frac{C_{wall}}{C_{bulk}}$; Equation 5.47b).

4) j = membrane flux.

   K'' = initial flux in absence of concentration polarisation.

   K' = mass transfer coefficient

   α = concentration factor.

5) See Figure 5.23. Plot membrane flux (j) against ln concentration factor (α). The mass transfer coefficient is then the slope of the line. The initial flux (in absence of concentration polarization) is intercept on the j-axis.

5.12

$\bar{\alpha} = 3.4$

The flux equation is $\bar{J} = K'' - K' \ln \alpha$ (Equation 5.48)

From Figure 5.23, K'' and K' are 50 and 20 $\text{l m}^{-2} \text{h}^{-1}$ respectively

Thus $\bar{J} = 50 - 20 \ln \bar{\alpha}$

$$\text{so } \ln \bar{\alpha} = \frac{50 - \bar{J}}{20}$$

From Table 5.5, $\bar{J} = 25.5$

$$\text{Thus } \ln \bar{\alpha} = \frac{50 - 25.5}{20} = 1.225$$

Thus $\bar{\alpha} = 3.4$

Responses to Chapter 6 SAQs

6.1 3 - 4 stages

To solve this problem we can either use Equation 6.7 or use the data in Figure 6.6.

Let us use Equation 6.7 first.

We can either use $\frac{C_N}{C_f} = \frac{S-1}{S^{N+1}-1}$ (Equation 6.7a)

or better still:

$$N = \frac{\ln\left[\frac{(S-1)}{(1-f)} + 1\right]}{\ln S} - 1 \quad \text{(Equation 6.7b)}$$

$$S = \alpha \frac{\phi_{v,e}}{\phi_{v,r}}$$

Thus $S = 5 \cdot \frac{2}{5} = 2$

and $1 - f = 1 - 0.95 = 0.05$

Therefore $N = \ln \frac{\left[\frac{2-1}{0.05} + 1\right]}{\ln 2} - 1 = 3.39$ stages

If we use the data on Figure 6.6, then for S = 2 and f = 0.95 (solid line), the number of stages can be read on the horizontal axis (ie about 3.4).

6.2

1) True.

2) False. Flow of raffinate is given by r(S + 1); rS describes flow of solvent.

3) False. The non-extracted fraction is given by $\frac{1}{1+S}$; $\frac{S}{S+1}$ described the extracted fraction.

4) True.

5) True.

6.3

1) The distribution coefficient for a base would increase with increasing pH because more of the base would be in an undissociated form and therefore available for extraction.

2) $\alpha^o = 9\alpha$. Since pH - pK_a = $\log(\frac{\alpha^o}{\alpha} - 1)$.

Then 5 - 4 = $\log(\frac{\alpha^o}{\alpha} - 1)$.

So 1 = $\log(\frac{\alpha^o}{\alpha} - 1)$.

Thus 10 = $(\frac{\alpha^o}{\alpha} - 1)$ and $\alpha^o = 9\alpha$

3) $\alpha^o = 99\alpha$

Because 2 = $\log(\frac{\alpha^o}{\alpha} - 1)$

$100 = \frac{\alpha^o}{\alpha} - 1$

Thus $\alpha^{o = 99}\alpha$

Obviously the extraction becomes less and less efficient as the acid becomes more ionised.

6.4

1) According to the Brønsted equation, the distribution coefficient given by $\alpha = \exp(M\lambda/KT)$ - see Equation 6.21.

2) If λ is positive, an increase in molecular weight will increase the distribution coefficient.

3) Aqueous two phase systems offer three major advantages, these are:

a) higher protein capacity.

b) non denaturing solvent environment.

c) high selectivity.

6.5

1) a) Proteins usually exhibit a net negative charge on their surface.

b) The Stern-layer comprises positive ions.

c) The Gouy-Chapman layer is a diffuse layer of mobile counter ions.

2) a) False

b) True

6.6

1) True - proteins have ionic surfaces. High dielectric constant decreases ionic attraction. Thus the protein molecules are less likely to associate and precipitate out. In other words they remain soluble.

2) True.

3) True.

4) True.

5) True.

Responses to Chapter 7 SAQs

7.1

1) Affinity adsorption is more specific than physical adsorption and ion exchange.

2) The main binding forces in physical adsorption are Van de Waals forces and polar (ionic) forces.

3) An adsorbent that is an inorganic material covered with a thin layer of organic material is termed a composite.

7.2

$4.15 \cdot 10^{-9}$m

Since $D_m = \dfrac{KT}{6\,\pi\,\eta\,r_m}$ (see Equation 7.10)

$$0.525.10^{-10} = \frac{1.38 \cdot 10^{-23} \cdot 298}{6\,\pi \cdot 10^{-3} \cdot r_m}$$

$r_m = 4.15 \cdot 10^{-9}$m

7.3

1) True.

2) False. The constant B in the Van Deemter equation represents longitudinal diffusion, not axial dispersion. Axial dipersion is also called Eddy diffusion (see 1 above).

3) False. Axial diffusion is virtually independent of the superficial velocity (flow rate). See Figure 7.13, curve A.

4) False. Dispersion due to incomplete mass transfer is a linear function of flow rates. See Figure 7.13 curve C.

5) True.

6) True, see Equation 7.8.

7.4

$v_{s,opt} = 1.08 \cdot 10^{-2}$ m s^{-1}, $H_{min} = 1.71 \times 10^{-3}$ m

Plotting $H = 0.2 \cdot 10^{-3} + \dfrac{0.82 \times 10^{-5}}{v_s} + 0.70 \cdot 10^{-1}\, v_s$ produces curve A in the following figure.

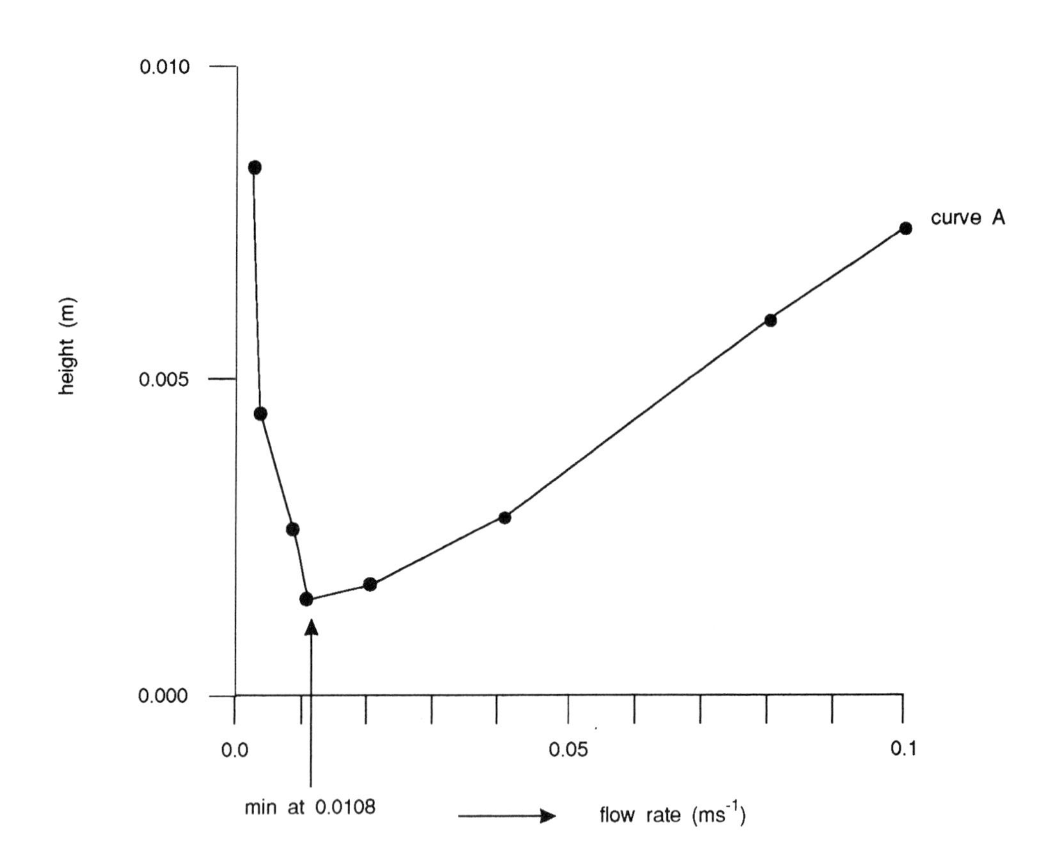

(Note we have plotted selected calculated values).

The optimum flow rate can be calcualted using Equation 7.13.

$v_{s,opt} = \sqrt{(0.82 \,.\, 10^{5}) \div (0.7 \,.\, 10^{-1})} = 1.08 \,.\, 10^{-2}\ m\ s^{-1}$ = approximately 1 cm s^{-1}

The minimum plate height H_{min} can be calcualted using Equation 7.14. Substition of the values given yields:

$H_{min} = 0.2 \,.\, 10^{-3} + 2\sqrt{0.82 \times 10^{-5} \,.\, 0.7 \times 10^{-1}}$

$= 1.61 \,.\, 10^{-3}$ m (1.6 mm).

7.5

1) L/v_s is the residence time of an empty column.

2) High capacity of adsorbent per unit volume is indicated by a high q_e value.

3) In a multiple stage adsorption process, high L_m/S_m ratio will increase the number of cycles needed to achieve the same final concentration. (See Figure 7.20, increasing

the slop of L_m/S_m will increase the number fo stages need to achieve the same final C_e value).

7.6

1) 8

From Equation 7.3b it follows that:

$$360 = \frac{0.5}{8.8 \times 10^{-3} \,.\, 0.6} \left\{ 0.6 + 0.4 \frac{\Delta q}{\Delta C} \right\}$$

$$\text{so } \frac{\Delta q}{\Delta C} = 8$$

2) 0.950 m^3

Feed treated in 360 s

$= A \,.\, v_s \,.\, t = 0.3 \times 8.8 \,.\, 10^{-3} \times 360 = 0.950\ m^3$.

3) 15.8

The net volume of the adsorbent = A L(1 - ε)

$= 0.3 \times 0.5 \times 0.4 = 6 \,.\, 10^{-2}\ m^3$ Thus the ratio of liquid feed volume (L_v) to solid adsorbent volume (S_v) = 15.83.

By increasing the adsorption characteristics of the solid phase (ie by changing $\frac{\Delta \bar{q}}{\Delta C}$) or by increasing the volume of adsorbent and by altering the superficial velocity U.

If for example we doubled both the length of the column and the superficial velocity, then the lapse time t would remain the same providing the other parameters were kept constant (see Equyation 7.3b). However, the volume of adsorbent (L . A . (1-ε)) would have doubled, so will the volume of feed since the volume of feed treated = t . U . A . m^2. Thus we could treat twice as much liquid feed per unit time we double the length of the column and double the superficial velocity of the liquid.

7.7

1) Doubling N (see Figure 7.24 or substitute values into Equation 7.18).

2) By a factor of 10

Since R is proportional to $\sqrt{N}$ (see Equation 7.18). If α and K remain constant:

R = 10(x) when N = 100 and R = 100(x) when N = 10000

3) a) column diameter, volumetric flow rate and sample load should be increased.

   b) bed height, linear flow and sample concentration should remain the same.

Responses to Chapter 8 SAQs

8.1

Novo	Gist-brocades
Use of glucose isomerase from *Streptomyces murines*	Use of glucose isomerase from *Actinoplanes missouriensis*
Use of glutaraldehyde as a cross-linking agent	Use of glutaraldehyde as a cross-linking agent
Fluidised bed dryer	Use of propylene glycol as a preservative
Use of glucose isomerase from *Bacillus coagulans*	Use of butylacetate
	Gelatin

8.2

1) True.

2) True. When the tablet is swallowed, the acid in the stomach will cause the bicarbonate to generate CO_2. This will disrupt the tablet and the acetylsalicylic acid will be dispersed. There are however a number of drawbacks. The generation of CO_2 in the stomach can cause problems. It is for this reason that manufacturers choose to add the citric acid to their formulation, thereby enabling CO_2 generation, tablet disruption and subsequent solution of acetylsalicylic acid to occur prior to ingestion.

3) False. Although the enzyme catalysed reaction may be initially speeded up by raising the temperature, this will also increase the rate of decay of the enzyme. Thus a high temperature might product a very short lived high productivity and then the productivity would fall dramatically as the enzyme because inactivated.

4) False. Generally small particles will increase productivity. But if they are too small and compressible then the reactor may become blocked.

Responses to Chapter 9 SAQs

9.1

1) $2.1 \times 10^{-8}\ m^2\ V^{-1}\ s^{-1}$

We used Equation 9.1.

$$\text{Thus } v_{emp} = \frac{1.6 \times 10^{-19} \times 5}{6 \times \pi \times 20 \times 10^{-10} \times 1 \times 10^{-3}}$$

$= 2.1 \times 10^{-8}\ m^{-2}\ V^{-1}\ s^{-1}$

2) $2.1 \times 10^{-5}\ m\ s^{-1}$

Since $v_{emp} = 2.1 \times 10^{-8}\ m^2\ V^{-1}\ s^{-1}$. this means that the molecules will move at a rate of $2.1 \times 10^{-8}\ m\ s^{-1}$ in a potential field of $1\ V\ m^{-1}$.

In the question set the field is $10\ v\ cm^{-1} = 1000\ v\ m^{-1}$.

Thus the molecules will move at a rate of

$2.1 \times 10^{-8} \times 1000\ m\ s^{-1}$

$= 2.1 \times 10^{-5}\ m\ s^{-1}$

3) v_{epm} will decrease.

As the pH approaches the isoelectric point of β-lactoglobulin, the net charge on the molecule will decrease. Thus the electrophoretic mobility will decrease (see Equation 9.1). At the isoelectric point, the electrophoretic mobility = 0 since the molecule will carry no net charge.

9.2

$$\sigma = \frac{(C_A/C_b)\text{ per}}{(C_A/C_B)\text{ feed}} = \frac{0.35/0.65}{0.8/0.2} = 0.135$$

Compared with the pervaporation process, distillation shows much less selectivity.

9.3

1) We anticipate that you will have suggested that the fixed protein would need to be split to release the desired product and that the desired product would have to be separated from the ther fussion products and reagents. We can represent this by the following scheme.

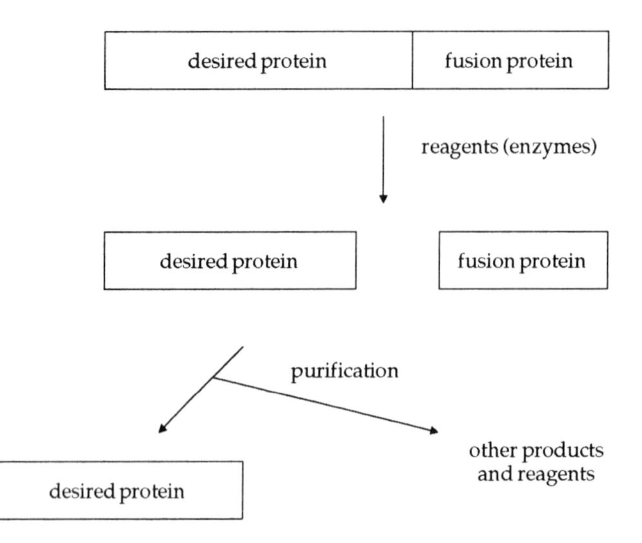

2) The proposed new gene product can be represented in the following way.

protein A	40 aspartate residue

Thus hybride protein has many features which may be used to help purification. For example, at neutral (or higher) pH, the aspartate residues will be ionised.

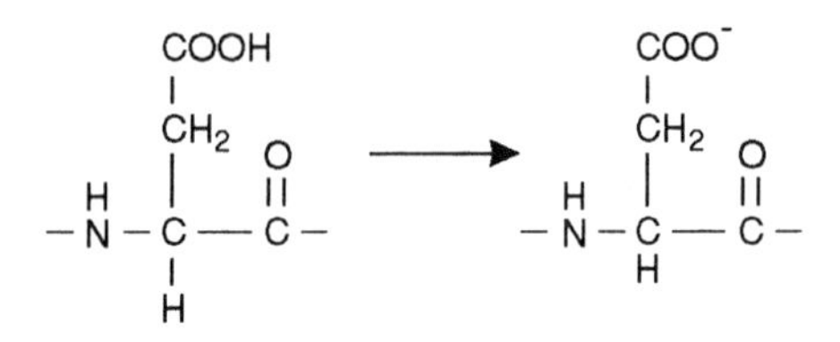

Thus the molecule would be much more water soluble than the original protein A.

It is probable that with so many aspairlate residues that the charge/mass ratio of this hybrid protein is much higher than many other proteins. This could aid separation by electrophoresis, although this is not used for large scale operations.

Alternatively the molecule by ion exchange chromatography or adsorption onto immobilised basic surfaces.

The point we are attempting to make with this SAQ is that by using genetic engineering we may design molecules which are easier to purify than their natural counter parts.

Appendix 1

Suggestions for further reading

Chapter 1

Krijgsman J (1990)
Increasing Interest in Bioseparations (in Dutch)
Dutch Chemical Industry (NCI), 7, 14-17,

Malby P S (1970)
Process Biochem 5, 22

Chapter 2

Schlegel H G (1988)
General Microbiology
Cambridge University Press, United Kingdom

Bird R B, Stewart E, Lightfoot N (1960)
Transport phenomena pp 10-13
John Wiley & Sons, New York, USA

Krijgsman J (1989)
Release of Intracellular Components
In: Advanced Course Downstream Processing, J Krijgsman, Ed, Chapter 3,
Delft Institute of Technology, The Netherlands

Chapter 3

Hetherington P J *et al* (1971)
Release of protein from baker's yeast by disruption in an industrial homogeniser
Trans Inst Chem Eng, 49 pp 142-148

Krijgsman J (1989)
Release of Intracellular Components
In: Advanced Course Downstream Processing, J Krijgsman, Ed, Chapter 3,
Delft Institute of Technology, The Netherlands

Limon-Lason J *et al* (1979)
Reactor Properties of a High-Speed Bead Mill for Microbial Cell Rupture
Biotech and Bioeng, Vol XXI, pp 745-774

Chapter 4

Purchas D B (1969)
Process Engineering Technique Evaluation: Filtration, pp 23-29, HK, Suttle Ed
Morgan-Grampion (Publ) Ltd, London, United Kingdom

Bosley R (1969)
Vacuum Filters in Process Engineering Technique Evaluation, pp 40-55
Morgan & Grampion (Publ) Ltd, London, United Kingdom

Svrovsky L (1981)
Filtration Fundamentals
In: Solid Liquid Separation, 2nd ed, pp 242-265, L Svarovsky Ed
Butterworth-Heinemann Ltd, UK

Shirato S, Esumi S (1963)
Filtration of a Cultured Broth of *Streptomyces griseus*
Part I, Journ of Ferm Tech Japan 41, pp 86-92

Brunner K H, Hemfort H (1988)
Centrifugal Separation in Biotechnological Processes
In: Downstream Processes; Equipment and Techniques, pp 1-50
Allan R Liss Inc

Kula M R (1985)
Recovery Operations
In: Biotechnology, H J Rehm, G Reed Eds, Vol 2, pp 726-760
VCH Verlag, Germany

Grimwood G L (1969)
Centrifuges
In: Process Engineering Technique Evaluation, pp 97-116
Morgan Grampion (Publ) Ltd, London, United Kingdom
Prospect Alfa-laval (Tumba, Sweden)
Centrifuges for the Pharmaceutical and Biotechnology Industries
No PB 41094E

Hemfort H, Kohlstette W (1988)
Centrifugal Clarifiers and Decanter for Biotechnology
Technical Scientific Documentation No 5
Westfalia Separator AG, Oelde, Germany

Chapter 5

Perry R H, Chilton C H (1973)
Chemical Engineers Handbook, 5th Edition
McGraw Hill, New York, USA

McCabe W L, Smith J C (1976)
Unit Operations of Chemical Engineering, 3rd Edition
McGraw Hill, New York, USA

Strathmann H (1979)
Separation of Molecular Mixtures using Synthetic Membranes (in German)
Stein Kopff Verlag, Darmstadt, Germany

Cheryan M (1986)
Ultrafiltration Handbook
Technomic Publ, Lancaster, USA

Lo T C, Baird M H I, Hanson C (1983)
Handbook of Solvent Extraction
John Wiley & Sons, New York, USA

Chapter 6

Lo T C, Baird M H I, Hanson C (1983)
Handbook of Solvent Extraction
John Wiley & Sons, New York, USA

Wesselingh J A, Kleizen H H (1990)
Separation Processes (in Dutch)
Dum, The Netherlands

Strathmann H (1979)
Separation of Molecular Mixtures using Synthetic Membranes (in German)
Stein Kopff Verlag, Darmstadt, Germany
Cheryan M (1986)

Ultrafiltration Handbook
Technomic Publ, Lancaster, USA
Albertson P A (1977)

Separation of Particles and Macromolecules by Phase Separation
Endeavour NS, vol 1, no 2
Abbot L, Hatton T A (1988)

Liquid-Liquid Extraction for Protein Separations
Chem Eng Progress, August, pp 31-41
Scopes R K (1982)

Springer Advanced Texts in Chemistry, Cantor C R Ed
Springer Verlag, New York, USA

Lyklema J (1978)
In: The Scientific Basic of Flocculation
Ives K J Ed
Sijthoff and Noordhof, Alphen a/d Rijn, The Netherlands

Kula R M (1985)
Recovery Operations in Biotechnology
vol 2, Rehm H J, Reed G Eds, pp 725-760
VHC-Verlag, Weinheim, Germany

Weast R C (1974)
Handbook of Chemistry and Physics 55th Edition
CRC Press, Cleveland (Ohio), USA

Foster P R, Dunhill P, Lilly M D (1976)
The Kinetics of Protein Salting Out
Biotech Bioeng 18, pp 545-580

Bell D J, Hoare M, Dunhill P (1983)
The Formation of Protein Precipitates and their Centrifugal Recovery

In: Advances in Biochemical Engineering/Biotechnology
vol 26, Giechter A Ed
Springer Verlag, New York, USA

Chapter 7

Dechow F J (1989)
Separation and Purification Techniques in biotechnology
Noyes Publ, Park Ridge (NY), USA

Ruthven D M (1984)
Principles of Adsorption and Adsorption Processes
John Wiley & Sons, New York, USA

Jagschies G (1988)
Process-scale Chromatography
Ullmann's Encyclopedia of Industrial Chemistry Vol B3, 10 1-44
VCH Verlag, Weinheim, Germany

Kula M R (1983)
Recovery Operations
In: Biotechnology Vol 2, Rehm H J, Reed G Eds, 726-760
VCH Verlag, Weinheim, Germany

Chapter 9

Brandt S, Randal A G, Kessler S B (1988)
Membrane based affinity technology for commercial scale purification
BIO/Technology, vol 6, July, pp 779-782

Mansoori G A,Schulz K, Martinelli E E (1988)
Bioseparation using supercritical fluid extraction/retrograde condensation
Biotechnology 6, April, pp 393-399

Weiss M D (1989)
Supercritical fluids offer potential solutions to bioprocessing problems
Genetic Engineering News 9, December

Enfors S O, Hellebust H, Kohler K, Strandberg L, Veide A (1990)
Impact of genetic engineering on downstream processing of proteins produced in *E. coli*
Adv Bioeng Biotechnol

Enfors S O, Kohler K, Veide A (1990)
Combined use of extraction and genetic engineering for protein purification
Bioseparation I, 305-310

Appendix 2

The major symbols used in the text

Chapter 2

K = consistency index ($Nm^{-2}\ s^{n}$)

n = power law index

$\mathring{\gamma}$ = shear rate (s^{-1})

η = dynamic viscosity ($Ns\ m^{-2}$; Pas)

τ = shear stress (Nm^{-2}; Pa)

τ_o = yield stress (Nm^{-2}; Pa)

ϕ = volume fraction

Chapter 3

C = concentration

c = dimensionless constant (for power input)

c_p = specific heat

C_r = concentration of product that has been released from biomass

C_r^{max} = concentration of product that can maximally be released from biomass

D = impeller diameter (m)

K_b = proportionally constant for bead mills (s^{-1})

K_h = proportionally constant for homogenous $(Nm^{-2})^{-n}$

L = length of bead mills

N = rotational speed of an impeller (s^{-1})

P = power input

p = pressure (bar; Nm^{-2})

T = cylinder diameter (bead mill) (m)

θ = temperature (K; °C)

ρ = density (kg m^{-3})

Chapter 4

A = filtration area (m^2)

A = clarification area (m^2)

C = concentration (kg m^{-3})

d_p = particle diameter (m)

F_b = buoyancy force

F_d = drag force

F_g = gravitational force

g = gravitational constant (m s^{-2})

H = height (m)

L = lenght (m)

m = mass

p = pressure (bar; Nm^{-2})

r = radius (m)

r_c = cake resistence (m^{-1})

r_m = medium resistance (m^{-1})

r_o = outer radius (centrifuge) (m)

r_i = inner radius (centrifuge) (m)

V = volume (m^3)

v_c = centrifugal sedimentation rates (ms^{-1})

v_g = gravitational sedimentation rate

v_{ax} = axial velocity

W = width (m)

w = mass of cake (filtration) per unit area (kg m^{-2})

α = specific cake resistence (m kg^{-1})

ξ = relative centrifugal force

η = dynamic viscosity (Pas; Ns m^{-2})

ρ_L = liquid density (kg m^{-3})

ρ_s = solid density (kg m^{-3})

Σ = equivalent clarification area

ϕ_v = volumetric flow rate (m^3 s^{-1})

ϕ''_v = volumetric flow rate per unit area (flux) (m s^{-1})

φ = cone angle (disc stack centrifuge)

ω = angular speed (rad s^{-1})

Chapter 5

A = area of a membrane (m^2)

A = heat transfer surface (m^2)

C_c = concentration of the solute in the concentrate (kg m^{-3})

C_f = concentration of the solute in the feed (kg m^{-3})

C_p = concentration in the permeate (mol m^{-3} or kg m^{-3})

$c_{p,c}$ = specific heat of concentrate (J kg K^{-1} or J kg $°C^{-1}$)

$c_{p,f}$ = specific heat of feed (J kg K^{-1} or J kg^{-1} $°C^{-1}$)

C_r = concentration in retentate (mol m^{-3} or kg m^{-3})

C_s = concentration of solute

C_w = molar density of a solution

C_{bulk} = concentration in the bulk liquid (kg m^{-3})

$C_{permeate}$ = concentration of the permeate (kg m^{-3})

C_{wall} = concentration at a wall (kg m^{-3})

d_m = membrane thickness (m)

E = energy

E_i (x) = exponential integral

h_c = enthalpy of concentrate (J)

h_f = enthalpy (heat content) of feed (J)

h_s = enthalpy of steam

h_v = enthalpy of vapour (J)

$\bar{J}$ = mean flux

j = flux (flux per unit area)

K = permeability coeffience (m)

K^i = mass transfer coefficient (membrane)

K^{ii} = initial flux through a membrane

N = number of effects (evaporators)

p = pressure (bar; $Nm^2 s^{-1}$)

R = gas constant (kJ mol^{-1} K^{-1})

R = membrane rejection coefficient

R_m = membrane resistence (m^{-1})

T = temperature (K)

U = overall heat transfer coefficient (kW m^{-2} K^{-1})

V = volume

V_f = volume of feed (m^3)

V_r = retentate volume (m^3)

V_w = volume of solution containing 1 kg of water

x_s = molar fraction of solute

x_w = molar fraction of water

α = concentration factor

γ_w = activity of water

Δ = permeate yield (recovery factor)

η = dynamic viscosity (Ns m^{-2}; Pas)

ε = membrane porosity

θ_c = temperature of condensate (K; °C)

θ_{eff} = effective temperature difference (K; °C)

θ_f = temperature of feed (K; °C)

λ = thermal conductivity ($Wm^{-1}K^{-1}$)

μ_w = thermodynamic potential of water in a solution

μ_w^o = thermodynamic potential of pure water

π = osmatic pressure (Pa)

τ = turtuosity

ϕ = flow rate

ϕ_c = concentration mass flow rate (kg s^{-1})

ϕ_f = feed mass flow rate (kg s^{-1})

ϕ_q = heat transfer rate (kW)

ϕ_s = steam mass flow rate (kg s^{-1})

$\phi_{s,c}$ = condensate mass flow rate (kg s^{-1})

ϕ_v = vapour mass flow rate (kg s^{-1})

$\phi_{V,f}$ = volumetric flow of feed (m^3 s^{-1})

$\phi_{V,p}$ = volumetric flow of permeate

$\phi_{V,r}$ = volumetric flow of retentate (m^3 s^{-1})

$\phi_{V,pore}$ = volumetric flow through a pore

ψ = solid yield

ψ_s = specific steam consumption

Chapter 6

a = specific area (m^2 m^{-3} = m^{-1})

C = concentration

C_e = concentration solute in extract (kg m^{-3})

C_f = concentration of solute in feed (kg m^{-3})

C_r = concentration of solute in raffinate (kg m^{-3})

C_s = concentration of solute in solvent (kg m^{-3})

C_{se} = concentration at the end of a process (kg m^{-3})

C_{s0} = concentration of solute in solvent phase at time t = 0

C_{st} = concentration of solute in solvent phase at time t (kg m^{-3})

C_s^* = concentration of solute in solvent phase in equilibrium with concentration in raffinate phase (kg m^{-3})

d = droplet diameter

D_ε = dielectric constant

E_b = fractional approach to equilibrium (batch system)

E_f = fractional approach to equilibrium (flow system)

f = extracted fraction

K = Boltzmann constant

K_a = dissociation constant of an acid

K_b = dissociation constant of a base

K_s = mass transfer coefficient based on solvent phase (s^{-1})

M = relative molecular mass

N = number of stages (solvent extraction)

S = separation factor

T = absolute temperature (K)

V_m = volume of mixer (m^3)

V_r = volume of raffinate (m^3)

V_s = volume of solvent (m^3)

α = effective distribution coefficient

α^o = intrinsic distribution coefficient

ε = volume fraction

Γ = ionic strength

ϕ_V = volumetric flow rate (m^3 s^{-1})

$\phi_{V,e}$ = volumetric flow rate of extract (m^3 s^{-1})

$\phi_{V,f}$ = volumetric flow rate of feed ($m^3\ s^{-1}$)

$\phi_{V,r}$ = volumetric flow rate of raffinate ($m^3\ s^{-1}$)

$\phi_{V,s}$ = volumetric flow rate of solvent ($m^3\ s^{-1}$)

Chapter 7

C = concentration (kg m^{-3}; mol l^{-1})

C_e = equilibrium concentration (kg kg^{-1}; kg m^{-3})

C_o = initial concentration (kg kg^{-1}; kg m^{-3})

D_m = diffusion coefficient in the mobile phase ($m^2 s^{-1}$)

D_s = diffusion coefficient in particle (solid) ($m^2\ s^{-1}$)

d_p = particle diameter

H = plate height (m)

K = Boltzmann's constant (J K^{-1}) also capacity factor

K_F = Freundlich equilibrium constant

K_L = Langmuir equilibrium constant

L = column length (m)

L_m = mass of liquid (kg)

L_v = volume of liquid feed (m^3)

N = number of plates

q = solid loading (kg m^{-3}; mol m^{-3})

q_e = solid loading at equilibrium (kg kg^{-1}; kg m^{-3}; mol m^{-3})

R = resolution factor

r_m = molecular (Stokes) radius (m)

S_m = mass of absorbent (kg)

S_v = volume of adsorbent (m^3)

T = absolute temperature (K)

t = time (s)

t_e = mean residence time of carrier or eluant (s)

t_r = residence time or retention time of peak (s)

V_e = volume of effluent (m^3)

V_o = interstitial or outer volume of a column

v_s = supreficial liquid velocity ($m\ s^{-1}$)

w_b = width at the base of the peak (chromatography)

α = selectivity

δ = boundary layer thickness (m)

ε = porosity or void fraction

η = dynamic viscosity (Pas; Nsm^{-2})

λ = particle ordering factor

σ = standard deviation

τ = obstruction factor

Chapter 8

D = diffusion coefficient

H = column height (m)

p = pressure (Pas)

P_t = productivity

R_p = radius of particle (m)

v_s = superficial flow velocity ($m\ s^{-1}$)

η = dynamic viscosity (Nsm^{-2}; Pas)

ϕ = Thiele Modulus

Chapter 9

C = concentration ($Kg\ m^{-3}$)

C_{crit} = density at critical point ($kg\ m^{-3}$)

e = charge on an electron

K = solubility constant

r_m = molecular (Stokes) radius (m)

v_{epm} = velocity in an electric field (electrophorietic mobility $m^2 V^{-1} s^{-1}$)

Z = net charge

η = dynamic viscosity (Nsm^{-2})

ρ = density

ϕ''_m = mass flux ($Kg\ m^{-2}\ s^{-1}$)

Appendix 3

Values of physical constants

Physical constant	Symbol	Value
acceleration due to gravity	g	$9.81\ m\ s^{-2}$
Avogadro constant	N_A	$6.022\ 52 \times 10^{23}\ mol^{-1}$
Boltzmann constant	k	$1.380\ 54 \times 10^{-23}\ J\ K^{-1}$
charge to mass ratio	e/m	$1.758\ 796 \times 10^{11}\ C\ kg^{-1}$
Curie	Ci	37.0×10^{9} disintegrations per second
electronic charge	e	$1.602\ 10 \times 10^{-19}\ C$
Faraday constant	F	$9.648\ 70 \times 10^{4}\ C\ mol^{-1}$
gas constant	R	$8.314\ 3\ J\ K^{-1}\ mol^{-1}$
ice-point temperature	T_{ice}	273.150 K
molar volume of ideal gas at stp	V_m	$2.241\ 36 \times 10^{-2}\ m^{3}\ mol^{-1}$
standard pressure, atmosphere	P	$101\ 325\ N\ m^{-2}$
unified atomic mass constant	m_u	$1.660\ 43 \times 10^{-27}\ kg$
velocity of light in a vacuum	c	$2.997\ 925 \times 10^{8}\ m\ s^{-1}$

Index

A

B

C

D

E

F

T

U

V

W

Y

Z